Construction Stakeholder Management

Edited by

Ezekiel Chinyio
School of Engineering and the Built Environment
University of Wolverhampton
Wolverhampton, UK

Paul Olomolaiye
Professor of Construction Engineering and Management
Pro Vice-Chancellor and
Executive Dean - Environment and Technology
University of the West of England
Bristol, UK

WILEY-BLACKWELL

A John Wiley & Sons, Ltd., Publication

This edition first published 2010
© Blackwell Publishing Ltd

Blackwell Publishing was acquired by John Wiley & Sons in February 2007. Blackwell's publishing programme has been merged with Wiley's global Scientific, Technical, and Medical business to form Wiley-Blackwell.

Registered office
John Wiley & Sons Ltd, The Atrium, Southern Gate, Chichester, West Sussex, PO19 8SQ, United Kingdom

Editorial office
9600 Garsington Road, Oxford, OX4 2DQ, United Kingdom
2121 State Avenue, Ames, Iowa 50014-8300, USA

For details of our global editorial offices, for customer services and for information about how to apply for permission to reuse the copyright material in this book please see our website at www.wiley.com/wiley-blackwell.

The right of the author to be identified as the author of this work has been asserted in accordance with the Copyright, Designs and Patents Act 1988.

Wiley also publishes its books in a variety of electronic formats. Some content that appears in print may not be available in electronic books.

Designations used by companies to distinguish their products are often claimed as trademarks. All brand names and product names used in this book are trade names, service marks, trademarks or registered trademarks of their respective owners. The publisher is not associated with any product or vendor mentioned in this book. This publication is designed to provide accurate and authoritative information in regard to the subject matter covered. It is sold on the understanding that the publisher is not engaged in rendering professional services. If professional advice or other expert assistance is required, the services of a competent professional should be sought.

Library of Congress Cataloging-in-Publication Data

Construction stakeholder management / edited by Ezekiel Chinyio, Paul Olomolaiye.
    p. cm.
  Includes bibliographical references and index.
  ISBN 978-1-4051-8098-6 (hardback : alk. paper)
  1. Strategic planning. 2. Organization. 3. Customer relations. 4. Consumer satisfaction. I. Chinyio, E. (Ezekiel) II. Olomolaiye, Paul O.
  HD30.28.C6616 2010
  624.068'8--dc22
                                                                                2009023944
A catalogue record for this book is available from the British Library.

Set in 9.5pt/12.5pt Palatino by Macmillan Publishing Solutions, Chennai, India
Printed and bound in Malaysia by Vivar Printing Sdn Bhd
1   2010

Contents

Contributors

Amir W. Al-Khafaji earned his BSCE in Civil Engineering and MSCE degree in Construction Management from Wayne State University, Detroit, Michigan. He received a second MS degree in Soil Dynamics and PhD in Geotechnical Engineering from Michigan State University in East Lansing, Michigan. Dr Al-Khafaji is the Executive Director of the Center for Emerging Technologies in Infrastructure at Bradley University. Through this center, he chaired many national and international conferences in construction, asphalt and transportation. He was selected as the Person of the Year by the Illinois Asphalt Pavement Association in 2006 and received the 2006 Award of Excellence from the Bradley University Parents' Association. He is the 1991 recipient of the Bradley University's Professional Excellence Award and the 1985 Spirit of Detroit Leadership Award. In 1998, he chaired a national ASCE Task Force to define scholarship in engineering. For the past 22 years, Dr Al-Khafaji has served as the chairman of the Department of Civil Engineering and Construction at Bradley University, Peoria, Illinois. He is the co-author of 11 engineering textbooks including Numerical Methods, Geotechnics, Statics and Dynamics and Software Application.

Kenneth Amaeshi is a Lecturer (Assistant Professor) in Strategy and Policy Innovation at the Doughty Centre of Corporate Responsibility, School of Management, Cranfield University, UK. He is also a Visiting Assistant Professor at the University of Warwick's Business School, UK, and a Research Associate of the Centre for the Study of Globalisation and Regionalisation at the University of Warwick. Previously he worked as a management consultant to multinational corporations in oil and gas, financial services, telecommunication and aviation industries in Africa and Europe and was a Visiting Scholar at the Said Business School, University of Oxford, United Kingdom (2007/2008 academic session). His PhD from the University of Warwick was on Comparative Political Economy of Corporate Social Responsibility in Germany and UK. Kenneth consults in the broad areas of strategic corporate social responsibility, private–public partnerships, financial innovation, sustainable development and corporate governance. His research focuses mainly on sustainable finance and innovation, comparative corporate social responsibility, commercialisation of intellectual property assets, governance of innovation networks and multinational corporations in developing economies. He is published in several leading journals wherein he is the 2007 winner of the International Award for Excellence in the field of interdisciplinary social sciences, awarded by the *International Journal of Interdisciplinary Social Sciences* (Australia/USA).

Brian L. Atkin holds a doctorate from the University of Reading, where he was formerly Professor of Construction Management, and has continued as visiting Professor to the present. He also holds a master's degree by research into the costs of major engineering infrastructure. Currently, he holds professorial appointments in Australia (QUT) and Sweden (Lund).

From 1996–2001, Brian was Professor of Construction Management and Economics at The Royal Institute of Technology, Stockholm, Sweden and from 2001 until the end of 2007 was Director of the Swedish national research and development programme, Competitive Building, which was funded by the Foundation for Strategic Research (SSF) and a group of industry-leading companies. He is a member of the Steering Committees for the Engineering and Physical Sciences Research Council (EPSRC) supported Innovative Construction Research Centre (ICRC) at the University of Reading and the Salford Centre for Research and Innovation (SCRI) at the University of Salford. He has enjoyed substantial support from research councils and the European Commission (EC) over many years, of which the FutureHome project under the global Intelligent Manufacturing Systems (IMS) initiative was the largest of its kind, involving 15 organisations across seven countries in Europe and a consortium of Japanese companies. For three years, Brian acted as coordinator of the EC targeted research action in Environmentally-friendly construction technologies with responsibility for the process area and, later, for E-CORE (European Construction Research Network) in examining technology transfer into construction from other industrial sectors.

He has three books in print and has published more than 100 papers, articles and reports on subjects spanning construction management and facility management in both the academic and practitioner press.

Wayne Baum is President and CEO of CORE Construction, an ENR Top 100 General Contracting firm, headquartered in Morton, Illinois. Mr Baum's companies have been a leader in risk management and safety. Under his leadership, the CORE Construction group has expanded its reach and across the United States and is now ranked 15th in the construction of schools and educational facilities. He is the past chairman, board of trustees, Progressive Health Systems; past chairman, board of trustees, Pekin Hospital; member, board of trustees, Eureka College; member, advisory board, Union Planters Bank, Bradley University board of trustees, Easter Seals, Central Illinois Memorial Kidney Fund. His list of public service is so extensive that he was named Citizen of the Year by the Boy Scouts of America.

Johan J. Bester is a Forestry Professional employed by the South African Government's national Department of Water Affairs and Forestry. He studied Forest Management and Conservation Forestry at University of Stellenbosch (Republic of South Africa) and obtained an MSc Forestry degree from the University of Wolverhampton (UK). Johan considers himself a generalist. His career in the forestry sector has involved a variety of work, including management of commercial forestry operations; forestry advisory services; community forestry and forestry policy development. Among the endeavours he has found most rewarding are the restoration of degraded arid savannah, exploring the connections between forests and climate change and involvement with stakeholder participation in forestry issues in Wales. The multiple benefits that people derive from trees and forests as well as the impacts that people have on forests remain areas of interest and concern to which he dedicates much of his efforts.

Colin A. Booth is a Senior Lecturer in Environmental Engineering at the School of Engineering and the Built Environment (SEBE), The University of Wolverhampton. He gained his Environmental Modelling PhD in 2002 from The University of Wolverhampton (UK). Before joining SEBE, he held several research fellowship and teaching posts (each with an environmental theme). More recently, he was a leading investigator in a 3-year EU-funded global research project 'The environmental and socio-economic contribution of palm geotextiles to sustainable development and soil conservation', directing activities in the United Kingdom, Belgium, Hungary, Lithuania, Brazil, South Africa, Gambia, China, Thailand and Vietnam. He has authored/co-authored ~50 peer-reviewed scientific research papers and book chapters and, furthermore, he is the supervisor of nine doctoral and two postdoctoral researchers. His research interests include: environmental magnetism, soil erosion and conservation, soil management, water engineering and management, urban pollution, coastal and estuarine science.

Lynda Bourne is Managing Director of Stakeholder Management Pty Ltd, based in Melbourne, Australia. She consults and runs training programs in 'Successful Stakeholder Management' based on the *Stakeholder Circle®* methodology and tools worldwide. Dr Bourne has over 20 years industry experience as Project Manager and Project Sponsor: this experience led to her research focused on improving the management of relationships within the project and organisational environment. Her doctoral research investigated improvements to existing stakeholder management practices and resulted in the development of the *Stakeholder Circle®* methodology and tools. Working with organisations to implement the *Stakeholder Circle®* methodology has subsequently led Dr Bourne to develop the concept of 'Stakeholder Relationship Management Maturity' (SRMM®). She has presented papers on the *Stakeholder Circle®* methodology since 2002, contributed to books on stakeholder management and risk management and has published her own book: *Stakeholder Relationship Management: A Maturity Model for Organisational Implementation* in 2009.

Todd Bridges is the Army's Senior Research Scientist for Environmental Science. His main areas of research activity at the U.S. Army Engineer Research and Development Center (ERDC) concern the 1) development of risk and decision analysis methods applied to water resources infrastructure and environmental management and 2) science and engineering related to sediment management. In addition to his research activity as Senior Scientist, Dr. Bridges also serves as the Director of the Center for Contaminated Sediments at ERDC and the Program Manager for the Dredging Operations Environmental Research (DOER) program, where he directs the execution of more than $6 million in research in one of the Corps' largest research programs. He has chaired international working groups within the Scientific Group of the London Convention and the International Navigation Association. Dr. Bridges serves as an editor for the journal *Integrated Environmental Assessment and Management* and is an active member of several professional organizations. Over the last 15 years Dr. Bridges has published numerous papers and book chapters in the fields of risk assessment, modeling and environmental management. He received his B.A. (1985) and M.A. (1988) in Biology/Zoology from California State

University, Fresno and his Ph.D. (1992) in Biological Oceanography at North Carolina State University.

Gail Charnley (PhD) is an internationally recognized scientist specializing in environmental health risk assessment and risk management science and policy. She has over 30 years of experience in the biological, chemical, and social policy aspects of environmental and public health protection. She has been executive director of the Presidential/Congressional Commission on Risk Assessment and Risk Management, director of the Toxicology and Risk Assessment Program at the National Academy of Sciences, and president of the international Society for Risk Analysis. She has published many papers, given many presentations, and served on numerous committees focused on various aspects of the relationships between environmental chemical exposures and public health protection.

Ezekiel Chinyio holds a BSc (Hons) degree in quantity surveying and MSc in construction management. He worked as a site quantity surveyor for 1 year in Africa before joining the academic sector in 1987. In the mid-nineties, Ezekiel spent 1 year as a visiting researcher at the University of Reading, UK and went on to study for his PhD at the University of Wolverhampton, UK. He has since then worked with Glasgow Caledonian University as a Research Fellow and with the University of Central England (now Birmingham City University) as a Senior Lecturer. He is now lecturing with the University of Wolverhampton. His research interests include procurement, risk, knowledge and operations management and organisational development. Ezekiel has published several articles in international academic journals as well as chapters in books and conference proceedings. He has also co-authored a previous book.

Abbas Ali Elmualim is a Lecturer and one of the leading academic of the £3.5million Engineering and Physical Science Research Council (EPSRC) funded Innovative Construction Research Centre (ICRC), University of Reading, UK where he coordinates the Sustainable Design, Construction and Facilities Management research group. He has a BSc (Hons) degree in Architecture from the University of Khartoum; MSc in the Environmental Design of Buildings from Cranfield University, UK and PhD in Sustainability and Renewable Energy from the University of Reading, UK. Dr Elmualim has acted as the coordinator for the Big Ideas research project on sustained competitiveness of the UK construction, a £1.7 million project in collaboration with IMRCs at Loughborough and Salford Universities. He was involved in the successful bid for the £3.8million Grand Challenge project in coordination with Bath and Cambridge Universities. He also acted as the Research Manager for the ICRC in developing a research community. Currently, Dr Elmualim is leading several collaborative projects with industry worth over £600K and funded by the UK Department of Trade and Industry as part of the Knowledge Transfer Partnership scheme and the Spearhead Technology Transfer programme. Dr Elmualim's research focuses on the development of integrative approaches to design, construction and facilities management with a particular perspective of broad socio-technical systems for sustainability and digital technologies.

João Alberto da Costa Ganzo Fernandez has degrees in Business Administration, Law and Architecture, as well as a PhD in Civil Engineering.

Over the last 10 years, he has concentrated on the real estate market in Florianópolis, the capital city of the State of Santa Catarina in Brazil. His 8 years as a researcher at the Federal University have been dedicated to the potential apartment-buyer segment. He provides consultancy services in market analysis and feasibility studies for multi-family projects. For 15 years, he has been full professor of Construction Planning and Architectural Design at the Federal Centre of Technological Education of Santa Catarina.

Felix Nikoi Hammond is a Senior Lecturer in real estate finance, investment, valuation and management. Felix is Head of the Land, Property and Flood Management research group of the School of Engineering and the Built Environment of the University of Wolverhampton and a Course Manager in charge of the BSc real estate course. His main research interest centres on economic analysis of real estate policies. Felix has published several peer-reviewed journal papers and has chaired a number of international conferences. Prior to becoming an academic, he worked with the Ghana Lands Commission as a Senior Lands Officer. He was also a land expert on the World Bank sponsored Land Administration project in Ghana from 2001 to 2003.

Frank Harris now retired, graduated in Civil Engineering during the 1960s and gained industrial experience with British and Australian contractors before setting out on an academic career in construction management and engineering at Wolverhampton and Loughborough Universities, returning to Wolverhampton in 1988 as Head/Dean of the School of Construction, Engineering and Technology. Research interests concentrated on the application of scientific and mathematical methods to construction management systems and engineering, demonstrated with projects funded by research councils, government departments and industry, extensively reported in international journals and textbooks.

Philip T. Harris has been an academic for over 22 years having cut his teeth in the FE sector for 3 years prior to moving to the University of Wolverhampton. Previously he had worked as a Site Manager on various projects in the midlands region of the United Kingdom. His areas of research interest relate to sustainability and vocational education. Phil holds a Degree in Construction Management and a Masters Degree in Construction Project Management from Loughborough University. Phil is heavily involved with Lifelong Learning Networks and Chairs the Birmingham Black Country and Solihull Construction Lifelong Learning Network group. In addition, he is responsible for his school's collaborative education programmes and the management of numerous UK and overseas collaborations. Phil is currently a Principal Lecturer at the University of Wolverhampton.

Bernie Koch is President of River City Construction, an ENR Top 400 General Contracting firm, headquartered in Peoria, Illinois. Mr Koch has 25 years experience in the management and administration of construction projects. His project experience includes correctional facilities, waste water and water treatment plants, healthcare facilities, military construction, as well as educational and university projects. Mr Koch is actively involved as an officer or director in a number of construction related organisations. Mr Koch holds a BS in Construction from Bradley University and has also attained his LEED AP certification.

Tas Yong Koh received a Diploma in Building from the Singapore Polytechnic in 1990, Bachelor of Building Construction Management from the University of New South Wales (UNSW) in Sydney, 1999, MSc (Building) from the National University of Singapore (NUS) in 2005 and had recently completed both stages of higher education teaching course leading to the Certificate of Teaching and Learning in Higher Education awarded by the University of Hong Kong (HKU). In between these times, he had worked in construction industry in Singapore. Tas had worked through various positions ranging from Supervisor to Project Manager and Quality Systems Manager. He is experienced in managing residential and industrial projects. Tas is also experienced in the quality, environmental and safety management of the construction project. From 2003, he has been active in construction management research, first in Singapore, and then in Hong Kong. He has been involved in teaching undergraduate courses in NUS in Construction Project Management and Productivity and Quality when in Singapore, and construction related management courses when in Hong Kong. Tas is currently with the Real Estate and Construction Department, the University of Hong Kong. As a researcher, he has published in *Construction Management and Economics*, and academic conferences.

Robert Kowalski is a Principal Lecturer with the Centre for International Development and Training at the University of Wolverhampton, UK. As a long-standing member of the Centre, he has a leadership role in training courses, consultancy work and research. He is the Director of Studies for CIDT programmes, developing the postgraduate portfolio and winning the prestigious Chevening Fellowship Programme on Governance and Environmental Democracy. He recently completed his role as Director of Studies for a piece of Action Research into institutional strengthening of a Rural University in Brazil. Originally with a technical background in Organic Agriculture, Bob has subsequently specialised in various aspects of International Development and change management. In this context, he has contributed to projects in SME development, conflict management and institutional strengthening. He has worked in Africa, Asia and South America, and has over the last 10 years concentrated on countries in economic transition: Poland, Czech Republic, former Soviet Union (including Central Asia), Bulgaria, Romania and former Yugoslavia. Bob has published widely, and successfully supervised a number of PhD students.

Mei-yung Leung has more than 15 years of practical/teaching experience in the construction industry/education and has participated in a number of prestigious construction projects in Hong Kong. Leung is a *Chartered Quantity Surveyor* in the RICS in the United Kingdom and the HKIS in Hong Kong and a *Charted Builder* in the CIOB in the United Kingdom and AIB in Australia. She received a BSc (Hons) with first class in Quantity Surveying at the University of Wolverhampton in the United Kingdom, a Degree in Religions Science at the Pontifical Urban University in Rome and a PhD degree at the University of Hong Kong in Hong Kong. Dr Leung is also a *Qualified facilitator* of the SAVE International 'The Value Society' in the United States and the HKIVM (List A) for facilitating governmental 'VM workshop' in the industry internationally. She is currently conducting a number of professional and research projects in the areas of project management, value management, stress management, facility management and construction education. All research projects are being investigated on the basis of a human behavioural paradigm and project management knowledge, which will assist and support the construction engineers throughout design, construction and operation stages. Over 70 refereed journal and conference papers in construction engineering and management have been published or accepted for publishing.

Igor Linkov is a Research Scientist at the US Army Engineer Research and Development Center and Adjunct Professor of Engineering and Public Policy at Carnegie Mellon University. He has a BS and MSc in Physics and Mathematics (Polytechnic Institute, Russia) and a PhD in Environmental, Occupational and Radiation Health (University of Pittsburgh). He completed his postdoctoral training in Biostatistics and Toxicology and Risk Assessment at Harvard University. Dr Linkov has managed ecological and human health risk assessments and risk management projects. Many of his projects have included application of the state-of-the-science modelling and software tools to highly complex sites (e.g. Hudson River, Dow Midland, Natick Soldier Systems Command, Elizabeth Mine and so on) and projects (e.g. insuring emerging risks, risk-based prioritisation of engineering projects, developing performance metrics for oil spill response). He was instrumental in developing an integrated risk assessment and multi-criteria decision analysis framework that is now being widely applied by the US Army Corps of Engineers, for example for the multi-billion dollar restoration planning for coastal Louisiana and Mississippi following hurricane Katrina. Dr Linkov is currently involved in several projects that examine factors responsible for nanotoxicology and nanomaterials risks. Dr Linkov has published widely on environmental policy, environmental modelling, and risk analysis, including 10 books and over 100 peer-reviewed papers and book chapters. Dr Linkov has organised more than a dozen national and international conferences and continuing education workshops. He has served on many review and advisory panels and other US and international agencies. He is DoD representative at the Interagency Working Group on Nanotechnology Environmental and Health Implications (NEHI). The Governor of Massachusetts has appointed Dr Linkov to serve as a Scientific Advisor to the Toxic Use Reduction Institute. He is the recipient of the prestigious Chauncey Starr Award for exceptional contribution to Risk Analysis.

Ektewan Manowong completed a PhD in Construction Engineering and Management at the Asian Institute of Technology, Thailand where he studied the effectiveness of construction stakeholder participation. He has been working in the areas of building structural design, civil engineering and infrastructure inspection and maintenance and construction business consultancy. He has a strong research interest in the area of good governance in construction projects. He recently worked as a Research Fellow at the Asian Institute of Technology conducting research on construction project stakeholder's participation in waste management.

Helder M. Moura has been in the construction field for over 20 years and is currently a Senior Project Manager in a Portuguese road concessionary, with special interests on claims management. He holds a degree in Civil Engineering and an MSc in Construction Management and for the last 5 years has been involved in research projects in topics related to construction management, claims and internationalisation of construction services. He also has 10 years' experience working as an expropriation expert and real estate valuator.

Amlan Mukherjee is an Assistant Professor at Michigan Technology University. He received his PhD from the University of Washington, MS from the University at Buffalo (SUNY) and BE from Birla Institute of Technology and Science, Pilani, India. Dr Mukherjee focuses his research and professional activities primarily in the area of planning and decision-making in civil and environmental engineering. His research goal is to study the impact of alternative planning and design of engineering operations and decision-making strategies on the fates of the economic, environmental and human systems that they are situated in. Dr Mukherjee combines research methods in system dynamics, discrete event simulations and agent based modelling with methods in AI planning, and data mining to further understanding of decision-making in complex systems. He is involved with the development of frameworks and models, and implementation of interactive simulations that can aid decision-makers assess risk, examine alternative sustainable designs and explore what-if scenarios. Dr Mukherjee advises graduate students and teaches interdisciplinary graduate and undergraduate classes. He has publications in journals and has presented at the American Education Research Association (AERA).

Douglas R. Oberhelman is a Group President of Caterpillar Inc., the world's largest manufacturer of construction and mining equipment, diesel and natural gas engines and industrial gas turbines. He oversees the company's human services and sustainable development functions, its growing remanufacturing business, machinery marketing operations in North America and worldwide manufacturing, marketing and support of industrial and large power systems. In his 30-plus years with Caterpillar, Mr Oberhelman has held a variety of positions on three continents, including 3 years as the company's Chief Financial Officer. He holds a Bachelor's degree from Millikin University and is a member of the board of directors of the Association of Equipment Manufacturers, the National Association of Manufacturers, the Manufacturing Institute and the Wetlands America Trust.

Stephen Ogunlana holds the chair of construction project management at the School of the Built Environment, Heriot-Watt University, UK. He was formerly the Professor of construction at the Asian Institute of Technology, Thailand. He is the joint coordinator of the International Council on Construction Innovation and Research's (CIB) working commission on construction in developing economies (W107). He is the editor of the book 'Profitable Partnering for Construction Procurement' published by Taylor and Francis (SPON). His research and consulting interests are in project management, public private partnerships, organisational learning, system dynamics simulation, human resources development and construction process improvement.

Stefan Olander is a Senior Lecturer at Lund University, Faculty of Engineering within the division of Construction Management. Since 1999, he has worked within research areas such as construction management, property development, real estate management, project management and industrialised construction. Where the main body of research has focused on how stakeholders affect the construction industry, which also was the topic of his PhD thesis in 2006. After achieving his PhD, he has continued to explore the concept of stakeholder management within the construction industry.

Roberto de Oliveira had been involved in the construction industry as a practitioner for over 16 years where he evolved from teaching to professorship and research 13 years ago. Firstly, he spent 16 years working as a site engineer and gearing an office of structural design. Then he moved to Canada where he completed his PhD at the University of Waterloo and has consolidated his position as Professor at UFSC (Universidade Federal de Santa Catarina), Department of Civil Engineering. The main focus of his research is the Management of the Design Process oriented to the production of buildings that meet customers' expectancies.

Paul Olomolaiye, Professor of Construction Engineering and Management, recently became the Pro Vice-Chancellor and Executive Dean for Environment and Technology at the University of West of England (UWE) in Bristol, UK. He holds a PhD degree in Civil Engineering from Loughborough University and gained his Professorship in 1999. He worked for some 20 years at the University of Wolverhampton and for 8 years was the Dean of the School of Engineering and the Built Environment where he achieved significant year on year business growth through a motivating and vision-driven leadership. Paul's School at Wolverhampton led the regeneration drive aimed at advancing products and business practices of more than 3000 Engineering and Construction Companies in Northern Birmingham by attracting research and knowledge exchange funds in excess of £20 million from Private Sector, Government and European Union to execute applied research projects which have helped to define Wolverhampton University as the regional leader in business interaction. With well published and ongoing research and active knowledge transfer initiatives, Paul has developed specific expertise in Productivity and Performance enhancement strategies to the benefit of a number of Governments, Agencies, Charities, Companies and Institutions worldwide who regularly consult him.

Adekunle Sabitu Oyegoke is a Research Fellow at the Salford Centre for Research and Innovation (SCRI) in the Built and Human Environment, in the University of Salford, UK. He obtained Doctor of Science in Technology in Construction Economics and Management with emphasis on procurement management from the Helsinki University of Technology, Finland, in 2007. Before commencing his academic career, he worked as a Quantity Surveyor in consulting and contracting firms. He holds a corporate membership of the Chartered Institute of Building (MCIOB) and the Association of Cost Engineers in the United Kingdom (MACostE).

Ritsuko Ozaki is Senior Research Fellow in the Innovation Studies Centre, Imperial College Business School, Imperial College London, UK. She is also Visiting Fellow of the University of Tokyo Institute of Industrial Science. Her background is sociology with a particular focus on housing and the built environment. Her research interests include: the adoption and consumption of innovation, and end-user perception and experience of innovative products and services. Her current research project focuses on pro-environmental technologies and services (e.g. hybrid cars, zero-carbon houses) and investigates consumers' adoption motivations, lifestyles and daily practices and use and appropriation of adopted innovations.

Ron Rosenhead is Chief Executive of Project Agency, a Consultancy practice dedicated to helping organisations deliver projects on time, to budget with the right results. He started his career in training and development and helped deliver several change management projects. It was during this period that he first realised that professional staff at all levels were getting involved in projects but had received little or no training to do this. He ran some basic project management events and he was hooked! Ron now runs Project Agency which has a large client base. He has worked with professionals at all levels from construction to social work, from engineers to doctors, from surveyors to managers. He has worked in many sectors and is the author of *Deliver that Project* and writes a regular Blog at www.ronrosenhead.co.uk. He coaches, trains and speaks all on the topic of project management.

Athena Roumboutsos is a Lecturer at the University of the Aegean, Department of Shipping, Trade, and Transport. She is a holder of an Engineering Diploma from the National Technical University of Athens (1986) and a PhD in Engineering (reservoir evaluation and business development) from Heriot-Watt University, UK (1988). With extensive experience in project management, she has held top management positions in the private sector (1989–2003). Her field of research concerns risk analysis and management, stakeholder management and change management, transport infrastructure evaluation, Public Private Partnerships, procurement and contract management, accountability and entrepreneurial responsibility, alliances and strategies where she has published/presented more than 20 papers and currently provides consultancy services to public and private companies.

Steve Rowlinson is a Civil Engineer by training, graduating from Nottingham University, and has worked for the past 22 years at the University of Hong Kong, first in the Department of Civil and Structural Engineering and (for the past 17 years) in the Department of Real Estate and Construction. During that time, he has been engaged as a consultant and expert witness for a number of companies, including the Hong Kong Housing Authority and Hong Kong Government Works Bureau. Steve has researched into the Hong Kong construction industry and has written and co-authored five books on that industry. His areas of specialisation are construction site safety, IT and procurement systems. In the latter area, he has been the co-coordinator of the international working commission W092 of the CIB (www.cibworld.nl) for 13 years and has extensive experience of how construction projects are procured worldwide. He has organised or co-organised seven conferences in the field of procurement systems and published two books and numerous journal articles and book chapters in the area. He is currently an Adjunct Professor at the QUT in Brisbane, Australia and was a project participant in the project *Value alignment in project delivery systems*; and project leader on the project *Value in project delivery systems: Facilitating a change in culture*. He has been an assessor for the professional examinations of both the HKIE and the ICE, UK for a number of years and is a keen golfer and skier.

Pantaleo D. Rwelamila is a Professor of Project Management at the Graduate School of Business Leadership (GSBL), University of South Africa (UNISA) and President of The South African Council for Project and Construction Management Professions (SACPCMP); Joint coordinator: International Council for Research & Innovation in Building & Construction (CIB) – W107: Construction in developing countries; Chairperson and non-executive director of Tau Pride Projects (Pty) Ltd, a construction project management consulting in South Africa; Chairperson and non-executive director of MSINGI Construction Project Management (Pty) Ltd, a Construction project management consulting firm based in Cape Town, South Africa and non-executive Director and joint owner, XPQ Consulting Services (Pty) Ltd, a project management consulting firm based in Pretoria, South Africa. Professor Rwelamila is a former Vice President (Education) of the Chartered Institute of Building (CIOB) – Southern Africa and a former Senior Partner of Quantum Consultants (Pty) Ltd – a Project Management and Quantity Surveying consulting firm in Botswana. P.D. Rwelamila graduated in Building Economics in Tanzania, has a Masters Degree in Construction Management from Brunel University, UK and a PhD in Project Management and Procurement systems from the University of Cape Town. He is a Fellow of the CIOB – UK, American Association of Cost Engineers (MAACE) and American Institute of Constructors (MAIC). He has worked in a number of countries, including Tanzania, Kenya, Uganda, Botswana, Zambia, Australia, United Kingdom and Sweden. Professor Rwelamila has more than 140 published and peer-reviewed journal papers, chapters in books and conference proceedings.

Michaela M. Schaffhauser-Linzatti has a tenure track position in teaching and research at the Accounting Department, University of Vienna, Austria. Having been Guest Lecturer at the Europa University Viadriana, Germany, for 4 years, she is currently also teaching MBA classes at the Vienna University of Technology, the Medical University of Vienna and the Polytechnical University Bucharest, Romania. Her research activities focus on efficiency and performance measurement of public organisations and hereby include topics on PPP, healthcare and university management and intangible asset accounting.

Nidhi Shah is currently a Project Officer with the University of Wolverhampton where she is working within the Engineering and Built Environment sector on several externally funded projects related to recruitment, widening participation and lifelong learning. She has recently completed a Masters in Human Resource Development and Consulting from Lancaster University and is keenly interested in areas such as Change Management, Knowledge Transfer and Life Long Learning.

Renato da Silva Solano's background is Civil Engineering and he teaches at the Federal University of Rio Grande do Sul as well as the Pontifícia University Católica of Rio Grande do Sul. His other major occupation relates to an office of Construction Design where he operates as Design Manager; doing works of Design Coordination and Compatibility. His current major field of research is focused on the influence of building forms on construction costs.

José Cardoso Teixeira is currently an Associate Professor with the University of Minho in Portugal. He holds a degree in Civil Engineering and a PhD in Construction Management from the University of Loughborough. His research interests are on project management and economics of construction. He has also worked as a project consultant in the last 25 years.

Michael Thompson has been working in construction for over 42 years, having spent his first 30 years with a major international consultancy in the water industry, in the United Kingdom and in the Near East, Middle East and Far East. He is a Fellow of the Institution of Civil Engineers. He is currently employed by the University of Wolverhampton as Executive Director of the West Midlands Centre for Constructing Excellence (WMCCE), which is part of the national Constructing Excellence 'family'. WMCCE, located in centre of the City of Birmingham, is one of the leading centres in Best Practice in the United Kingdom, helping to improve the performance of the Construction Industry in various ways, so that the Industry not only becomes more efficient, but also becomes more competitive in an increasingly difficult market. In 1996, Michael set up his own business, helping the construction industry to understand how they can work in a collaborative environment, and once in that environment, helping it to understand how it can take maximum advantage of that environment, for mutual benefit with other stakeholders and for the benefit of the project. He specialises in partnering, risk and value management, and believes that 'value' is still underrated.

Martin Morgan Tuuli is currently Senior Lecturer in Construction Commercial Management at the University of Westminster, London-UK. He holds a Ph.D. in Construction Project Management from the University of Hong Kong. Martin also holds a B.Sc. in Building Technology (First Class Honours) from Kwame Nkrumah University of Science and Technology (KNUST), Kumasi-Ghana. His research interests are in Empowerment and Control Dynamics, People Management, Ethics, Diversity, Equality and Human and Social Sustainability in Project Organizations. He was previously Head/Quantity Surveyor, Batching Plant/Pre-cast Department of Taysec Construction Ltd (a then Taylor Woodrow plc), overseeing the production and supply of a wide range of pre-cast concrete products and ready-mix concrete to the company's portfolio of projects. He also worked as Teaching Assistant at the Department of Building Technology, KNUST. Martin is an Incorporate Member of the Chartered Institute of Building (CIOB) and a Probationer of the Ghana Institution of Surveyors (GhIS).

Izak J. Van der Merwe is Assistant Director: Woodlands at the Department of Water Affairs and Forestry (South Africa), where he is responsible for biodiversity and conservation planning for woodlands and natural forests among other. He studied Town and Regional Planning and Geography at the University of Pretoria, followed by an extensive career within the Environmental and Forestry Administrations of South African Government and as environmental consultant. He has contributed to several government publications of various kinds, and is the author of *The Knysna and Tsitsikamma Forests: Their History, Ecology and Management*. He is co-author and co-editor of five scientific publications and conference proceedings, and author of more than 40 environmental articles and papers for environmental and popular magazines.

Jeroen Warner is a Dutch political scientist. He has researched, published and lectured on issues of water governance, environmental conflict and multi-stakeholder participation as a Wageningen University associate and is currently co-ordinating work on the Joint Planning Approach for the EU-funded Freude am Fluss project at Radboud University, Nijmegen. He published two edited volumes: Conflictos y Participación with Nordan, Montevideo (with Alejandra Moreyra) in 2004 and Multi-Stakeholder Platforms for Integrated Water Management with Ashgate in 2007. His PhD research is to appear with IB Tauris as Flood planning: the politics of water security.

Patrick Weaver is the Managing Director of Mosaic Project Services Pty Ltd, an Australian project management consultancy business specialising in project scheduling and project management training. His consultancy work encompasses: developing and advising on project schedules, forensic analysis of project data, developing and presenting project management training courses, acting as an expert witness and assisting with dispute resolution and claims management, he is a qualified Arbitrator and accredited Mediator.

Matthew Wood is a graduate student in the Cognitive Psychology program at Carnegie Mellon University and a recipient of the Science, Mathematics and Research for Transformation Scholarship (part of the National Defense Education Program). He has an MS in General Psychology from Villanova University and a BA in Psychology from Elizabethtown College. His primary interest is in the area of cognitive and social processes related to creative problem-solving and expert performance.

Foreword

This is an important book. At last the issue of stakeholder management has been written about in a way which relieves the term of the burden of the facile sloganeering so beloved of those charged with directing the construction industry for the last 10 years. At one time, stakeholder management was a term used to indicate that a wider set of interests needed to be satisfied; certainly beyond the narrow and dominant interests of the clients that have so frequently trumped those of contractors, sub-contractors, suppliers and construction workers.

This book brings a fresh perspective. It does so because it has recruited a cast of authors, many of whom are drawn neither directly from the ranks of construction industry academics nor from construction consultants. As such, a fresh eye has been cast over the subject and authors from different nationalities offer refreshing perspectives. This enables a comprehensive coverage to be given to the pedagogy of the subject; Theory and Practice: Cases and Codification: Formal and Informal Processes: Strategic Tactical and Operational Concerns: Research and Speculation. All are given space in this book which enables the editors to launch a *cri de coeur* to put stakeholder management at the heart of the project process and embedded in the procurement procedures used within the construction industry.

An important point made in the book is that stakeholder management does not eliminate differences between the parties to a project but that these differences have different sources, causes and effects. The dominant theme in the book is that stakeholder management is about recognising that self-interest has been nurtured, both politically and socially over a long period of time and consequently, differences between the stakeholders is inevitable. The insight is that different styles of leadership and negotiation practices are required to get the best from a project team and beyond. The emphasis upon negotiated outcomes of stakeholder interest offers another new dimension. 'Partnering', the 'buzz-word' of the 1990s and the early 2000s has often been a subterfuge under which clients have imposed, often unilateral and unfair conditions, including 'Dutch Auctions' to tie in contractors and suppliers through the power of near monopolistic authority. Stakeholder Management as promoted in the book provides a more egalitarian context for 'partnering'.

The range of coverage is also helpful. The book recognises that stakeholders go beyond the people and organisations that populate projects. The wider social and physical aspects of the environment may be important, if intangible and, at times, incorporeal stakeholders. This opens the door for the issues of morality and ethics to be brought out into the open. Stakeholders may have legal, contractual, financial and political rights and obligations. These same stakeholders and others may have both moral ethical and functional stakes in a project. The book attracts the eye because it addresses, for the first time (of which I am aware) for the construction industry these essential moral arguments which can be marshalled to the discussion of stakeholder management.

It has been both a pleasure and a privilege to provide this Foreword and the Editors are to be congratulated for putting together such an important volume. I should like to wish this book every good fortune in both its critical and commercial futures.

Dave Langford

Professor *Emeritus*
School of the Built and Natural Environment
Glasgow Caledonian University.

Honorary Life President of the Association of Researchers in Construction Management (ARCOM).

Preface

The book employs case studies in several places to explain and advocate for stakeholder management. Each chapter emphasises one or more aspects of stakeholder management.

A preview of the contents of the book is provided in the following paragraphs.

Chapter 1 sets the scene by explaining what stakes, stakeholders and stakeholder management are. The need for stakeholder management is discussed. The transient nature of stakes is highlighted as well as an attempt to group stakeholders into different types. The position of firms regarding stakeholder management is touched upon wherein the role of shareholders is clarified.

Chapter 2 covers the history of the stakeholder perspective to business and organisational studies, and discusses the different underpinning philosophies of this perspective, and how they relate to corporate social responsibility, corporate governance and accountability. The perspectives and paradigms of stakeholding are discussed via a micro-level theorisations of corporate stakeholding practice. In this regard, the chapter highlights the significance of examining the roles and influences of stakeholders, their interactions with managers as well as the institutional configurations in which they act, and are acted upon, in order to further explain variations in behaviour of firms in different institutional contexts. This chapter essentially identifies alternative modes of theorising corporate stakeholding practices, drawing theories from disciplines such as politics, economics and sociology to explain the behaviours of firms at the multi-levels. In all this, the external environment identified as a significant explanatory factor of behaviour of firms.

Chapter 3 uses formal contracts (and forms of procurement) to examine stakeholder interaction. In this chapter, the evolution of different contractual arrangements is highlighted with their associated implications for the duties of primary stakeholders. Many stakeholders can be found in a construction project and this chapter casts them on the two dimensions of primary players and secondary players. Each of these two categories is further examined at three sub-levels. Given the myriad of stakeholders in a project, the chapter winds-up by advocating stakeholder management as an antidote to conflicts. The chapter also recommends that the implementation of stakeholder management should embrace a moral, political, technological and economic non-obligatory tone. The process requires a champion to drive it.

Chapter 4 discusses the extent to which construction practice has embraced stakeholder management. The turn of the 21st century is used as a landmark for evaluating construction stakeholder management practice. Significantly, Sir Michael Latham reviewed the construction industry in a report in 1994 and offered suggestions for improvement. Sir John Egan ensued in 1998 and 2002. These two reviews advocated partnering and teamwork as panacea for improving practice, outputs and benefits. Partnering augurs well for stakeholder management. Whilst supporting partnering and stakeholder management, the chapter uses two case studies to highlight its importance.

Chapter 5 provides a project overview as involving several stakeholders, viz: internal, external, social, environmental, etc. These stakeholders interact in the course of a project with potential conflicts on interest. A prudent approach is to manage these diverse stakes to avoid conflicts. The Finnish approach to stakeholder management is then covered in the second half of the chapter. This approach involves six steps and is applicable in the five project phases of identification, programming, appraisal, implementation and facilities management.

Chapter 6 starts by considering the broad and narrow definitions of stakeholders; and follows this with a consideration of internal and external stakeholders in the context of construction projects. The stakeholders often found in a construction project are established along with the roles they play as a project unfolds. Inherent in the functions of stakeholders are risks that can manifest. A risk management concept is thus needed to manage stakeholders so as to avoid their associated risks.

Chapter 7 is about stakeholder mapping. The chapter explains the objective of every stakeholder mapping process, that is to develop a useful list of stakeholders, assess some of their key characteristics and present these assessments in a way that helps the project team develop insight and understanding to support their implementation of planned stakeholder management initiatives. It follows this discussion with a review of the evolution of stakeholder mapping and the methods and techniques being used today. After that, a specific mapping method and technique is discussed and demonstrated, that is the *Stakeholder Circle®*. This methodology provides guidance to knowing who the right stakeholders are at any stage of the project. It also provides guidance for implementing targeted communication with stakeholders as well as evaluating the dynamism of stakeholders. The tool helps its users in tracking their stakeholders and their priority standing.

Chapter 8 provides an insight on how to strategically manage and deal with project stakeholders; provides a framework for analysing and understanding construction stakeholders and their interrelations because strategies and tactics for managing stakeholders start with defining and identifying stakeholders. The chapter explains that, having analysed its stakeholders, an organisation can evolve strategies that befit the unique requirements of each, for example whether to approach each stakeholder directly or indirectly and whether to inform, consult, involve or partner with stakeholders. In terms of tactics, stakeholders are classified and prioritised and relationships and communication with them are based on this analysis; likewise the quest to improve their satisfaction.

Chapter 9 provides an exposition to the principles of negotiations. Two or more individuals, groups or organisations can negotiate on one or more issues. The tactics to use in negotiation are discussed such as making unilateral concessions, being (un)cooperative, etc. The principles of negotiation are also examined such as involving a facilitator, considering several options, clarity and openness, respecting the other party, etc.

Chapter 10 is on communication. The concept and patterns of communication are introduced and used as a basis for discussing how internal stakeholders should communicate more effectively and efficiently. Three case studies, one each from Caterpillar, CORE and RCC are used to embellish the relevance and applicability of communication.

Chapter 11 provides a broad and deep exposition on the role and need for leadership in construction project delivery. This leadership should champion the function of stakeholder management in the construction projects' process. Construction is largely viewed as a fragmented and diverse industry. Individual actors in construction are often adversarial and act with the objective of achieving their self-interests. The industry is commonly criticised of being lagging behind other industry, as of low trust and with adversarial relationships as the main strands of its current culture that exacerbated the problem of late and over budget delivery of projects. Sometimes, collaboration is employed as a means of actualising these self-interests; but fundamentally the self-interest mission is a cultural norm. So in a construction project, it is not unexpected if stakeholders do not act altruistically. The chapter thus advocates an overarching concept for managing stakeholders' interests – this is described as a 'psycho and socio-cultural approach'. This approach should be managed by leaders/managers for maximum impact and contextualised with the industry cultural terrain with the view of bringing the required change.

Chapter 12 draws from the results of a number of research projects to demonstrate how various construction project procurement systems have impacted on the management of stakeholders. An appropriate construction project procurement system is necessary for the project manager in order to balance the project parameters, allocate risk appropriately and consequently meet stakeholders' requirements. The choice of a procurement system for a project is a strategic decision that should be made after assessing and interpreting the stakeholders requirements as its characteristics define the structure and interaction of stakeholders. Against this background, the chapter discusses how different types of procurement systems impact, implicitly, on different project stakeholders. The chapter concludes that the management of stakeholders is conditioned by the type of procurement system, and vice versa.

Chapter 13 discusses the impact of procurement on stakeholder management – using Hong Kong as a base country. Hong Kong was administered by British colonial rule for 150 years until 1997. Part of the influence of this colonial rule was the adoption of traditional procurement in construction practice and this mode of procurement is still prevalent in Hong Kong today. Coupled with Hong Kong (Chinese) culture, the stakeholders in Hong Kong are predisposed to keeping to contractual principles. Although modern projects are introducing 'partnering' in their procurement; the implementation of partnering ethos is not yet fully entrenched in practice. Two elaborate case studies are used to show the steady progress of partnering and stakeholder management in Hong Kong construction. While there is a willingness to manage all stakeholders' concerns, there are challenges that are being overcome.

Chapter 14 concentrates on decision-making in a multi-stakeholder setting. The mental models applied in decision-making are particularly discussed. In this regard, previously acquired knowledge is often used in making current decisions. However, the mental mechanism involved in this process is very complex. The 'mental model theory' which is discussed in this chapter captures this complexity and explains different approaches to cognition, decision-making and expression. Significantly, these different approaches can manifest in multi-stakeholder scenarios. Stakeholders need to be aware of these different models and to accommodate these in their interactions.

Chapter 15 reviews the stakeholder management process through the lens of theory and medium of case studies, to show where and how the prospects for success can be enhanced. The benefits of engaging with stakeholders are reinforced in these discussions, as well as the consequences of not doing so. Stakeholders can delay or thwart a project, hence the need to engage with them in the course of an undertaking. Trust and communication are some of the features which can be used to endear stakeholders. The chapter argues that the concerns of stakeholders should never be taken for granted but treated with utmost respect. Examples of project delays and cost overruns are cited to demonstrate the pains of stakeholder management. Consequently, organisations are encouraged to manage their stakeholders towards avoiding such consequences while exploiting the upsides.

Chapter 16 Electronic archives help save costs and time, reduce procedures and bureaucracy, increase security to data, improve accessibility and the satisfaction of citizens. Austria is the first country worldwide to realise high-security archives by introducing her so-called 'original fiction' under which electronically submitted and archived documents have the same legal status as the original paper documents. Through this achievement, Austria has been the champion of e-Government in Europe on multiple occasions. The authenticity of these documents is proven by lawyers and notaries who import the documents into the archives by their secure digital signatures. This chapter presents the judicial, technical and organisational aspects of Austria's successful electronic archiving solution by especially focusing on stakeholder aspects. It further discusses the general lessons learnt from the best practice example 'Archivium Dokumentenarchiv GmbH'.

Chapter 17 is about conflict management and explains the sources, nature and types of conflicts. Consequently, conflicts can be associated with legal, political and institutional framework, economic constrains and pressures, people's culture, social structure, stakeholder interests, technical knowledge, environment and history, etc. Conflicts may involve any two or more stakeholders. Some conflicts will be between external stakeholders, some between internal stakeholders while some will be between a combination of stakeholders. The management of conflicts between stakeholders is underpinned by a framework. The chapter proposes a set of options for resolving conflicts, arguing for a need to build bridges in view of stakeholder management.

Chapter 18 The environment is discussed as a distinctive stakeholder in this chapter as well as whether managers of construction and property enterprises manage the environment as a primary stakeholder albeit voluntarily and without government intervention. Aspects of the environment which project it as an important stakeholder are discussed: atmosphere, land, biodiversity and water. Notably, the environment is non-human and not represented in person in the course of business decisions and actions that affect it or even those that it affects. Someone or some people must take the initiative to manage this stakeholder else its interests could be undermined in projects.

As a stakeholder can impact and be impacted upon, the chapter argues that a fair distribution of the gains and burdens of construction and property enterprises with the natural environment must be a comparison between the external benefits and costs associated with the undertaken.

Chapter 19 discusses how to implement change in an organisation; explaining that it is a sensitive issue that needs to be handled carefully. The external environment in which businesses operate can trigger the need for change. Evidence from literature indicates that although change is imperative, when people are confronted with it, they usually resist. In fact, studies show that nearly two in every three change initiatives in large-scale corporations tend to fail. If a transformation plan fails, imperfect implementation and resistance to change are usually identified as the reasons for it. Organisations use leadership, communication and consultation as drivers of successful change. In addition to these, the chapter discusses other aspects of change management which will assist an implementation endeavour.

Chapter 20 provides 10 case studies to supplement the main discussions in the book. The case studies illustrate the consequences of dealing with stakeholder issues in an inadequate manner. The challenge of stakeholder management is to monitor the changing profile of one's stakeholders and to be ready to address their demands, else they could affect you negatively. In this regard, the case studies emphasise the potential downsides of stakeholders. However, the main concept of stakeholder management is to maximise the benefits which can be derived from stakeholders while minimising the downsides. In this regard, Chapter 20 flags up the downsides as a call to action towards planning for stakeholder issues upfront.

Chapter 21 concludes with a call to excellence in stakeholder management. The expectation is that stakeholder management will be a core competence in construction. The chapter reiterates the benefits of stakeholder management and downsides of inaction. While stakeholder management can be implemented informally, it is worthwhile to adopt a formalised approach. That way some people will be tasked with the responsibility of implementing it and their effort can be supported and sustained. In addition, the need for preparation is important. Training may be used while preparing so that each organisation can be in a position to attend to its stakeholders when they come calling.

1 Introducing Stakeholder Management

Ezekiel Chinyio and Paul Olomolaiye

1.1 Introduction

This book discusses stakeholder management as it is particularly applicable to construction practice. This chapter sets the scene by defining stakeholders and the concept of stakeholder management. Readers who are fairly conversant with stakeholder management may wish to skip this introductory chapter.

The generic origins and nature of stakeholders' interests are identified. The pluralistic nature of stakes and the need to manage these in an undertaking is portrayed. In most construction projects, there will be many stakeholders and their presence in itself is a potential conflict of interests. There is a need to manage this diversity proactively to forestall conflicts and even disputes. Therefore, most of the stakeholders that are visible in construction are flagged up in this chapter.

1.1.1 Stakes and stakeholders

A stake is an interest or a share in an undertaking while a stakeholder is an individual with a stake (Weiss, 2006). Moloney (2006) argues that stakeholders are individuals or groups that benefit from an organisation. Further, stakeholders can be harmed or have their rights affected by an organisation. Fundamentally, stakeholders affect and are affected by an organisation and its activities. Stakeholders can affect an organisation's functioning, goals, development and even survival. Stakeholders are beneficial when they help you achieve your goals and they are antagonistic when they oppose your mission. In effect, stakeholders have power to be either a threat or a benefit to an organisation (Gibson, 2000).

Sometimes stakeholders will trigger project schemes in other organisations (Orndoff, 2005) and can support or obstruct an ongoing project (Vogwell, 2002). Their influence can be small or great and can be exerted either deliberately or incidentally. Individuals and organisations need to be wary of their stakeholders and their influences.

Diverse sources can trigger stakes, e.g. stakes can be influenced by economic and other considerations. Mintzberg (1995) reckons that stakes can have cultural or political origins too. Shareholders constitute a stakeholder group, and often have a vested interest in the profits their organisation will make. To them, if keeping other stakeholders happy will yield more profit, then so be it.

If stakeholders can have a negative influence on us, why should we deal with or bother about them? The reason is that most often you cannot do without them. Organisations often depend on external stakeholders for resources, services, information, etc. Our operations make us interact with several stakeholders. Most often, an organisation would depend on others for something and this can give the latter some leverage (Frooman, 1999). The argument is that stakeholders have claims, rights and expectations that ought to be honoured and not taken lightly (Carroll and Buchholtz, 2006). Thus, stakeholders must be managed in each undertaking to avoid any of their negative influences, especially those that could be contrary to a firm's objectives. Conversely, business endeavours and indeed construction projects affect stakeholders. So it is a tit-for-tat affair. Businesses must recognise their stakeholders and manage them and vice versa.

1.1.2   Government as a stakeholder

Governments can be counted as stakeholders in some way as they certainly affect organisations and groups through their fiscal and regulatory policies (Moloney, 2006). Governments could have an interest in the operations and existence of certain organisations. For instance, in terms of operations, some governments have regulated the tobacco industry obliging the latter to warn that its product can kill. Another instance of the influence of government as a stakeholder can be seen in the food industry wherein manufacturers have been compelled through regulations to label their food products in certain ways. However, the relationship between organisations and the government is on a different dimension. According to Moloney (2006):

> Non-government stakeholders can threaten organisational existence but rarely in the sovereign way of governments against which there is ultimately no constitutional defence. A business can be threatened by another in a hostile takeover bid; employees can strike against their firm; a cause group can interfere with building a by-pass road. These actions, however, do not have the sustained potential for continuous damage that a government with executive and legislative power can have, albeit a liberal, democratic one.

Government is a type of stakeholder with unique interests. Their involvement with firms is on a different level and scale. For instance; in terms of the survival of organisations, the UK government recently came to the rescue of 'Northern Rock' when it faced adverse financial problems. In addition, the UK Government has injected billions of pounds (£) into its finance sector to forestall some banks like HBOS and Royal Bank of Scotland from collapsing. Similarly, some financial institutions in the United States faced serious crises and the US government intervened to avoid these banks going under.

1.1.3   Stakeholders in construction

There are stakeholders in construction undertakings, just as there are stakeholders in other endeavours. The checklist of stakeholders in a construction project is often large and would include the owners and users of facilities, project managers, facilities managers, designers, shareholders, legal authorities, employees, subcontractors, suppliers, process and service providers, competitors, banks, insurance companies, media, community representatives, neighbours, general public, government establishments, visitors, customers, regional development agencies, the natural environment, the press, pressure groups, civic institutions, etc. (Newcombe, 2003; Smith and Love, 2004). Each of these would influence the course of a project at some stage. Some bring their influence to bear more often than others. If diverse stakeholders are present in construction undertakings, then the construction industry should be able to manage its stakeholders.

1.2   Types of stakeholders

Stakeholder management involves identifying and classifying stakeholders, thus facilitating both initial and subsequent engagement with them in a timely, planned and coordinated manner. This engagement involves identifying different categories of stakeholders; gathering information about them; identifying their missions in a project; determining their strengths and weaknesses; identifying their strategies; predicting their behaviour and developing and implementing a strategy for managing these stakeholders (Cleland, 2002). Stakeholders have been grouped in several ways and some of these are discussed below.

The stakeholders in a project can be divided into (Calvert 1995; Winch and Bonke 2002):

- Internal stakeholders, that is those who are members of the project coalition or who provide finance.
- External stakeholders, that is those affected by the project in a significant way.

Stakeholders can be internal or external to the project team or project scope (Sutterfield *et al.*, 2006).

Similar classifications are inside and outside stakeholders (Newcombe, 2003), and direct and indirect stakeholders (Smith and Love, 2004).

Another delineation considers primary versus secondary stakeholders (Carroll and Buchholtz, 2006). A primary stakeholder group is one without whose continuing participation the corporation cannot survive as a going concern, whereas secondary stakeholders are those who influence or are influenced by the firm, but who are not essential to its survival (Clarkson, 1995; Pajunen, 2006). Some stakeholders can be very critical to an organisation and others less critical (Calvert, 1995; Winch and Bonke, 2002).

Stakeholders could also be contrasted between those that are contracted to provide services (e.g. contractors, subcontractors, consultants) that is in a primary or direct relationship with an organisation; in contrast to those that have no contracted responsibility or formal redress, but are in an indirect or secondary relationship with an organisation (Smith and Love, 2004; Carroll and Buchholtz, 2006). The un-contracted stakeholders (e.g. members of the community and potential end users who are committed to occupy/use the facilities) can have power to disrupt projects through their actions, which can be political, but are not easily liable for their actions.

Some stakeholders could be viewed as financial developers and regulatory authorities. Carroll and Buchholtz (2006) have also considered the categories of social versus non-social, and core, strategic or environmental stakeholders.

Given the several dimensions on which stakeholders can be interpreted, some stakeholders may be members of two or more types; so a multidimensional plot is really needed to capture the full complexity of stakeholders and their often large number.

In terms of decision-making, it is worthwhile to consider stakeholders as being supportive, neutral, or anti (Chinyio and Akintoye, 2008). The anti's are often in the minority but can be very vocal. The idea is to endeavour to shift stakeholders from the neutral and especially anti side of the fulcrum to the supportive side.

1.2.1   The legitimacy and power of stakeholders

Stakeholders and their associated stakes will manifest the attributes of legitimacy and power (Carroll and Buchholtz, 2006). Legitimacy is the perceived validity of a claim to a stake. Power is the capacity to induce, persuade or coerce the actions of others and is displayed when one part in a relationship is able to impose its will on the other part (Johnson *et al.*, 2005). Power may be displayed through (Ihlen and Berntzen, 2007):

- force (coercive power),
- material or financial resources (utilitarian power), or
- symbolic resources (normative power).

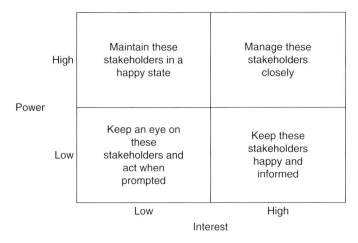

Figure 1.1   A Power-interest matrix.

As stakeholders have claims, rights and expectations, they must be managed in each project to avoid any of their influences that could be contrary to a firm's objectives. Conversely, business endeavours and indeed construction projects affect stakeholders. So businesses must recognise their stakeholders and manage them, and vice versa. The ideal is to optimise by maximising the benefits that are derivable from stakeholders while minimising their potential negative impacts.

Templates are useful for mapping stakeholders. Figure 1.1 is a simple and popular tool for mapping an organisation's stakeholders. The power differential between a firm and its stakeholders will inform the strategies and tactics for dealing with each other (Frooman, 1999; Kolk and Pinkse, 2006). This book discusses different ways of relating with stakeholders especially Chapters 7 and 8.

In addition to the power-interest dimension Newcombe (2003) also considered a power-predictability matrix. A risk perspective overshadows the predictability of stakeholders. In this regard, an organisation should be able to identify those stakeholders that can spring a surprise to them in terms of making a demand on or exercising power in the project. When things are progressing well with an organisation and its stakeholders, it does not necessarily mean that a stakeholder cannot place a sudden and unnecessary demand on the project. Thus in running projects, organisations may often act in a tolerance zone which is a performance band in which the firm is satisfying the interests of all its key stakeholder groups (Doyle and Stern, 2006). As projects can swing out of this tolerance band, there is a need to monitor their progress continuously.

1.2.2   *The saliency and dynamism of stakeholders*

Saliency (or urgency) is the intensity of claim, attention and priority attached to a stake (Mitchell *et al.*, 1997). It is also the degree to which a claim demands immediate attention (Gago and Antolín, 2004). It seems thus that urgency influences the manner and extent to which power is exercised in stakeholder engagement. A tri-axial template can be used to map stakeholders. Chapter 7 considers stakeholder mapping in more detail.

The urgency associated with stakes is often not static but dynamic. Thus, stakes must be monitored regularly for any (major) changes. The levels of power and saliency of stakeholders may change with the passage of time. Also, coalitions of stakeholders may emerge in the course of a project and their compositions could change with prevalent circumstances (Freeman,

1984). In each project therefore, there is a need to monitor the stakeholders and their stakes and respond to their dynamism in order to avoid any negative effects. Being proactive is worthwhile in this regard. The continuous assessment of stakeholders' power and urgency informs the choice of approaches for engaging with them. As stakeholders are dynamic, the choice and use of any engagement approach or combination of approaches is always circumspect. It is thus worthwhile for stakeholders to be familiar with the array of engagement approaches, including their strengths and weaknesses, and be able to use these effectively and circumspectly. When the differing expectations of all stakeholders cannot be achieved at the same time, compromises become worthwhile (Johnson *et al.*, 2005). In this regard, an organisation may sometimes have to trade-off the needs of one stakeholder against another (Thompson, 2002).

Any matrix used for mapping stakeholders should be updated regularly to track their positions, especially regarding those stakeholders that are critical to the operations or survival of an organisation. A dynamic matrix is essentially not an answer to everything, rather it helps to (Vogwell, 2002):

- bring order to a very complex situation,
- bring collective understanding if compiled by a group,
- suggest up-to-date strategies for management and communication between the various groups,
- manage resources and time and use these where most benefit will be derived.

1.3 Stakeholder management

Stakeholder management, the subject of this book, is about relationships between an organisation and its stakeholders. These relationships impact on individuals and organisations both positively and negatively. Stakeholders need to be managed in order to minimise their negative impacts and ensure that they do not hinder the achievement of goals by individuals and organisations.

An organisation can be influenced on several dimensions and in different ways, as the checklist of stakeholders in most undertakings is often long and their differing stakes can also become a major source of conflict. It is thus worthwhile to manage stakes in most undertakings. As stakes are not static but dynamic, there is a need to manage the constantly shifting balance between the interests of stakeholders (Goodijk, 2003).

Stakeholder management dictates that an organisation should relate with many constituent groups and should engender and maintain the support of these groups by considering and balancing their relevant interests (Goodpaster, 1991; Freeman, 1994; Logsdon and Wood, 2000). Stakeholding is thus a form of social inclusion and so it diminishes barriers to the expertise that is flowing into and out of organisations and groups (Moloney, 2006).

Differing stakes can become a major source of conflict between stakeholders and hence it is worthwhile to manage stakeholders in most undertakings. Stakeholders' influences are varied (Lynch, 2006) hence the need to respond to different stakeholders in different ways. Even if all stakeholders have good intentions, and they often do, their large number in a given project warrants their management because the pursuit of their individual objectives may not necessarily be congruent. A proactive approach is needed in dealing with stakeholders as opposed to being reactive. While minimal effort is required in satisfying stakeholders with low levels of interest, greater effort is required in keeping those with high levels of interest happy (Carter, 2006).

The differing claims, rights and expectations of stakeholders can exert tangential forces in different directions. This effect must be countered by managing stakeholders collectively in accordance with the objectives of a given cause (Gibson, 2000). Firstly, each stakeholder should be managed uniquely on the basis of their disposition. That way, the missions, strengths, weaknesses, strategies and behaviour of the different stakeholders will be engaged circumspectly (Cleland, 2002) to avoid any threats they may pose to projects and corporate governance, processes and outcomes (Freeman, 1984; Logsdon and Wood, 2000). Secondly, each project-based set of stakeholders must be managed as a cohort. This activity extends beyond the construction phase of a project. Users of facilities, members of the public, etc. may exert their interests after the construction phase and so stakeholder management stretches in consonance with the life of a facility.

1.3.1   Need for construction stakeholder management

Modern construction clients tend to manifest as dynamic configurations of stakeholders who engage with a multifaceted market (Newcombe, 2003). In projects involving multifaceted clients, large project teams and many other stakeholders, there is a dire need for effective coordination and general management of the different stakes, and this warrants effective client leadership. This role of the client is underachieved (Latham, 1994; Egan, 1998, 2002; Boyd and Chinyio, 2006). Stakeholder management enhances greater competency in relational issues and minimises risks therein.

In order to achieve a successful project outcome, the project manager must be adept at managing the interests of multiple stakeholders throughout the entire project management process (Sutterfield *et al.*, 2006). Although principles can be adopted across boundaries, construction has its peculiarity, hence the need to evolve principles of construction stakeholder management based on empirical research.

1.4   Aspects of stakeholder theory

The concept of stakeholder management is accepted as theory, especially in academic discourse. Stakeholder management theory evolved from business management and aims to describe, understand, analyse and manage stakeholders. Many scholars identify the book by Freeman (1984) as a pacesetter; thus some scholars attribute the introduction of modern stakeholder theory to Freeman. Stakeholder management evolved as a tenet of 'corporate social responsibility' and is underpinned on ethical, social and economic considerations. Socially responsible organisations endeavour to employ ethical behaviour in their conduct (Moloney, 2006) and this informs their management of diverse stakes.

Stakeholder management involves managing relationships in order to motivate stakeholders to behave in ways that support the objectives of a firm. The theory posits that businesses, causes, interests and pressure groups have to manage their relationships with those external entities that can influence the achievement of their goals (Moloney, 2006). Stakeholder management is about creating the most positive environment in which to develop a project (Vogwell, 2002).

A stakeholder management approach takes many factors into account, for example moral, political, technological and economic interests (Weiss, 2006). Three approaches are useful for dealing with stakeholders (Goodpaster, 1991):

1. *Strategic approach* – This approach allots shareholders' profit a greater priority above the interests of other stakeholders.
2. *Multifiduciary approach* – This assumes a fiduciary responsibility to stakeholders, allotting them equal stakes with shareholders.
3. *Stakeholder synthesis approach* – This approach assumes a moral but non-obligatory responsibility to stakeholders, e.g. dealing with them ethically.

To an organisation, the key considerations in practical stakeholder management should include the following (Caroll and Buchholtz, 2006):

1. Who are our stakeholders?
2. What are their stakes?
3. What opportunities do they present?
4. What challenges or threats do they present?
5. What responsibilities do we have towards our stakeholders?
6. What strategies or actions should we use to engage our stakeholders?

Caroll and Buchholtz (2006) elaborated on their sixth point above. This significant and loaded point considers such aspects as:

- Should we deal directly or indirectly with our stakeholders?
- Should we be aggressive or defensive in dealing with stakeholders?
- How and when should we accommodate, negotiate, manipulate or resist the overtures of our stakeholder?
- How and when should we employ a combination of the above strategies or pursue a singular course of action?

The foregoing are decisions which each organisation has to make. Ideally, there should be an organisational approach to stakeholder management. This approach should be coordinated within the organisation and guide a company and its employees or representatives while engaging their stakeholders. Individuals may bring their personal attributes to bear in stakeholder engagement, for example some will be more dramatic in their communication, some will be more formal than others and so on. However, each character outlook in the course of corporate stakeholder management should be based on a pre-defined approach.

As organisations nurture and sustain stakeholder management, they should endeavour to specify their approach to this activity. An organisational approach could be pitched at two levels: macro and micro. At the macro level will be the broad guidelines while the micro-level will concern operational tactics. Chinyio and Akintoye (2008) discussed some practical approaches in this regard.

Freeman (1984) explicitly linked stakeholder interests to corporate strategy (Goodpaster, 1991; Freeman, 1994; Logsdon and Wood, 2000). A major purpose of stakeholder theory is to help corporate managers understand their stakeholder environments and manage them more effectively. A larger purpose is to help corporate managers improve the value of the outcomes of their actions while minimising any harm to stakeholders. In essence, stakeholder theory concerns relationships between corporations and their stakeholders (Logsdon and Wood, 2000).

Stakeholder management is useful to procurement in general and has been fostered in several disciplines, for example land remediation, forestry; business marketing; IT; electronics industry; hospitals; automotive industry and so on. Construction practice and procurement in particular has embraced stakeholder management and is now promoting its full implementation. Chapter 3 provides an evolution of construction practice and the conjoint growth of stakeholder management while Chapter 2 discusses other dimensions of stakeholder theory. Stakeholder management is now being researched in both construction and other disciplines (Kolk and Pinkse, 2006).

1.4.1   *The principles of stakeholder management*

The key principles of stakeholder management (Table 1.1) were first proposed by Max Clarkson who became legendary for his early support of the stakeholder concept.

Table 1.1   Principles of stakeholder management

Principle	Stipulation – that managers should:
No. 1	*Acknowledge* and actively *monitor* the concerns of all legitimate stakeholders, and should take their interests appropriately into account in decision-making and operations.
No. 2	*Listen* and openly *communicate* with stakeholders about their respective concerns and contributions, and about the risks that they assume because of their involvement with the corporation.
No. 3	*Adopt* processes and modes of behaviour that are sensitive to the concerns and capabilities of each stakeholder constituency.
No. 4	*Recognise the interdependence* of efforts and rewards among stakeholders, and should attempt to achieve a fair distribution of the benefits and burdens of corporate activity among them, taking into account their respective risks and vulnerabilities.
No. 5	*Work cooperatively* with other entities, both public and private, to ensure that risks and harms arising from corporate activities are minimised and, where they cannot be avoided, appropriately compensated.
No. 6	*Avoid altogether* activities that might jeopardise inalienable human rights (e.g. the right to life) or give rise to risks that, if clearly understood, would be patently unacceptable to relevant stakeholders.
No. 7	*Acknowledge the potential conflicts* between (a) their known roles as corporate stakeholders and (b) their legal and moral responsibilities for the interests of stakeholders, and should address such conflicts through open communication, appropriate reporting, incentive systems and, where necessary, third-party review.

Source: http://www.mgmt.utoronto.ca/~stake/Principles.htm

The principles emerged organically after an international group of management scholars, including Clarkson himself, explored the role of the large corporation in modern, highly interdependent economies. Their goal was to develop a broad concept of the firm as a vehicle for advancing the interests of multiple and diverse stakeholders. The principles thus incorporate a variety of perspectives and provide a template and guide to organisations for managing their stakeholders (Donaldson, 2002).

Although principles can be adopted across boundaries, construction has its peculiarity, hence the need to adopt, evolve or refine principles of construction stakeholder management based on empirical research.

1.4.2   Engaging with stakeholders

An organisation has to engage with its stakeholders during normal and difficult times. There are different avenues for engaging with stakeholders and these include consultation, dialogue, partnership and regular supply of information. These avenues can be exploited at corporate events, exhibitions and meetings. Channels of communication could also be exploited, such as uses of posters, websites, newsletters and emails. The idea is to use an approach and tactics that are effective.

1.5   The firm and corporate social responsibility

Corporate social responsibility (CSR) can be perceived as the voluntary integration of social and environmental concerns into business operations and interactions with stakeholders (Enquist *et al.*, 2006). From this perspective, stakeholding can be said to have its origins in the theory of the firm (Freeman 1984; Moloney, 2006). Although the term CSR gained earlier and more widespread use in the United States, the European business environment has long been associated with the presumption that corporations have societal obligations that go beyond their responsibilities to shareholders (Doh and Guay, 2006). An argument thus emerged from a CSR perspective that corporations have a moral obligation to their stakeholders.

It seems worthwhile that in corporate affairs, stakeholder management should be pursued and this activity should be backed by policy. That way, its implementation will be empowered directly and certain individuals will be tasked with its responsibility. It has been argued that stakeholder management should be driven from board and executive level and be auditable (Wheeler and Sillanpää, 1997). At least, one board member should have the responsibility of overseeing an organisation's stakeholder management practice.

Corporate existence and activities are conceivable in the three domains of governance, processes and outcomes (Logsdon and Wood, 2000). These three domains have a bearing on stakeholder management. Governance and processes have a direct relationship with the way a firm engages its stakeholders. The achievement of outcomes is then influenced by an organisation's stakeholders. This reinforces the need for stakeholder management in organisations.

Good corporate governance deals in part with how to manage the involvement of an organisation with its stakeholders and to balance their interests. This objective is arranged in such a way as to enable the optimum weighing of stakeholders' diverse interests (Goodijk, 2003). The top-management duties of issuing press releases, press conferences and advertising – referred to generally as reputation management activities – should in part influence the perception of a firm's reputation by its stakeholders' (Carter, 2006).

With firms also, customer-retention management is in place in order to create distinctive, long-lasting relationships with customers (Normann, 2001). This is a blending of neo-institutional and stakeholder theory (Doh and Guay, 2006). Neo-institutional theory has suggested that organisations and their strategies are substantially influenced by the broader institutional settings in which they operate, and are also shaped by the institutional legacies that reflect the culture, history and polity of the particular country or region where they are located (Doh and Guay, 2006). The prime merit of the stakeholder concept is that it points out the important relational aspects of organisations, and it functions as a useful heuristic in this sense (Ihlen and Berntzen, 2007).

Relationship marketing has not fully replaced conventional transactional marketing as the dominant paradigm but the concept is often used where appropriate (Knox and Gruar, 2007). Relations with stakeholders are also evidence of organisations and groups using cooperation as well as competition while searching for means of gaining advantage over each other (Moloney, 2006).

Is stakeholder theory sensitive to the differences between employees, contractually bound to the organisation in a servant–master relationship, and customers who are free agents in the market and often called 'sovereign'? (Moloney, 2006). Ideally yes. However, what stakeholding does is to seek to strike a balance that yields optimal benefit. It looks at that which is good and bad for everybody and identifies the optimum in the given circumstances. It is about looking at the wider picture.

Corporations are cognisant of stakeholder management and implement it. Some do it implicitly while others do it proactively and explicitly. Each organisation should implement it in a way they deem fit and employees should be made aware of the practice that is acceptable.

1.5.1   Shareholders

Shareholders constitute one set of internal stakeholders who have invested in a firm in expectation of rewards in terms of dividends, share appreciation and capital repayments

(Doyle and Stern, 2006). Shareholders have an influence on the objectives of an organisation. Likewise, other stakeholders can influence the pursuit of an organisation's objectives (Freeman, 1984). While shareholders can influence an organisation from within, most other stakeholders often influence it from without. The management drive of a company is in someway sandwiched by its shareholders and stakeholders. Both shareholders and stakeholders place demands on the management of a company and it is worthwhile to address both sets of demands.

An organisation needs to satisfy its shareholders as well as its other stakeholders. Enquist *et al.* (2006) described this as a balance between shareholder strategy and the social-harmony strategy. The former focuses on satisfying the desires of shareholders (e.g. profitability and return on capital), while the latter emphasises the balancing of the various stakeholders' demands. The ideal is to strike a balance where the objectives of a business are achieved while satisfying stakeholders, that is, a win–win approach (Carroll and Buchholtz, 2006).

1.5.2   Acceptability of stakeholder management

It should be noted that not all researchers agree on the importance of stakeholders, and that stakeholder theory has been criticised on both theoretical and empirical grounds (Doh and Guay, 2006). However, the need for customer retention has made companies to esteem and implement stakeholder management either formally or informally, or on both counts. Employees in construction and other disciplines will thus find themselves having to manage stakeholders in their endeavours. A useful approach to this activity is to be prepared. This book brings a comprehensive understanding to the subject matter and should provide a useful guide for managing stakeholders.

It is reckoned that over 95% of organisations in construction practice are either small- or medium-sized enterprises. To some of these firms, maintaining or loosing a customer can be very critical to their continued existence. Stakeholder management will enable organisations to understand their stakeholders better, manage them properly and enhance repeat business opportunities. Those who can manage their stakeholders better will reap the rewards while those who cannot may live to bear the pains. There are gains and pains in stakeholder management and it is worthwhile to aim for the gains. Chapter 15 explores this aspect further.

1.6   Summary

This chapter has defined stakes and stakeholders. The power, urgency and saliency of stakeholders have been highlighted. The concept and theory of stakeholder management have been introduced as well. In doing so the usefulness of stakeholder management to organisations (including those in the construction sector) has been demonstrated. Stakeholder management was discussed as a corporate function that should be supported by top-level management. Employees should be empowered to carry it out, and clear lines of responsibility for its effective implementation should be drawn.

The scene is now set to discuss stakeholder management in more detail. The following chapters discuss various aspects of the concept and its theory.

References

Boyd, D. and Chinyio, E. (2006) *Understanding the Construction Client*. London: Blackwell Science.
Calvert, S. (1995) *Managing Stakeholders: The Commercial Project Manager*. New York: McGraw-Hill.

Carroll, A.B. and Buchholtz, A.K. (2006) *Business & Society: Ethics and Stakeholder Management* (6th edn). Mason: Thomson South-Western.

Carter, S.M. (2006) The interaction of top management group, stakeholder, and situational factors on certain corporate reputation management activities. *Journal of Management Studies*, **43**(5):1145–1176.

Chinyio, E. and Akintoye, A. (2008) Practical approaches for engaging stakeholders: Findings from the UK. *Construction Management and Economics*, **26**(6):591–599.

Clarkson, M.B.E. (1995) A stakeholder framework for analyzing and evaluating corporate social performance. *Academy of Management Review*, **20**(1):92–117.

Cleland, D.I. (2002) *Project Management: Strategic Design and Implementation* (4th edn). London: McGraw-Hill.

Doh, J.P. and Guay, T.R. (2006) Corporate social responsibility, public policy, and NGO activism in Europe and the United States: An institutional-stakeholder perspective. *Journal of Management Studies*, **43**(1):47–73.

Donaldson, T. (2002) The stakeholder revolution and the Clarkson principles. *Business Ethics Quarterly*, **12**(2):107–111.

Doyle, P. and Stern, P. (2006) *Marketing Management and Strategy* (4th edn). Harlow-Essex: Pearson Education Limited.

Egan, J. (1998) *Rethinking Construction*: The report of the Construction Task Force to the Deputy Prime Minister, John Prescott, on the scope for improving the quality and efficiency of UK construction. London: Department of the Environment, Transport and the Regions.

Egan, J. (2002) *Accelerating Change*: A report by the Strategic Forum for Construction, chaired by Sir John Egan. London: Rethinking Construction.

Enquist, B., Johnson, M. and Skale'n, P. (2006) Adoption of corporate social responsibility – Incorporating a stakeholder perspective. *Qualitative Research in Accounting and Management*, **3**(3):188–207.

Freeman, R.E. (1984) *Strategic Management: A Stakeholder Approach*. Boston: Pittman Publishing Company.

Freeman, R.E. (1994) A stakeholder theory of the modern corporation. In: Beauchamp, T.L. and Bowie, N.E. (eds) *Ethical Theory and Business*. Englewood Cliffs, NJ: Prentice-Hall, pp. 66–76.

Frooman, J. (1999) Stakeholder influence strategies. *Academy of Management Review*, **24**(2):191–205.

Gago, R.F. and Antolìn, M.N. (2004) Stakeholder salience in corporate environmental strategy. *Corporate Governance*, **4**(3):65–76.

Gibson, K. (2000) The moral basis of stakeholder theory. *Journal of Business Ethics*, **26**(3):245–257.

Goodijk, R. (2003) Partnership at corporate level: The meaning of the stakeholder model. *Journal of Change Management*, **3**(3):225–241.

Goodpaster, K.E. (1991) Business ethics and stakeholder analysis. *Business Ethics Quarterly*, **1**(1):53–73.

http://www.mgmt.utoronto.ca/~stake/Principles.htm (2008) The Clarkson Principles of Stakeholder Management – Based on the 'Clarkson Centre for Business Ethics and Board Effectiveness' (Accessed on 20 November 2008).

Ihlen, Ø and Berntzen, Ø (2007) When lobbying backfires: Balancing lobby efforts with insights from stakeholder theory. *Journal of Communication Management*, **11**(3):235–246.

Johnson, G., Scholes, K. and Whittington, R. (2005) *Exploring Corporate Strategy: Texts and Cases* (7th edn). Harlow: Financial Times Prentice-Hall.

Knox, S. and Gruar, C. (2007) The application of stakeholder theory to relationship marketing strategy development in a non-profit organization. *Journal of Business Ethics*, **75**(2):115–135.

Kolk, A. and Pinkse, J. (2006) Stakeholder mismanagement and corporate social responsibility crises. *European Management Journal*, **24**(1):59–72.

Latham, M. (1994) *Constructing the Team: Final Report of the Government/Industry Review of Procurement and Contractual Arrangements in the UK Construction Industry*. London: HMSO.

Logsdon, J.M. and Wood, D.J. (2000) Introduction. In: Logsdon, J.M., Wood, D.J. and Benson, L.E. (eds) *Research in Stakeholder Theory, 1997–1998: The Sloan Foundation Minigrant Project*. Toronto: Clarkson Centre for Business Ethics, pp. 1–4.

Lynch, R. (2006) *Corporate Strategy* (4th edn). Harlow: Pearson Education Limited.

Mintzberg, H. (1995) Beyond configuration: forces and forms in effective organisations. In: Mintzberg, H., Quinn, J.B. and Ghoshal, S. (eds) *The Strategy Process*, (European edn, 3rd revision). London: Prentice Hall.

Mitchell, R.K., Agle, B.R. and Wood, D.J. (1997) Toward a theory of stakeholder identification and salience: Defining the principle of who and what really counts. *Academy of Management Review*, **22**(4):853–886.

Moloney, K. (2006) *Rethinking Public Relations: PR Propaganda and Democracy* (2nd edn). London: Routledge.

Newcombe, R. (2003) From client to project stakeholders: A stakeholder mapping approach. *Construction Management and Economics*, **21**(8):841–848.

Normann, R. (2001) *Reframing Business – When the Map Changes the Landscape*. New York: Wiley.

Orndoff, C.J.W. (2005) Promising new tool for stakeholder interaction. *Journal of Architectural Engineering*, **11**(4):139–146.

Pajunen, K. (2006) Stakeholder influences in organizational survival. *Journal of Management Studies*, **43**(6):1261–1288.

Smith, J. and Love, P.E.D. (2004) Stakeholder management during project inception: Strategic needs analysis. *Journal of Architectural Engineering*, **10**(1):22–33.

Sutterfield, J.S., Friday-Stroud, S.S. and Shivers-Blackwell, S.L. (2006) A case study of project and stakeholder management failures: Lessons learned. *Project Management Journal*, **37**(5):26–35.

Thompson, J.L. (2002) *Strategic Management* (4th edn). London: Thomson.

Vogwell, D. (2002) Stakeholder Management. *Davis Langdon Everest*, http://www.davislangdon.com/pdf/EME/Publications/RM_VM/Stakeholder%20Management.pdf#search=%22stakeholder%20management%20vogwell%22 (Accessed on 20 June 2006).

Weiss, J.W. (2006) *Business Ethics – A Stakeholder and Issues Management Approach* (4th edn). Mason: Thomson Higher Education.

Wheeler, D. and Sillanpää, M. (1997) *The Stakeholder Corporation: A Blueprint for Maximizing Stakeholder Value*. London: Pitman Publishing.

Winch, G. and Bonke, S. (2002) Project stakeholder mapping: Analysing the interests of project stakeholders. In: Project Management Institute (PMI) *The Frontiers of Project Management Research*. Pennsylvania: PMI (Chapter 23), pp. 385–403.

2 Stakeholder Management: Theoretical Perspectives and Implications

Kenneth Amaeshi

2.1 Introduction

There is a growing interest on the institutional embeddedness and variations of corporate social responsibility (CSR) (Matten and Moon, 2008), accountability (Gray, 2002; Lounsbury, 2008) and corporate governance (Jackson, 2005; Aguilera *et al.*, 2006) practices. This interest is mainly driven by the understanding that organisational practices are not only determined by managerial rationality, but are also constrained and enabled by their institutional configurations and social conditions (DiMaggio and Powell, 1983; Granovetter, 1985; Hollingsworth and Boyer, 1997; Whitley, 1998; Hall and Soskice, 2001; Amable, 2003; Crouch, 2005). Despite the increasing attention paid to the institutional embeddedness of corporate behaviour and performance in business and management literature, the application of institutional theory to account for corporate stakeholder practices is rather scarce in the extant CSR literature.

Hitherto, the literature has presented decision-making on corporate stakeholding practices (CSPs) as something solely internal to the firm and under managerial perception and bounded rationality (Mitchell *et al.*, 1997; Agle *et al.*, 1999). In order to provide a complementary view to the notion of managerial capture of CSP (Owen *et al.*, 2000; Swift, 2001), we accept the logic that CSP could be shaped differently not only by managers, but also by the institutional configurations in which they are enacted. In other words, the chapter argues that CSPs are not only expressions of managerial choice and rationality, as often presented in the literature, but are also products of their institutional contexts. This view does not claim any superiority to the managerial view but rather complements it. It suggests that equal attention should be paid to both institutional contexts and managerial discretions in CSP discourses. This is in recognition of the fact that 'Firms are not simple "institution-takers"; firm strategies interact with the institutional framework, which can lead to institutional reconfigurations, especially in the process of adjustment' (Borsch, 2004:370).

The discussion here is largely exploratory and does not present or adopt any normative stance (or 'best practice' approach) towards corporate stakeholding, governance and accountability. These are rather examined as neutral business practices (Amaeshi and Adi, 2007). It is the position of this study that CSP, as a firm behaviour, is implicated in dynamic interactions between managerial discretion and rationality, on one hand, and institutional contexts, on the other (Greening and Gray, 1994; Jones, 1999; Sharma, 2000; Matten and Moon, 2008). This chapter will briefly cover the history of the stakeholder perspective to business and organisational studies, discuss the different underpinning philosophies of this perspective and how they relate to the CSR movement, corporate governance and accountability. It will also highlight the significance of examining the roles and influences of stakeholders, their interactions with managers as well as the institutional configurations in which they act, and are acted upon, in order to further explain variations in behaviour of firms in different institutional contexts.

2.2   Corporate stakeholding practices

2.2.1   Brief history of stakeholder salience in management literature

The stakeholder perspective to organising and managing firms is one of the major management paradigm shifts in the late last century. The theory, in its present form, is traceable to Freeman (1984:246) who defined stakeholders as '. . . those groups and individuals who can affect, or are affected by the achievement of an organization's purpose'. More recently, Freeman provides a more instrumental definition of stakeholders as '. . . those groups who are vital to the survival and success of the corporation' (Freeman, 2004:58). The use of the term stakeholder in an economic sense could be traced to the works of the Stanford Research Institute (now SRI) in the 1960s (Freeman, 1984; Slinger, 1999; Freeman and McVea, 2001). It was then used as a metaphor to encourage an inclusive approach – that takes wider perspectives – to adapting to the turbulent business environment. From the start, the stakeholder approach grew out of management practice (Freeman and McVea, 2001:190) to such areas as corporate planning, systems theory, organisational theory and CSR. It took another 20 years from the 1960s for the stakeholder approach to crystallise. It was in pulling together of these perspectives that Edward Freeman in the early 1980s articulated the stakeholder approach as a framework for strategic management (Freeman and McVea, 2001) in his classic – *Strategic Management: A Stakeholder Approach*. Since then, the concept has been embedded in management scholarship and in managerial practice (Mitchell *et al.*, 1997).

2.2.2   Definitions of stakeholding

Despite its contemporary popularity and proliferation since Freeman (1984), the term 'stakeholder' has joined the league of words (e.g. the word 'strategy') that are overused and yet not clearly understood, or at best are still contested (Friedman and Miles, 2006). This is more so, especially as the word easily lends itself to multiple applications in support of what could be considered as 'fashionable constructs' of the time. For example, it has featured in such combinations as stakeholder society (Ackerman and Alstott, 1999), stakeholder capitalism (Kelly *et al.*, 1997; Jones, 1999), stakeholder corporation (Kay, 1997) and so on, which renders any attempts to provide a succinct definition for stakeholding cumbersome. Nonetheless, many authors have gone round to define stakeholders by the nature of 'stakes' they hold. Mitchell *et al.* (1997:858) provided a list of such definitions.

2.2.3   Broad and narrow perspectives to stakeholding

The definitions and use of stakeholding in management literature range from broad, inclusive definitions to narrow views of the firm's stakeholder environment. The narrow view of stakeholding refers specifically to those stakeholders that have vested 'stakes' in the firm. Carroll (1993:22) defines them as: 'individuals or groups with which business interacts who have a "stake", or vested interest, in the firm'. These could be employees, shareholders, management, government, society, etc. as long as they have explicit stakes and or vested interests in the firm. The broad view of stakeholding goes beyond those stakeholders that have explicit stakes and vested interests in the firm to extend to those stakeholders that could affect and/or affected by the activities of the firm (Starik, 1994, 1995). Starik (1994:92), for instance, suggests that these stakeholders, from a broad view of the concept, could be 'any naturally occurring entity which affects or is affected by organizational performance'. In essence these stakeholders will include living entities such as plants and animals as well as non-living beings such as the natural environment and ecology. It could also be stretched to include such groups as 'unborn generations' (Freeman and Miles, 2006), as often articulated in sustainable development discourse. The 'stakes' and vested interests associated with stakeholding are further categorised into primary and secondary stakeholding. Primary stakeholding involves fiduciary obligations from the firm while secondary stakeholding does not involve such obligations. In this regard, examples of primary stakeholders will include shareholders, employees and managers, while those of secondary stakeholders might be local communities, environmental groups, suppliers, etc. This notion of fiduciary and non-fiduciary stakeholding underpins most of the existing corporate governance frameworks and typologies (Slinger, 1999; Aguilera and Jackson, 2003; Aguilera, 2005). Commenting on this from a comparative governance perspective, Aguilera and Jackson (2003:454) write:

> The corporate governance literature largely neglects employees. . . . This omission partly reflects weak employee participation in the United States relative to that in economies such as Germany or Japan, where labor participation is politically important and often a source of competitive advantage. . . . In addition, a major assumption of agency theorists is that shareholders are the only bearers of ex post residual risk, and, thus, employee interests are treated only as an exogenous parameter.

In addition, some inclusive definitions are driven by the understanding that stakeholders have intrinsic value and managers have a moral duty to be responsible to a variety of stakeholders (Evan and Freeman, 1988; Freeman and Evan, 1990; Donaldson and Preston, 1995). More narrow strategic or instrumental perspectives (Donaldson and Preston, 1995), on the other hand, define stakeholders as those groups or individuals that are in a mutually dependent, risk-based or exchange relationship with a firm (Clarkson, 1995; Mitchell *et al.*, 1997). The oscillation between the narrow and broad interpretations often leads to multiple views and practices of stakeholding in management literature.

2.2.4 *Varieties of stakeholding paradigms*

Donaldson and Preston's (1995) major contribution to the stakeholder theory in management studies is identifying that the stakeholder concept is largely theorised in the literature from three main perspectives: (a) descriptive (b) instrumental and (c) normative.

Descriptive perspective

The descriptive aspect of the stakeholder theory, as the name suggests, merely describes what the corporation is – '. . . a constellation of co-operative and competitive interests possessing intrinsic value' (Donaldson and Preston, 1995:66) and who they consider as possible stakeholders. It also highlights the interactions between firms and their stakeholders with the aim of contributing to knowledge, theory and practice. Its justifications are to show that theory corresponds to observed reality. It is neither judgmental nor prescriptive. However, it is difficult to claim that it is value neutral, as research and researchers are often and even inadvertently value laden and value driven (Darke *et al.*, 1998; Appadurai, 1999; Hardy *et al.*, 2001; Ritchie and Lewis, 2003; Johnson and Onwuegbuzie, 2004).

Instrumental perspective

This perspective theorises stakeholding by examining the consequences of corporate stakeholding. It is underpinned by the paradigm that meeting stakeholder needs could be driven by instrumentalist goals and objectives (Jones *et al.*, 2007). According to Jones *et al.*, '. . . instrumentalist firms place preeminent value on the pursuit of corporate self-interest with *guile*.

Other terms used to convey this orientation are *enlightened self-interest, pragmatic morality*, and *strategic morality'* (emphasis, theirs, p. 000). The instrumentalist form of stakeholder relationship does not necessarily give voice to stakeholders and is often characterised by a one-way communication and unequal balance of power (Crane and Livesey, 2003). A more critical view of it suggests that it is not genuine; it is selfish and firms involve in it because ' . . . it makes good business sense . . . (and) . . . helps companies to mitigate risk, protect corporate brand, and gain competitive advantage . . .' (Deloitte Touche Tohmatsu, 2002:2 cited in Brown and Fraser, 2006). From a sort of neutral stance, it may be better aligned to the language of contemporary capitalism than most other perspectives (Amaeshi and Adi, 2007).

Normative perspective

Stakeholder theory can, to a large extent, be argued to be fundamentally and originally rooted in norms and mores. The normative perspective to stakeholding is largely prescriptive of 'who' ought to be considered as stakeholders and what is 'right' or 'wrong' to do in relation to stakeholders. It draws its legitimacy from its inclination towards some moral standards. According to Donaldson and Preston (1995:72), '. . . a normative theory attempts to interpret the function of, and offer guidance about, the investor-owned corporation on the basis of some underlying moral or philosophical principles'. Morality in this case might be seen as not merely a matter of rules, but also of principles – general standards for evaluating conduct, standards that we apply to all behaviours and rules. According to Freeman and Miles (2006:36), normative can refer to:

- The norms or standard practices of society as it exists.
- The way one would live in an ideal 'good' society.
- What we ought to do, either in order to achieve a good society or unconnected with any notion of the 'good'.

At the individual level these principles include that of utility, also, known as the principle of the greatest happiness. It tells us to produce the greatest balance of happiness over unhappiness, making sure that we give equal consideration to the happiness and unhappiness of everyone who stands to be affected by our actions. Morality also includes the principle of fairness founded on the golden rule that states, 'Do unto others as we should have them do unto us' and which is, basically founded on the need to respect the other person. In addition, there is the ethics of care and virtue ethics views that are gaining popularity today (Gilligan, 1977, 1982). These principles of morality at the individual level are also thought to be applicable to institutional and social morality (Olen and Barry, 1992). In relation to firms and stakeholders, therefore, it could be argued that firms with high moral standards will undertake genuine stakeholder engagement, which is thought to be characterised by genuine intentions, dialogue, engagement, trust and fairness (Phillips, 1997; Swift, 2001). In other words, '. . . moralist firms have a genuine concern for stakeholder interests, making legitimacy the *primary* driver of salience for their managers' (emphasis, theirs, Jones *et al.*, 2007:152).

Firms get involved in it, because they know that doing so is good in itself. Donaldson and Preston (1995:67) argue that the fundamental basis of the stakeholder theory is normative and therefore involves acceptance of the following ideas:

a. Stakeholders are persons or groups with legitimate interests in procedural and/or substantive aspects of corporate activity. Stakeholders are identified by *their* interests in the corporation whether the corporation has any corresponding functional interest in *them* (emphasis in original).

b. The interests of all stakeholders are of *intrinsic value*. That is, each group of stakeholders merits consideration for its own sake and not merely because of its ability to further the interests of some other group, such as the shareowners (emphasis in original).

Leveraging these ideas, Phillips (2003) makes a distinction between normative and derivative stakeholder legitimacy arguing that normative stakeholders are those stakeholders to whom the organisation has a moral obligation, and the answer to the seminal stakeholder question 'for whose benefit . . . should the firm be managed?' (Freeman, 1984); while '. . . [d]erivative stakeholders are those groups whose actions and claims must be accounted for by managers due to their potential effects upon the organization and its normative stakeholders' (Phillips, 2003).

2.2.5   Nesting of paradigms

While these stakeholding paradigms have been presented as if they are independent of each other, Donaldson and Preston (1995:66) conclude that the three approaches to stakeholder theory – i.e. descriptive, instrumental and normative – '. . . are mutually supportive and that the *normative* base serves as the critical underpinning for the theory in all its forms' (emphasis in original). They argue that the different aspects of the stakeholder theory – i.e. normative, instrumental and descriptive – are rather nested. They explain the nested nature of these aspects diagrammatically (see Figure 2.1) and in the following words:

> . . . the external shell of the theory is its descriptive aspect; the theory presents and explains relationships that are observed in the external world. The theory's descriptive accuracy is supported, at the second level, by its instrumental and predictive value; *if* certain practices are carried out, *then* certain results will be obtained. The central core of the theory is, however, normative. The descriptive accuracy of the theory presumes that managers and other agents act as *if* all stakeholders' interests have intrinsic value. In turn, recognition of these ultimate moral values and obligations gives stakeholder management its fundamental normative base (p. 74, emphasis in original),

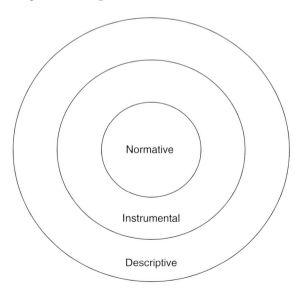

Figure 2.1   Three aspects stakeholder theory – schematic nesting of paradigms.
Source: Donaldson and Preston, 1995:74

2.3   Micro and macro level theorisations of corporate stakeholding behaviour

The stakeholder concept lies at the heart of contemporary interests in corporate accountability, governance and social responsibility. Stretching the moralist bent of the argument further, some authors have argued that the stakeholder perspective of CSR ought to extend to the concept of accountability. However, exposition of the stakeholder theory in the literature has mainly been at the micro level – i.e. managerialist (Freeman, 1984; Mitchell *et al.*, 1997; Agle *et al.*, 1999) and organisational level perspectives (Jawahar and McLaughlin, 2001; Jones *et al.*, 2007). Donaldson and Preston's conception of stakeholder theory, for example, is heavily managerialistic and they maintain that managers have an essential role in the identification of stakeholders (Donaldson and Preston, 1995: 86). The literature on managerialist view emphasises the centrality of managers in stakeholder-related decisions, while the organisational level theorists place emphasis on such constructs as stakeholder culture (Jones *et al.*, 2007) and organisational life cycle (Jawahar and McLaughlin, 2001) as drivers of corporate stakeholder salience. While the stakeholder theory is evolving in management literature, there is parallel stream in the broader domain of social sciences – e.g. political economy, politics, international relations and economic sociology that studies firm behaviour more from the meso and macro levels. An example of such streams of studies will include the national business system literature (Whitley, 1998), Varieties of Capitalism (Hall and Soskice, 2001; Crouch, 2005, 2006) and the national system of innovation (Lundvall, 1988) to mention but a few. These studies could be grouped together under the comparative business system umbrella – i.e. '. . . that institutionalist tradition of research that emphasizes that the way in which economic activities are coordinated and controlled (between holders of capital and managers, between managers and employees) is crucially affected by national institutional contexts. . . . These contexts (of the financial system, the political system, the educational and training system and the cultural system) set the rules of the game embedded in specific historically emergent social practices such as how capital is made available to entrepreneurs and firms, the types of skills and knowledge possessed by managers and workers, and the mechanisms of coordination and control utilized by managers' (Morgan, 2001:114). These studies tend to categorise firms in relation to their institutional contexts; and one thing they all share in common is an interest in the influences institutional contexts exert on firms and industries either in relation to stability (i.e. institutional isomorphism) or dynamism (i.e. institutional change). Unfortunately, both streams of literature have continued to run in parallel with the chances of converging ever diverging.

   This section first provides a further review of the stakeholder theory in management literature, with emphasis on the implications of its managerial and organisational level perspectives. The implications of these micro level views are reviewed in relation to the growing interest in broad CSR – and particularly on stakeholder accountability – which is currently dominated by managerialist views and in dire need to transcend this perspective (Gray, 2002; Lounsbury, 2007). The section then examines this interest in stakeholder accountability through the developments in comparative corporate governance studies (Aguilera and Jackson, 2003; Aguilera, 2005). The interest in comparative corporate governance studies is particularly helpful because these studies have been successful in going beyond manager-centric views to incorporate the role of meso and macro level variables (i.e. industry and institutional levels, respectively) on corporate governance and accountability. Given that the stakeholder theory is a precursor to both CSR and stakeholder accountability, studies in comparative corporate governance will provide a smooth connection to introduce meso and macro perspectives to stakeholder theory in management.

2.4   Micro level theorisation of corporate stakeholder salience in management literature

Micro level theorisation of corporate stakeholder salience in this case refers to those literatures that have examined stakeholder salience either from the managerialist view or from an organisational (corporate) level perspective. The managerialist view of stakeholding suggests that the firm is a nexus of contracts between stakeholders and managers, who are at a central node, where it is the responsibility of managers to reconcile divergent interests by making strategic decisions and allocating strategic resources in a manner that is most consistent with the claims of the other stakeholder groups (Hill and Jones, 1992:134). Therefore, 'the stakeholders that receive priority from management will be those whom managers – especially CEOs – perceive as highly salient' (Agle *et al.*, 1999:510). The organisational or corporate aspect of the micro level theorisation of stakeholder salience on the other hand, places emphasis on the role of firms (as opposed to managers) in shaping stakeholder-related decisions. However, in the same approach as the managerialist view, it places the firm at the centre of stakeholding, from which the firm exercises power and maintains legitimacy.

2.4.1   Corporate stakeholder salience: a managerialist view

It could be argued that contemporary interest in stakeholder theory in management research and practice has been overly managerialist in focus. And the same could be said of it right from its earlier conceptualisation (Freeman, 1984). The managerialist view is driven by its emphasis and dependency on the centrality of manager's perceptions in stakeholder-related decisions (as shown in the hub and spoke schematic Figure 2.2). According to this school

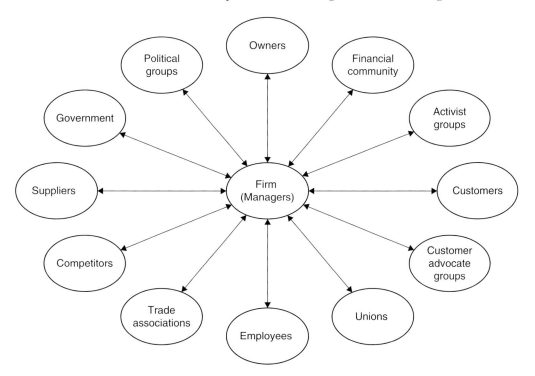

Figure 2.2   Stakeholder wheel (*Source:* Adapted from Freeman, 1984:55), managers added to emphasise the managerialist view (*Source:* Hill and Jones, 1992)

of thought, stakeholder salience is the degree to which managers give priority to competing stakeholder claims (Mitchell *et al.*, 1997; Agle *et al.*, 1999).

Mitchell *et al.* (1997) and Agle *et al.* (1999) are amongst the key contributors in espousing this managerialist view of stakeholder salience. The stakeholder theory of management, according to Mitchell *et al.*, involves identifying and prioritising stakeholder issues based on managerial perceptions of stakeholder salience. Mitchell *et al.* identified these stakeholder salience variables as: power, legitimacy and urgency. It could be said that a stakeholder has power when it can impose its will on the firm. Legitimacy implies that stakeholder demands comply with prevailing norms and beliefs. In other words, power accrues to those who control resources needed by the firm (Pfeffer, 1981) and legitimacy is achieved if patterns of organisational practice are in congruence with the wider social system (Scott and Meyer, 1983; Scott 1987, 1995; Powell and DiMaggio, 1991). However, power and legitimacy can appear together, giving authority to those who have both, but they can also appear independently. Finally, urgency is a concept sustained on two elements: (1) the importance stakeholders accord their own demands; and (2) their sensitivity to how long it takes managers to deal with their demands (Gago and Antolin, 2004). These salient variables according to Mitchell *et al.* will determine how managers respond to stakeholders.

Drawing from social cognition theory (Fiske and Taylor, 1984), Agle *et al.* (1999:509) explain that '. . . as the stakeholder attributes of power, legitimacy, and urgency cumulate in the mind of a manager, selectivity is enhanced, intensity is increased, and higher salience of the stakeholder group is the likely result'. Agle *et al.* (1999) also tested Mitchell *et al.*'s (1997) theoretical model of stakeholder salience and confirm this model. They found that in the minds of CEOs, 'the stakeholder attributes of power, legitimacy, and urgency are individually . . . and cumulatively . . . related to stakeholder salience across all groups; (which) . . . suggests that these stakeholder attributes (of power, legitimacy and urgency) do affect the degree to which top managers give priority to competing stakeholders' (p. 521).

While '. . . the stakeholder approach to management can be considered a knowledge structure that determines how a manager selectively perceives, evaluates, and interprets attributes of the environment' (Wolfe and Putler, 2002:65), some have criticised the stakeholder theory of management on the grounds that it provides unscrupulous managers with a ready excuse to act in their own self-interest thus resurrecting the agency problem that the shareholder wealth maximisation imperative was designed to overcome (Phillips *et al.*, 2003). Opportunistic managers can more easily act in their own self-interest by claiming that the action actually benefits some stakeholder group or other (Jensen, 2000; Marcoux, 2000). In this regard, Marcoux (2000:97) wrote: 'All but the most egregious self-serving managerial behavior will doubtless serve the interests of some stakeholder constituencies and work against the interests of others'. In the same trend, Sternberg (2000:51f) argues that stakeholder theory, 'effectively destroys business accountability . . . because a business that is accountable to all, is actually accountable to none'.

In response to this criticism of opportunistic self-interest on the part of managers, Phillips *et al.* (2003) argue that no small measure of managerial opportunism has occurred in the name of shareholder wealth maximisation, as well. While this sounds like a *tu quoque* (and you too!) fallacy, Phillips *et al.* simply describe this criticism as a version of the evil genie argument – '. . . one that is no more (or less) problematic for any one theory or idea than only of the extant alternatives' (p. 482). Continuing, they argue that although managerial opportunism is a problem, it is no more a problem for stakeholder theory than the alternatives. On the criticism of multiple master service (i.e. accountability to all), Phillips *et al.*, citing examples from Hill and Jones (1992) stakeholder-agency theory, argue that managers' interest in organisational growth runs contrary not only to the interests of stockholders, but also contrary to the interests of stakeholders. As such, the claims of different groups may conflict, however, on a more general level; each group can be seen as having a stake in the continued existence of the firm (Hill and Jones, 1992:145). Stakeholder theory, therefore, does not advocate the service of two masters. Rather, '. . . managers serve the interest of one master: the organisation' (Phillip *et al.*, 2003:484).

However, Phillip *et al.*'s response does not take away from the fact that dominant corporate stakeholder salience theorisation is largely managerialist in approach. Although both Mitchell *et al.* (1997) and Agle *et al.* (1999) linked stakeholder salience to legitimacy, which is an attribute of the wider social system (Scott 1987; Powell and DiMaggio, 1991), they seem to promote managerial perception in such decisions and under-emphasise this institutional dimension of stakeholder salience. As a result, they fail to account for how the wider social system enables and/or constrains corporate stakeholder salience decisions. In summary then, it could be said that the managerialist view of corporate stakeholder salience championed by such dominant views as those of Mitchell *et al.* (1997) and Agle *et al.* (1999) exhibit the following characteristics in common:

1. *Managers are placed at the centre of stakeholding:* The starting point of the stakeholder theory was around managers: '. . . the impetus behind stakeholder management was to try and build a framework that was responsive to the concerns of mangers who were being buffeted by unprecedented levels of environmental turbulence and change. Traditional strategy frameworks were neither helping managers to develop new strategic directions nor were they helping them to understand how to create new opportunities in the midst of so much change' (Freeman and McVea, 2005:189).
2. *Managers are framed and positioned as autonomous independent actors:* 'A stakeholder approach emphasizes *active* management of the business environment, relationships and the promotion of shared interests' (Freeman and McVea, 2005:192 – emphasis in original).
3. *Managerial perceptions are emphasised more than institutional influences:* The influence of wider social system on stakeholder salience (i.e. the institutional context in which stakeholder salience is embedded and enacted) is under-emphasised. On the contrary, stakeholders and stakeholder salience are theorised as subject to managerial perceptions, constructions and choices.

2.4.2   Corporate stakeholder salience: an organisational level view

Firm level theorisation of stakeholder salience goes beyond the managerialist perspective to emphasise the role of organisational context on stakeholder-related decisions. In this section, we review two major contributions to this perspective. One is Jawahar and McLaughlin (2001) who proposed that decisions on stakeholder salience are influenced by where a firm is on its organisational life cycle and the other is Jones *et al.* (2007) who argue that organisational stakeholder culture influences stakeholder salience decisions.

Organisational life cycle (Jawahar and McLaughlin, 2001)

In opposition to the managerialist perspective of stakeholder salience and drawing from resource dependence theory (Pfeffer and Salancik, 1978), Jawahar and McLaughlin (2001:401) argue that '. . . managers do not have unbridled strategic choice . . . but must make strategic choices within constraints'. One of these constraints include where the organisation is in its life-cycle development – i.e. where it is in one of the four overlapping phases comprising of start-up, emerging growth, maturity and revival (p. 404). The organisation strives to survive and as such is very likely to naturally gravitate towards those stakeholders that provide essential resources to its survival and sustenance.

. . . organizations in start-up or decline/revival stages are likely to favor certain stakeholders . . ., depending on the extent to which they are dependent on those stakeholders for
resources critical to organizational survival.

Organizations are unlikely to fulfil all the responsibilities they have toward each primary
stakeholder group. Instead, they are likely to fulfil economic and all noneconomic responsibilities of some primary stakeholders but not others and, over time, to fulfil responsibilities relative to each stakeholder to varying extents. This variation is how organizations
deal with different stakeholders, simultaneously and over time (i.e., across life cycle stages)
. . . (p. 397).

This dependency on specific resourceful stakeholders is the source of power over the firm
on the part of the stakeholders (Mitchell *et al.*, 1997). Continuing, Jawahar and McLaughlin
(2001:405) in their study confirm that:

1. At any given organisational life-cycle stage, certain stakeholders, because of their potential to satisfy critical organisational needs, will be more important than others.
2. Specific stakeholders are likely to become more or less important as an organisation
   evolves from one stage to the next.
3. The strategy an organisation uses to deal with each stakeholder will depend on the
   importance of that stakeholder to the organisation relative to other stakeholders.

The dependency discussed by Jawahar and McLaughlin (2001) is summarised in Table 2.1
and explained further below. Four organisational life phases are used for this elaboration:
start-up, emerging growth, mature and decline/transition stages.

Start-up phase: In this phase, the organisation is desperate to survive and as such requires
access to finance and market. Given these required resources, the organisation is likely to
be inclined to such stakeholder groups as shareholders and creditors for finance and to
customers for market share. Jawahar and McLaughlin argue that the other stakeholder

Table 2.1   Stakeholder salience and organisational life cycle

Phases	Pressing needs	Important stakeholders
Start-up	Access to finance, market share	Shareholders, creditors, customers
Emerging growth	Need to build a quality workforce and products and to obtain resources to accommodate rapid growth and expansion (p. 408)	Suppliers, employees
Mature stage	Often characterised by 'tempered overconfidence' of success and attended by strong cash flows, without particularly attractive investment opportunities (p. 408)	Likely to deal with all primary stakeholders in a proactive manner
Decline/transition stage	Dwindling patronage, loss of market share and or efforts to build a new market or rebuild market share	Main stakeholder focus will be customers and creditors. Unless government, community, trade associations, etc. are essential for survival, the organisation is very likely to adopt defensive strategies towards these latter groups.

groups (e.g. government, employees and suppliers) would only be considered if they were thought to be critical to the survival of the organisation at this stage.

Emerging growth stage: This stage is mainly characterised by the need to build quality brand, workforce and products and to obtain resources to accommodate rapid growth in expansion (Jawahar and McLaughlin, 2001:408). Employees and suppliers are considered very important at this stage because they provide the quality of workforce and material inputs to production development, respectively, needed to sustain the organisation.

Mature stage: This stage is often characterised by 'tempered overconfidence' of success and attended by strong cash flows, without particularly attractive investment opportunities (p. 408). The organisation is likely to deal with all primary stakeholders in a proactive manner at this stage. Jawahar and McLaughlin (2001), in this case, borrow Clarkson's articulation of primary stakeholders as groups typically comprised of shareholders and investors, employees, customers and suppliers, together with what is defined as the public stakeholder group: the government and communities that provide infrastructures and markers, whose laws and regulations must be obeyed, and to whom taxes and other obligations may be due (Clarkson, 1995:106).

Decline/transition stage: At this stage, the organisation is likely to experience dwindling patronage, loss of market share and/or make efforts to build new markets or rebuild existing market share. Main stakeholder focus will be customers and creditors. Unless government, community, trade associations, etc. are essential for survival, the organisation is very likely to adopt defensive strategies towards these latter groups.

One of the limitations of this framework, amongst others, is that it does not explicitly address differences in stakeholder salience arising from industry of the organisation. For instance, most firms in such sectors as the chemical and/or oil/gas might be constraint by government policies or environmental pressures to take on environmental and community issues earlier in their life cycle (for details on industry driven differences in corporate stakeholding, see Baucus and Near, 1991; Beliveau *et al.*, 1994; Greening and Gray, 1994; Jones, 1999). However, several scholars have suggested that an organisation can adopt different approaches to deal with its stakeholders, including proaction, accommodation, defence and reaction (Carroll, 1979; Gatewood and Carroll, 1981; Wartick and Cochran, 1985; Clarkson, 1995).

Organisational Stakeholder Culture (Jones *et al.*, 2007)

Jones *et al.* (2007) started from the point that '. . . whereas the focus of attention in stakeholder theory mainly has been on top managers, understood as relatively autonomous decision makers, these managers are often profoundly influenced by the organisational context in which they are embedded' and suggests a need to '. . . identify *organization-level* factors that could help us predict how firms manage stakeholder relationships' (p. 137, emphasis in original). This is a radical departure from the view that stakeholder-related decisions are functions of managerial choice. They recognise that stakeholder relationships are often fraught with tensions and note that managers often feel these tensions between meeting narrow demands of stakeholding based on self-interest and the broad demands based on 'a concern for the interests of others' (p. 137). These tensions are further exacerbated by the continuous pull on managers between what Hendry (2004) regards as traditional morality (obligation and duty, honesty and respect, fairness and equity, care and assistance) and market morality (self-interest). To resolve these tensions, Jones *et al.* introduce stakeholder culture as an organisational level construct that helps managers go through stakeholder-related decisions with less stress. Emphasising the mediating role of stakeholder culture on managerial stakeholding decisions, Jones *et al.* posit that: '. . . stakeholder culture is a potent organizational factor, profoundly influencing the way in which managers understand, prioritize, and respond to stakeholder issues and, as an example, how they establish stakeholder salience' (pp. 140–141).

Jones *et al.* (2007) define the stakeholder culture as '. . . the beliefs, values, and practices that have evolved for solving stakeholder-related problems and otherwise managing relationships with stakeholders' (p. 142). It is articulated as a central facet of organisational culture as well as an organisational memory of how moral tensions between self-interest and other-regarding interests were resolved in the past. It is as well a simultaneous outcome of '. . . employee sentiments and reified "social facts" that have an independent effect on managerial decisions making' (p. 143). According to Jones *et al.* (2007), the stakeholder culture influences managerial stakeholder-related decisions in two related ways: '(1) by constituting a common interpretive frame on the basis of which information about stakeholder attributes and issues is collected, screened, and evaluated and (2) by motivating behaviours and practices – and, by extension, organizational routines – that preserve, enhance, or otherwise support the organization's culture' (p. 143). However, this stakeholder culture is 'grounded in ethics and based on a continuum of concern for others that runs from self-regarding to other-regarding' (p. 143). In other words, the beginning part of the continuum has a narrow stakeholder orientation while the latter stages are broadly oriented. Based on this continuum and combination of narrow and broad stakeholder orientations, respectively, Jones *et al.* (2007) come up with five categories of corporate stakeholder cultures, which are further subdivided into three typologies: amoral (i.e. agency culture or managerial egoism), limited morality (i.e. corporate egoism and instrumentalism) and broad morality (i.e. morality and altruism).

1. *Agency culture:* This is characterised by managerial egoism, and is '. . . the pursuit of self-interest at the individual level, even if the interests of the corporation and its shareholders, for whom managers nominally work, must be sacrificed' (p. 144). This culture is dominated by self-centredness among the managers of the firm. In other words, managers work entirely for their self-interests. This sort of culture lies at the heart of the old 'agency problem' between managers and shareholders.
2. *Corporate egoist:* In firms characterised by corporate egoism, the predominant culture is pursuit of short-term profit maximisation. This kind of culture is primarily geared towards shareholders' wealth maximisation. Adherence to law is only done when the costs of law breaking is considered to outweigh the gains.
3. *Instrumentalist:* Managers in instrumentalist cultures recognise that moral behaviour could be beneficial to the firm, and practice morality as a strategic device for increasing profitability (Lantos, 2001). Behaviour of managers appears morally to people, but the underlying motive of managers is to advance economic interest of shareholders. In this case, stakeholders are seen more as means or impediments to the goals of the firm (p. 146).
4. *Moralist:* This is a broadly moral culture where the focus is to adhere to principles irrespective of economic pressures. Moral standards are only violated if there is a threat to the survival of the firm (p. 149).
5. *Altruistic:* In altruistic cultures, the concern for others dominates. Adherence to rules irrespective of the implications to the firm dominates and there is also emphasis in treating all the stakeholders fairly and with respect.

A summary of these stakeholder cultures and their orientations are presented in Table 2.2.

Table 2.2   Varieties of corporate stakeholder cultures and orientations

Alternative descriptions	• Amoral management • Management egoism	• Short-term profit maximisation • Short-term self-interest of the corporate level	• Enlightened self-interest • Instrumental or strategic morality	• Genuine concern for welfare of normative stakeholders • Moral pragmatism	• Pure intrinsic morality • Moral purism
Moral orientation; self- versus other-regarding	• Pure egoism • Purely self-regarding	• Regard for others extends to shareholders; belief in efficiency of the market	• Same as corporate egoist	• Morally based regard for normative stakeholders	• Morally based regard for normative stakeholders only
Relevant stakeholders	• None	• Shareholders only	• Shareholders only, but other stakeholders as means to shareholder ends	• All normative and derivative stakeholders	• Normative stakeholders only

Source: Adapted from Jones *et al.* (2007:145).

2.4.3   Summary of organisational level view of corporate stakeholder salience

Drawing from accounts of corporate stakeholding of Jawahar and McLaughlin, (2001) – organisational life-cycle approach – and Jones *et al.*, (2007) – organisational stakeholder culture view – presented above, it could be said that the organisational level view of corporate stakeholder salience also shares some characteristics in common with the managerialist view – albeit from non-atomised actor perspective. In this case, instead of placing managers at the centre of corporate stakeholding, the organisational level view places the firm at the centre. It fundamentally shifts from the individual cognition central to the managerialist view to a form of organisational collective cognition and action, whilst still theorising the firm as powerful and legitimate, and under-emphasising the influence of the wider social system in corporate stakeholding process.

2.5   Influences of micro level theorisation of stakeholding on CSR literature

In the sections below, we draw from insights from the micro level theorisations of stakeholding presented in previous sections to X-ray the different paradigms underpinning the contemporary stakeholder approach to CSR – where firms and managers are encouraged to prioritise other stakeholders in addition to shareholders. However, this prioritisation of CSR agenda is still largely assumed to be solely dependant on managerial discretion and organisational strategic choices (Child, 1972, 1997). In such cases, the institutional influences on both managerial and organisational choices are under-emphasised.

2.5.1   Micro level theorisations of CSR

The stakeholder theory of the firm is central to the current CSR movement. Without the stakeholder theory, CSR loses its fundamental structure and crumbles. As a precursor to CSR, the micro level theorisation of corporate stakeholding also finds an expression through contemporary CSR discourse. In this case, however, managers assume central positions in shaping and setting CSR agenda. In their work on 'Managers' Personal Values as Drivers of Corporate Social Responsibility', Hemingway and Maclagan (2004:34) propose two key dimensions for the analysis of CSR in practice, as shown in Figure 2.3. The motivational basis of the framework, which lies on the vertical axis, asks if the CSR practice analysed is driven primarily by commercial interests or is it just idealistic, even altruistic. In this case, one can see some similarities between commercial interests and instrumental dimension of stakeholding, and idealistic/altruistic interests as related to normative dimensions of stakeholding proposed by Donaldson and Preston (1995).

The second aspect of the framework is what Hemingway and Maclagan (2004:34) call the 'locus of responsibility'. And by this, they mean if the CSR practice in question could be said to be primarily driven by corporate or individual interests – i.e. whether they reflect organisational level and managerialist views of stakeholding, respectively. This thesis adopts these theoretical dimensions, which the authors conclude '. . . point towards a framework for analysis of corporate social responsibility' (p. 34). Given that stakeholder theory has for a long time focused on micro level theorising, a large number of the literature on CSR has also been at this micro level[1]. A couple of examples are highlighted below as typical of this literature.

Managerialist views of CSR

The CSR literature is dotted with accounts emphasising the role of individual actors in promoting (or inhibiting) CSR practices in organisations. These cases tend to draw inspiration from Giddens's (1984) structuration theory that attributes equivalent importance to both agency and structure in understanding social change – which could be extended to understanding firm behaviour. In such instances, powerful personalities within organisations are constructed as moral change agents who leverage their legitimacy and personal values to sway organisation level agenda and actions (Visser, 2007). CEOs and business leaders are often considered to be such personalities (Agle *et al.*, 1999), although Hemmingway (2005)

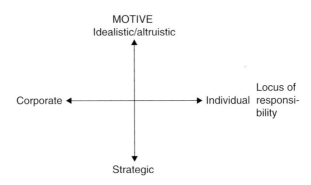

Figure 2.3   A framework for analysing CSR.
(*Source*: Hemingway and Maclagan, 2004:34)

has argued that this form of 'corporate social entrepreneurship' could '. . . operate at a *variety of levels* within the organization: from manual workers or clerical staff to junior management through to directors. They may not *necessarily* be the most senior executives at the top of the organizational hierarchy setting the moral tone of the corporation' (p. 236, emphasis in original). This exhibition of managerial or employee heroism is well documented in the corporate greening (Fineman and Clarke, 1996; Walley and Stubbs, 1999; Crane, 2000a, b, 2001) and ethical leadership literatures (Dukerich *et al.*, 1990; Sims and Brinkman, 2002; Sivanathan and Fekken, 2002). And a key theme central to these is the emphasis they place on the centrality of the 'manager' in shaping firm behaviour, often at the expense (under-emphasis) of the institutional influences.

Organisational views of CSR

Carroll, through his numerous works, is one of the major figures that have contributed significantly to shaping the organisational level CSR agenda since the late last century. Standing out amongst his works is his classic on the pyramid metaphor of CSR (Carroll, 1991), which he orchestrated recently (Carroll, 2004). In these works, Carroll argued that CSR is made up of the following components in a bottom-up order: (1) economic responsibility – 'be profitable' (2) legal responsibility – 'obey the law' (3) ethical responsibility – 'be ethical' (4) philanthropic responsibility –'be a good global corporate citizen'. Much of the CSR literature and practices have been greatly influenced by Carroll's typology of CSR.

In line with this organisational level theorising, Lantos (2001) identified the following strands of CSR: (a) ethical CSR, (b) altruistic CSR and (c) strategic CSR. According to him, ethical CSR is a firm's mandatory fulfilment of economic, legal and ethical responsibilities. It is akin to the first three components of Carroll's typology. Altruistic CSR is the same as philanthropic responsibility of Carroll's typology but differed from it in the sense that Lantos (2001) argued that it would only be possible for private firms to be philanthropic and irresponsibility on the part of public corporations since they do not have the rights to use the funds of shareholders (who might also be involved in private philanthropy) for public philanthropy. Non-instrumental CSR practices transcend (and often defy) rational economic principles underlying most organisational decisions (Korhonen, 2002) and are, thus, informed and governed by trans-material *ratio* of emotion (Fineman, 1996, 2001). Finally, strategic CSR is '. . . good works that are also good for the business'. Lantos (2001), therefore, proposes that ethical CSR, grounded in the concept of ethical duties and responsibilities, is mandatory; concludes that strategic CSR is good for business and society and advises that marketing take a lead role in strategic CSR activities.

This is not an entirely new venture. A number of scholars (Burke and Logsdon, 1996; Greening and Turban, 2000; Maignan and Ferrell, 2001; Moir, 2001; Zairi and Peters, 2002) have advocated for CSR to be solely used to support business objectives, but they are still in the minority camp. Drawing from concepts and practices within strategy, as a management domain, Burke and Logsdon (1996), for instance, argued that the probable contributions of CSR activities to value creation could be assessed from the following dimensions (pp. 496–499):

- *Centrality*: a measure of the closeness of fit between a CSR policy or programme and the firm's mission and objectives.
- *Specificity*: the firm's ability to capture or internalise the benefits of a CSR programme, rather than simply creating collective goods which can be shared by others in the industry, community or society at large.

- *Proactivity:* the degree to which CSR activities are planned in anticipation of emerging economic, technological, social or political trends and in the absence of crisis conditions.[2]
- *Voluntarism:* the scope of discretionary decision-making by the firm and the absence of externally imposed compliance requirements.
- *Visibility:* the observability of a business activity and the firm's ability to gain recognition from internal and external stakeholders.

The visibility dimension of value creation through CSR lends credence to the importance of pursuit of positive corporate reputation, which has been acknowledged in both theory and practice (Swift, 2001). According to Roberts and Dowling (2002), good corporate reputations are critical not only because of their potential for value creation, but also because their intangible character makes replication by competing firms considerably more difficult. In a similar vein, good corporate reputation has been argued to attract good job applicants (Greening and Turban, 2000; Maignan and Ferrell, 2001).

In what has become a classic, Baron (1995) proposed that robust corporate strategies should incorporate elements of the market and non-market environments, respectively. According to Baron (1995:47), '. . . the market environment includes those interactions between the firm and other parties that are intermediated by markers or private agreements. These interactions typically are voluntary and involve economic transactions and the exchange of property'. On the other hand, the non-market environment is characterised by interactions that are '. . . intermediated by the public, stakeholders, government, the media, and public institutions'; and these interactions may be voluntary, such as when the firm adopts a policy of developing relationships with government officials, or involuntary when government regulates an activity or activist groups organise a boycott of a firm's product. Going further, Baron (1995:48) outlined the following as the major components of the non-market environment: issues, institutions, interests and information. The non-market strategies address issues, by seeking to influence institutions (such as regulatory bodies) and interests (e.g. activists, individuals and groups) that drive these issues. The non-market strategies, also, seek to ascertain the information available to these different drivers through environmental scanning.

2.6   Influences of micro level theorisation of stakeholding on corporate accountability

The stakeholder perspective to organising and managing firms is one of the major management paradigm shifts in the late last century. The theory, in its present form traceable to Freeman (1984:246), broadly and loosely defines stakeholders as '. . . those groups and individuals who can affect, or are affected by the achievement of an organization's purpose' – e.g. shareholders, employees, suppliers, government, competitors, local communities and the environment. One of the popular propositions of the stakeholder theory is the view that firms exist at the nexus of series of interdependent relationships with groups that affect or are affected by them (Crane and Livesey, 2003). Given the infinite network of relationships, a firm could be entangled in, this proposition, however, poses some fundamental managerial challenges such as defining the boundaries of stakeholder-ship and effectively managing these relationships that often come with conflicting interests and goals. This challenge tends to polarise views on stakeholder approach to management into three broad camps: descriptive, normative and instrumental (Donaldson and Preston, 1995). The descriptive paradigm explains who a stakeholder is, the normative view prescribes who a stakeholder ought to be, while the instrumental view highlights the consequences of considering a stakeholder or not and suggests that stakeholders could be prioritised based on their salience (importance) (Freeman, 1999:233).

Freeman (1999) acknowledged that his 1984 stakeholder theory is instrumental and pragmatic. As such, he suggested that: '. . . if organizations want to be effective, they will pay attention to all and only those relationships that can affect or be affected by the achievement of the organization's purposes' (p. 234). In addition, it is necessary for an effective firm to manage the relationships that are important, irrespective of the purpose of the firm. Extending the instrumental view, Mitchell *et al.* (1997) theorised that stakeholder salience is a combination of the following factors: power, legitimacy and urgency. A stakeholder group has power when it can impose its will on the firm, especially when it controls resources needed by the firm (Pfeffer, 1981); while legitimacy implies that stakeholder demands comply with prevailing norms and beliefs. Legitimacy is achieved if patterns of organisational practice are in congruence with the wider social system (Scott 1987; Powell and DiMaggio, 1991). However, power and legitimacy can appear together, giving authority to those who have both (Weber, 1947), but they can also appear independently. Finally, urgency is a concept sustained on two elements: (1) the importance stakeholders accord their own demands; and (2) their sensitivity to how long it takes managers to deal with their demands (Gago and Antolin, 2004). These salient variables according to Mitchell *et al.* will determine how a firm responds to its stakeholders. Optimal strategic stakeholder management is, therefore, dependant on the ability of firms to identify and be responsive to salient stakeholders within their business environment.

Stakeholder salience is a precursor to stakeholder accountability and both are interdependent. Roberts and Scapens (1985:447) define accountability as 'the giving and demanding of reasons for conduct'. It is an art of '. . . making the invisible visible' (Munro, 1996:5) through the '. . . provision of information . . . where the one accountable, explains or justifies actions to the one to whom the account is owed' (Gray *et al.*, 1997). Traditionally, under the principal agent dispensation, firms have limited their accountability to shareholders as economic and legal owners of the firm. Friedman (1962) reinforced this form of accountability when he argued that the primary responsibility of firms is to pursue profits within the limits of the law. The economic logic of accountability leans heavily on what Korhonen (2002) called the 'dominant social paradigm' (DSP) of profit maximisation for the owners of the firm. The DSP emphasises such issues as competitive advantage, cost minimisation, equilibrium, market efficiency, optimal returns on investments (including labour) and market dominance. Shareholder accountability is the bedrock of modern capitalism. Adherence to this culture of capitalism often comes with its rewards in terms of increase in shareholders wealth and firm growth; although it sometimes leads to market failures (i.e. monopolies, pollutions, etc.). Stakeholder accountability has emerged, towards the end of the last century, as complement to shareholder accountability (Gray *et al.*, 1988; Owen *et al.*, 2000; Gray, 2002).

Drawing from the works of other academics (Roberts and Scapens, 1985; Williams 1987; Gray *et al.*, 1988), Swift (2001:17) broadly describes accountability as '. . . the requirement or duty to provide an account or justification for one's actions to whomever one is answerable' and narrowly as '. . . being pertinent to contractual arrangements only, . . . where accountability is not contractually bound there can be no act of accountability'. Borrowing from a later work of Gray *et al.* (1997), Swift notes that '. . . essentially accountability is about the provision of information between two parties where the one is accountable, explains or justifies actions to the one to whom the account is owed'. This form of accountability can easily be glimpsed from that characteristic of principal–agent relationship, which is central to the firm as an economic and legal entity. But no matter the side taken, and however defined, one factor that is central to the notion of accountability is the *duty to account*, which connotes institution of rights and as such, should hurt (Owen *et al.*, 2000). In the same line of thought, Gray *et al.* (1988) sought to explain the firm's accountability to the wider society as inherent in a social contract between the society and the business – the idea that business derives its existence from the society. This accountability inherent in the form of social contract is enforced through the market forces that punish or reward corporate behaviour (Donaldson and Preston, 1995; Swift, 2001). Korten (2004) argues that the market by necessity needs information to be effective – as such, corporations should be demanded to produce the necessary and complete information required by the market to punish or reward – this will constitute accountability to the market, which cannot be achieved through self-regulation.

According to Gray *et al.* (1988), the underlying principles of stakeholder accountability derive from a firm's accountability to the wider society as inherent in a social contract between the society and business – i.e. the idea that business derives its existence from the society. This accountability inherent in the form of social contract is enforced through the market forces that punish or reward corporate behaviour (Donaldson and Preston, 1995; Swift, 2001). In this regard, Korten (2004) argues that the market by necessity needs information to be effective – as such, corporations should be demanded to produce the necessary and complete information required by the market to punish or reward – this will constitute accountability to the market, which cannot be achieved through self-regulation. Accountability, therefore, in turn connotes some level of transparency; and transparency carries with it some risks of disclosure that could hurt (Owen *et al.*, 2000; Gray, 2002).

This perspective of stakeholder accountability seems to be driving the current surge of interests in social reports. Interest in and demand for stakeholder accountability has been on the increase. The 1970s enjoyed a boom in social accounting which disappeared in the 1980s and has reappeared since the 1990s. In addition, the accounting and governance travesties of such firms as Enron and WorldCom in the United States and Parmalat in Italy, to mention but a few, have made such demands for corporate accountability and social reports even more pertinent. Within these social reports, firms aim to signal accountability towards, and willingness to be held accountable by, their different stakeholder groups on such issues as their environmental footprints, poverty reduction, labour and employment conditions, gender and equality, community and consumer welfare, corporate governance and ethics. It is also argued that firms use corporate social reports as subtle strategies to reaffirm their legitimacy (Brown and Deegan, 1998; Neu *et al.*, 1998), and appeal to salient stakeholders (Hooghiemstra, 2000; Gray, 2002).

Unfortunately, the target of social reports has been one of the vexed issues about these reports in recent times. Unlike corporate annual reports that are specifically addressed to shareholders, corporate social reports often start with such diffused salutations as 'Dear Readers' or 'Dear Stakeholders'. This diffused and non-specific addressee approach tends to demean social reports as mere 'talks to all, but to none'. Some critics have even gone as far as describing social reports as artefacts of managerial capture (Owen *et al.*, 2000) '. . . used by a privileged part of the socio-economic-political system (capitalist elites) to protect and advance their sectional interests' (Unerman, 2003:429). This line of argument, which has dominated stakeholder accountability thinking for a long time now, tends to assume that managerial actions are largely rational and thus discretional. It is within this discretional rationality, it is argued, that managers as representatives of firms exercise power and dominion over different stakeholder groups.

Over the years, stakeholder management discourse and practice has also been anchored on managerial discretion. In other words, stakeholders that receive priority from management will be those whom managers perceive as highly salient (Agle *et al.*, 1999). This managerial elitism has, in the main, continued to dominate stakeholder management discourse, with little or no emphasis placed on the contextual embeddedness of managerial thoughts and actions in stakeholder management practice and discourse. This situation, which is arguably a manifestation of the rational choice school of thought, could be, borrowing from Granovetter (1985), described as an under-socialised account of stakeholder management practice. Theorists have recently begun to challenge this managerialist view and to interpret firms' interactions with their stakeholder from a much broader perspective that incorporates institutional, cultural and societal contexts, into the debate.

2.6.1 Summary of micro level theorisations of CSP

All the examples presented above tend to suggest that CSR agenda and actions are largely subject to firms' strategic choices. This suggestion rubs-off from organisational level theorisation of CSP. Notwithstanding, the view that firms exist at the nexus of series of interdependent relationships with groups that can affect or are affected by them (Freeman, 1984; Crane and Livesey, 2003) poses some fundamental managerial challenges such as defining the boundaries of stakeholder-ship and effectively managing these relationships that often come with conflicting interests and goals. However, central to this stakeholder approach is the principle of who or what really counts (Freeman, 1994). That is, who (or what) are the stakeholders of the firm? And to whom (or what) do managers pay attention? (Mitchell *et al.*, 1997). These come with a burden of defining the boundaries of stakeholder-ship, and establishing appropriate mechanisms for stakeholder consultation and involvement in strategic development. They, as well, constitute great challenges for managers and decision-makers, which ultimately impact on CSR agenda setting and actions – both at the managerial and organisational levels. It is the argument of this research study that in addition to managerial and organisational level influences, CSP, as a precursor to CSR, governance and accountability, is equally constrained and enabled by the institutional contexts in which it is enacted. Unfortunately, this institutional dimension to accounting for CSPs is marginalised (or under-emphasised) in the extant CSR literature.

2.7 Towards a 'new paradigm': theorising corporate stakeholder salience from an institutionalist perspective

Despite the under-emphasis of institutional embeddedness of CSPs, there is an emerging literature on variations of CSR and governance across national institutional contexts (Aguilera and Jackson, 2003; Chapple and Moon, 2005; Amaeshi *et al.*, 2006). Following its normative underpinnings, for instance, it is expected that stakeholder salience will differ according to industry and country since ethics have been found to differ along those lines, as well. In their large-scale survey among senior executives in the United States, the United Kingdom, Germany and Austria, Schlegelmilch and Robertson (1995) found that both country and industry have strong influence on perception of ethical issues (and that firm size does not). In another study, Thelen and Zhuplev (2001) present a comparative analysis of attitudes between Russian and US undergraduate students on ethical issues in managing Russian small firms engaged in business transactions with US firms. Based on the real-life situations, Russian and American respondents were asked to select decision alternatives dealing with ethical dilemmas. Significant differences were found between the two groups. Russians do not recognise significant differences between various alternatives, despite the disparity in the severity of these alternatives for resolving business problems. Russians, compared to Americans, tend to prefer more forceful decision alternatives resorting to business practices that would be considered unethical in the United States. This is attributable to differences in the countries' history, political, legal and cultural environment. The transitional nature of the Russian economy affects decision-making and business ethics.

Robertson *et al.* (2002) also queried 210 financial services managers from Australia, Chile, Ecuador and the United States about their ethical beliefs when faced with four diverse dilemmas. In addition, the situational context was altered so the respondent viewed each dilemma from a top management position and from a position of economic hardship. Results suggest a complex interaction of situation, culture and issue when individuals make ethical judgments. Specifically, Chileans were found to have different beliefs about sex discrimination and child labour dilemmas when compared to their colleagues from the other three nations. Chileans and Australians also disagreed on the bribery dilemma. Anglo managers were more likely than Latin American managers to change their ethical responses when the situation was altered. In a situation like this where interpretations and manifestations of ethical beliefs are determined by cultural differences, what should be the criteria for making ethical decisions? Robertson *et al.* (2002) suggested that multinational firms interested in maintaining healthy ethical climates, should consider adapting culturally contingent ethical guidelines, or policies to the local customs. If this suggestion should be adhered to, what happens in a situation where board members from different cultures and beliefs need to take ethical decisions that are not location specific?

Hooghiemstra and van Manen's (2002) research among 2500 of the largest companies in The Netherlands reveals the growing importance of social and ethical issues in the corporate governance debate. Such issues can place non-executive directors in a dilemma when his point of view is neither shared by the management board nor by the other supervisory board members: Should he resign or should he try to influence the others of his opinion? That is, in terms of Hirschman's (1970) classical work, should he 'exit' or 'voice'. The chapter reports the findings regarding non-executive directors' choice based upon a qualitative and a quantitative study conducted among almost 300 Dutch supervisory directors. Regarding bribing civil servants, non-executives seem to make a distinction based upon location. While a bribe in a third world country seemed acceptable to approximately half of the responding outside directors, it was considered unacceptable (and would lead to repercussions) in the case where the bribe involved either a Dutch civil servant or another company's employee. Indeed, in the qualitative study many of the non-executive directors remarked that bribing people is sometimes necessary to do business, although it is not a good thing to do. Furthermore, they also commented that ethical behaviour is a dependant variable and has its limits. For example, whereas bribing was considered unacceptable only in The Netherlands, non-executive directors did not make a distinction based upon location in case of environmental pollution – the same rules applied irrespective of whether it concerned a third world country or The Netherlands.

Institutionalists (Scott and Meyer, 1983; Powell and DiMaggio, 1991, Whitley, 1992, 1998) argue that organisations are products of their external environments, which '. . . are characterized by the elaboration of rules and requirements to which individual organizations must conform if they are to receive support and legitimacy . . .' (Scott and Meyer, 1983:149). In this case, Kondra and Hinings (1998) state that conformity to norms is facilitated by normative, coercive and mimetic processes (DiMaggio and Powell, 1983) and compliance may be for pragmatic reasons or due to paradigm stasis (Oliver, 1991). Continuing, Kondra and Hinings argue that organisations within an organisational field may conform to these rules and requirements, not necessarily for reasons of efficiency, but rather for increasing their legitimacy, resources and survival capabilities (Meyer and Rowan, 1977; DiMaggio and Powell, 1983); and these organisations that conform to institutional norms become 'optimal', if not efficient, in the sense that they increase their survival capabilities by conforming to those norms, thereby minimising the risk of organisational death (Baum and Oliver, 1991). As a result, values and beliefs external to the organisation play a significant role in determining organisational norms.

The same line of thought is shared by culturalists; necessitating Carroll and Gannon (1997) to argue that the ethical orientation of a particular culture may have a significant impact on internal organisational activities such as human resource management. Similarly, cultures may not extend their beliefs concerning ethical conduct to individuals deemed members of an out-group (Katzenstein, 1989; Pratt, 1991). However, whether or not ethics are 'relative' or contingent on national culture is indeed controversial. Donaldson and Dunfee (1994), for example, present a convincing argument for a superordinate set of normative ethical principles. Yet, multinational organisations often confront serious human resource management issues when operating in cultures with values different from their own (Carroll and Gannon, 1997). Firms that fail to consider the values and ethics of their host culture by appropriately aligning their human resource management policies may be perceived of as opportunistic and potentially unethical. Given this scenario, it is possible for firms to display different ethical orientations in relation to different target audiences (external and internal).

2.7.1   *Research gaps in the literature*

In summary, then, if stakeholder salience, as a matter of managerial perceptions, is a reality constructed over time rather than an objective reality (Agle *et al.*, 1999:508–509), then it could be argued that these constructions are likely to draw from (or are functions of) the sedimented broader social constructions within the institutional contexts in which the managerial perceptions are crafted and enacted. Moreover, legitimacy is a function of social context (Suchman, 1995) and '. . . bounded by cultural norms and behaviour' (Agle *et al.*, 1999:509). Therefore, following discussions so far, it could be argued that corporate stakeholder salience patterns are not only shaped by managerial influences but are also implicated in series of multilevel influences varying between micro and macro variables as shown in Figure 2.4.

Figure 2.4   Multilevel influences on corporate stakeholder salience.

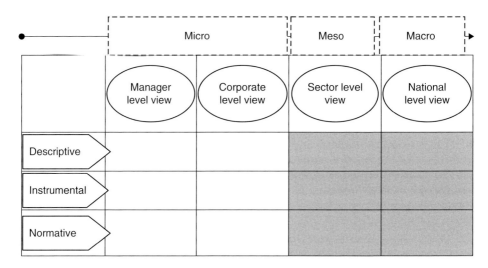

Figure 2.5   **Multilevel dimensions of studying corporate stakeholder salience.**

These influences, which could be bottom-up – e.g. through some form of institutional entrepreneurship[3] (Crouch, 2005; Lawrence and Suddaby, 2007) or top-down – e.g. through government/transnational influences (Djelic and Quack, 2008) interact to shape corporate stakeholder salience practices. In this regard, these interactions could either enable or constrain corporate stakeholding activities. In line with Jones (1999), the intention here is neither to imply that each of these levels exerts equal influences on determining corporate stakeholder salience patterns nor is it to unpack the intensity of each of these influences, but to '. . . stress the inter-relatedness of these levels, particularly with their most proximate counterparts, and their combined impact on determining the necessary and sufficient conditions for the practice of stakeholder management' (p. 165).

Following this line of thinking, Jones and Fleming (2003) criticise conventional theory of stakeholding for its failure '. . . to consider the underlying structural linkages that may exist between various stakeholders along with complex and deeply embedded (institutionalized) processes that constitute stakeholders' materiality, identity and even forms of rationality' (p. 433). The literature in the main, surprisingly, takes these interactions for granted and assumes that managers and firms can easily select or deselect stakeholders. Contrary to this common view are situations where these interactions constrain corporate stakeholding and ultimately, CSR agenda. A good example of the latter is the current European Union (EU) regulation on procurement, which constrains the EU Utilities from enforcing green procurement policies and practices across their supply-chain (see Arrowsmith, 2000, 2006 for details). This leaves a gap in the literature, which needs to be filled. Figure 2.5 helps to summarise where the literature on stakeholding is and shows where the gaps are (the shades).

2.8   Conclusion

This chapter has identified alternative modes of theorising CSPs. It is recognised that such disciplines as politics, economics and sociology have robust theories that deal with firm behaviours at the identified multilevels. Some of these include, but not limited to, the following: systems theory (Ackoff and Churchman, 1947; Pfeffer and Salancik, 1978), organisational theory (Katz and Kahn, 1966), national business systems, neo-institutionalism, varieties of capitalism, political economies, corporate governance frameworks and recently the explicit–implicit models (Matten and Moon, 2008). These approaches emphasise the external environment as a significant explanatory factor of the organisation of the firm (Pfeffer and Salancik, 1978).

Finally, the chapter problematises CSPs and theorisations as a precursor to contemporary pursuits of CSR, governance and accountability. It also highlights the predominance of the managerialist view on the practice and theorisation of corporate stakeholding, which appears to under-emphasise the institutional influences on CSP. Drawing insights from emergent comparative studies of business practices (particularly those on business ethics, CSR and corporate governance) across national borders, the chapter identifies possible research gaps in the literature – albeit at the macro level – that could complement the micro level managerialist accounts of CSPs. The goal of this chapter is to re-introduce the institutionalist perspective to the understanding and interpretation of firms' behaviours; and, more so, to emphasise the relevance of this perspective to the practice of corporate stakeholding, which is hereby theorised as a corporate practice.

Endnotes

1. It is only recently that corporate social responsibility theorisation is attempting to incorporate macro-theorisation – e.g. Matten and Moon (2008), Maignan and Ferrell (2001), which this book will cover in subsequent sections.
2. An example of proactivity in the CSR context, according to Burke and Logdson (1996), is a manufacturer monitoring emerging social trends and regulatory initiatives regarding pollution control (p. 498).
3. Institutional entrepreneurs are '. . . organized actors who skilfully use institutional logics to create or change institutions, in order to realize an interest that they value highly' (Leca and Naccache, 2006:634). Further discussions on institutional entrepreneurship and the link between micro and macro institutionalisation are presented later on in Chapter 8 of this book.

References

Ackerman, B.A. and Alstott, A. (1999) *The Stakeholder Society*. New Haven, CT: Yale University Press.

Ackoff, R. and Churchman, C. (1947) An experimental definition of personality. *Philosophy of Science*, **14**:304–332.

Agle, B.R., Mitchell, R.K. and Sonnenfeld, J.A. (1999) Who matters to CEOs? An investigation of stakeholder attributes and salience, corporate performance, and CEO values. *The Academy of Management Journal*, **42**(5):507–525.

Aguilera, R. (2005) Corporate governance and director accountability: An institutional comparative perspective. *British Journal of Management*, **16**(1):39–53.

Aguilera, R. and Jackson, G. (2003) The cross-national diversity of corporate governance: Dimensions and determinants. *Academy of Management Review*, **28**(3): 447–465.

Aguilera, R.V., Williams, C.A., Conley, J.M. and Rupp, D.E. (2006) Corporate governance and social responsibility: A comparative analysis of the UK and the US. *Corporate Governance: An International Review*, **14**(3):147–158.

Amable, B. (2003) *The Diversity of Modern Capitalism*. Oxford: Oxford University Press.

Amaeshi, K. and Adi, B.C. (2007) Reconstructing corporate social responsibility construct in utilish. *Business Ethics European Review*, **16**(1):3–18.

Amaeshi, K., Bongo C.A., Chris O. and Olufemi O.A. (2006) Corporate social responsibility in Nigeria: Western mimicry or indigenous influences. *The Journal of Corporate Citizenship*, 24(Winter).

Appadurai, A. (1999) Globalization and the research imagination. *International Social Science Journal*, **51**(160):229–238.

Arrowsmith, S. (2000) Public private partnerships and the European procurement rules: EU policies in conflict? *Common Market Law Review*, **37**(3):709–737.

Arrowsmith, S. (2006) Past and future evolution of EC procurement law: From framework to common code? *Public Contract Law Journal*, **35**(3):337–383.

Baron, D.P. (1995) Integrated strategy: Market and nonmarket components. *California Management Review*, **37**(2):47–65.

Baucus, M. and Near, J. (1991) Can illegal corporate behaviour be predicted? *Academy of Management Journal*, 34:9–36.

Baum, J.A.C. and Oliver, C. (1991) Institutional linkages and organizational mortality. *Administrative Science Quarterly*, **36**:187–218.

Beliveau, B., Cottrill, M. and O'Neill, H. (1994) Predicting corporate social responsiveness. *Journal of Business Ethics*, **13**(9):731–738.

Borsch, A. (2004) Globalisation, shareholder value, restructuring: The (non)-transformation of Siemens. *New Political Economy*, **9**(3):365–387.

Brown, N. and Deegan, C. (1998) The public disclosure of environmental performance information – a dual test of media agenda setting theory and legitimacy theory. *Accounting and Business Research*, **29**(1):21–41.

Brown, J. and Fraser, M. (2006) Approaches and perspectives in social and environmental accounting: An overview of the conceptual landscape. *Business Strategy and the Environment*, **15**:103–117.

Burke, L. and Logsdon, J.M. (1996) How corporate social responsibility pays off. *Long Range Planning*, **29**(4):495–502.

Carroll, A.B. (1979) A three dimensional conceptual model of corporate social performance. *Academy of Management Review*, 4:497–505.

Carroll, A.B. (1991) The pyramid of corporate social responsibility: Toward the moral management of organizational stakeholders. *Business Horizons*, **34**(4):39–48.

Carroll, A.B. (1993) *Business and Society: Ethics and Stakeholder Management*. Cincinnati, OH: South-Western.

Carroll, A.B. (2004) Managing ethically with global stakeholders: A present and future challenge. *Academy of Management Executive*, **18**(2):114–119.

Carroll, S.J. and Gannon, M.J. (1997) *Ethical dimensions of international management*. Thousand Oaks, CA: Sage.

Chapple, W. and Moon, J. (2005) Corporate social responsibility (CSR) in Asia: A seven-country study of CSR web site reporting. *Business and Society*, **44**(4):415–439.

Child, J. (1972) Organizational structure, environment and performance: The role of strategic choice. *Sociology*, **6**(1):1–22.

Child, J. (1997) Strategic choice in the analysis of action, structure, organizations and environment: Retrospect and prospect. *Organization Studies*, **18**(1):43–76.

Clarkson, M.B.E. (1995) A stakeholder framework for analyzing and evaluating corporate social performance. *Academy of Management Review*, **20**:92–117.

Crane, A. (2000a) Facing the backlash: Green marketing and strategic reorientation in the 1990s. *Journal of Strategic Marketing*, **8**(3):277–296.

Crane, A. (2000b) *Marketing, Morality and the Natural Environment*. London: Routledge.

Crane, A. (2001) Corporate greening as amoralization. *Organization Studies*, **21**(4):673–696.

Crane, A. and Livesey, S. (2003) Are you talking to me? Stakeholder communication and the risks and rewards of dialogue. In: Andriof, J., Waddock, S., Rahman, S.S. and Husted, B. (eds)

Unfolding Stakeholder Thinking Vol II: Relationships, Communication, Reporting and Performance. Sheffield: Greenleaf, pp. 39–52.

Crouch, C. (2005) *Capitalist Diversity and Change: Recombinant Governance and Institutional Entrepreneurs.* Oxford: Oxford University Press.

Crouch, C. (2006) Modelling the firm in its market and organizational environment: Methodologies for studying corporate social responsibility. *Organization Studies*, **27**:1533–1551.

Darke, P., Shanks, G. and Broadbent, M. (1998) Successfully completing case study research: Combining rigour, relevance and pragmatism. *Information Systems Journal*, **8**(4):273–289.

DiMaggio, P.J. and Powell, W.W. (1983) The iron cage revisited: Institutional isomorphism and collective rationality in organizational fields. *American Sociological Review*, **48**(2):147–160.

Djelic, M.L. and Quack, S. (2008) Institutions and transnationalization. In: Greenwood, R.,Suddaby, R., Oliver, C. and Sahlin-Andersson, K. (eds) *The Handbook on Institutional Theory.* London: Sage.

Donaldson, T. and Dunfee, T. (1994) *Ties That Bind: A Social Contracts Approach to Business Ethics.* Boston, MA.: McGraw-Hill.

Donaldson, T. and Preston, L.E. (1995) The stakeholder theory of the corporation: Concepts, evidence and implications. *Academy of management review*, **20**(1):65–91.

Dukerich, J.M., Nichols, M.L., Elm, D.R. and Vollrath, D.A. (1990) Moral reasoning in groups: Leaders make a difference. *Human Relations*, **43**(5):473–493.

Evan, W. and Freeman, R.E. (1988) A stakeholder theory of the modern corporation: Kantian capitalism. In: Beauchamp, T. and Bowie, N. (eds) *Ethical Theory and Business.* Englewood Cliffs, NJ: Prentice-Hall, pp. 75–93.

Fineman, S. (1996) Emotional subtexts in corporate greening. *Organization Studies*, **17**(3):479–500.

Fineman, S. (2001) Fashioning the environment. *Organization*, **8**(1):17–31.

Fineman, S. and Clarke, K. (1996) Green stakeholders: Industry interpretations and response. *Journal of Management Studies* **33**(6):715–731.

Fiske, S.T. and Taylor, S.E. (1984) *Social Cognition.* Reading, MA: Addison-Wesley.

Freeman, R.E. (1984) *Strategic Management: A Stakeholder Approach.* Boston: Pitman.

Freeman, R.E. and Evan, W.M. (1990) Corporate governance: A stakeholder interpretation. *The Journal of Behavioral Economics*, **19**(4):337–359.

Freeman, R.E. (1994) The politics of stakeholder theory: some future directions. *Business Ethics Quarterly*, **4**(4):409–421.

Freeman, R.E. (1999) Response: divergent stakeholder theory. *The Academy of Management Review*, **24**(2):233–236.

Freeman, R.E. and McVea, J. (2001) A stakeholder approach to strategic management. In: Hitt, M., Freeman, E. and Harrison, J. (eds) *The Blackwell Handbook of Strategic Management.* Oxford: Oxford University Press.

Friedman, A.L. and Miles, S. (2006) *Stakeholders: Theory and Practice.* Oxford: Oxford University Press.

Friedman, M. (1962) *Capitalism and Freedom.* Chicago IL, University of Chicago Press.

Gago, R.F. and Antolin, M.N. (2004) Stakeholder salience in corporate environmental strategy. *Corporate Governance*, **4**(3):65–76.

Gatewood, E. and Carroll, A.B. (1981) The anatomy of corporate social response. *Business Horizons*, **24**(1):9–16.

Giddens, A. (1984) *The Constitution of Society: Outline of the Theory of Structuration.* Cambridge, UK: Polity Press.

Gilligan, C. (1977) In a different voice: Women's conception of the self and of morality. *Harvard Educational Review*, **47**:481–517.

Gilligan, C. (1982). *In a Different Voice: Psychological Theory and Women's Development.* Cambridge, MA: Harvard University Press.

Granovetter, M. (1985) Economic action and social structure: A theory of embeddedness. *American Journal of Sociology*, **91**(3):481–510.

Gray, R., Dey, C., Owen, D., Evans, R. and Zadek, S. (1997) Struggling with the praxis of social accounting. *Accounting, Auditing and Accountability Journal*, **10**(3):325–364.

Gray, R. (2002) The social accounting project and *Accounting, Organizations and Society*: Privileging engagement, imaginings, new accountings and pragmatism over critique? *Accounting, Organizations and Society*, **27**:687–708.

Gray, R., Owen, D. and Maunders, K. (1988) Corporate social reporting: Emerging trends in accountability and the social contract. *Accounting, Auditing & Accountability Journal*, **1**(1):6–20.

Greening, D.W. and Gray, B. (1994) Testing a model of organizational response to social and political issues. *Academy of Management Journal*, **37**(3):467–498.

Greening, D.W. and Turban, D.B. (2000) Corporate social performance as a competitive advantage in attracting a quality workforce. *Business & Society*, **39**(3):254–280.

Hall, P.A. and Soskice, D. (eds) (2001) *Varieties of Capitalism – The Institutional Foundations of Comparative Advantage*. Oxford: Oxford University Press.

Hardy, C., Phillips, N. and Clegg, S.R. (2001) Reflexivity in organization and management theory: A study of the production of the research 'subject'. *Human Relations*, **54**(5):531–560.

Hemingway, C.A. and Maclagan, P.W. (2004) Managers' personal values as drivers of corporate social responsibility. *Journal of Business Ethics*, **50**(1):33–44.

Hemingway, C.A. (2005) Personal values as a catalyst for corporate social entrepreneurship. *Journal of Business Ethics*, **60**(3):233–249.

Hendry, J. (2004) *Between Enterprise and Ethics: Business and Management in a Bimoral Society*. Oxford: Oxford University Press.

Hill, C.W.L. and Jones, T.M. (1992) Stakeholder-agency theory. *Journal of Management Studies*, **29**:131–154.

Hirschman, A.O. (1970) *Exit, Voice, and Loyalty. Responses to Decline in Firms, Organisations and States*. Cambridge, MA: Harvard University Press.

Hollingsworth, R. and Boyer, R. (eds) (1997) *Contemporary Capitalism. The Embeddedness of Institutions*. Cambridge: Cambridge University Press.

Hooghiemstra, R. (2000) Corporate communication and impression management – New perspectives why companies engage in corporate social reporting. *Journal of Business Ethics*, **27**:55–68.

Hooghiemstra, R. and van Manen, J. (2002) Supervisory directors and ethical dilemmas: Exit or voice? *European Management Journal*, **20**(1):1–9.

Jackson, G. (2005) Stakeholders under pressure: corporate governance and labour management in Germany and Japan. *Corporate Governance: An International Review*, **13**(3):419–428.

Jawahar, I.M. and McLaughlin, G.L. (2001) Toward a descriptive stakeholder theory: An organizational life cycle approach. *Academy of Management Review*, **26**(3):397–414.

Jensen, M.C. (2000) Value maximization, stakeholder theory, and the corporate objective function. *Business Ethics Quarterly*, **12**(2):235–256.

Johnson, R.B. and Onwuegbuzie, A.J. (2004) Mixed methods research: A research paradigm whose time has come. *Educational Researcher*, **33**(7):14–26.

Jones, M.T. (1999) The institutional determinants of social responsibility. *Journal of Business Ethics*, **20**:163–179.

Jones, M.T. and Fleming, P. (2003) Unpacking complexity through critical stakeholder analysis: The case of globalization. *Business Society*, **42**:430–454.

Jones, T.M., Felps, W. and Bigley, G.A. (2007) Ethical theory and stakeholder-related decisions: The role of stakeholder culture. *Academy of Management Review*, **32**(1):137–155.

Katz, D. and Kahn, R.L. (1966) *The Social Psychology of Organizations*. New York: Wiley.

Katzenstein, G. (1989) *Funny Business: An Outsider's Year in Japan*. New York: Prentice Hal.l

Kay, J. (1997) The stakeholder corporation. In: Kelly, G., Kelly, D. and Gamble, A. (eds) *Stakeholder Capitalism*. London: Macmillan.

Kelly, G., Kelly, D. and Gamble, A. (eds) (1997) *Stakeholder Capitalism*. Houndmills, Basignstoke: Macmillan.

Kondra, A.Z. and Hinings, C.R. (1998) Organizational diversity and change in institutional theory. *Organization Studies*, **19**(5):743–767.

Korhonen, J. (2002) The dominant economics paradigm and corporate social responsibility. *Corporate Social Responsibility and Environmental Management*, **9**:67–80.

Korten, D.C. (2004) The responsibility of business to the whole http://www.flora.org/library/mai/responsibility.html visited 10 May 2004.

Lantos, G.P. (2001) The boundaries of strategic corporate social responsibility. *Journal of Consumer Marketing*, **18**(7):595–630.

Lawrence, T.B. and Suddaby, R. (2007) Institutions and institutional work. In: Clegg, S.R., Hardy, C., Nord, W.R. and Lawrence, T. (eds) *Handbook of Organization Studies* (2nd edn). London: Sage Publications, pp. 215–254.

Lounsbury, M. (2008) Institutional rationality and practice variation: New directions in the institutional analysis of practice. *Accounting, Organizations and Society*, **33**(4–5):349–361.

Lundvall, B.A. (1988) Innovation as an inter-active process; from user–producer interaction to the national system of innovation. In: Dosi, G. *et al.* (eds) *Technical Change and Economic Theory*. Pinter, London.

Maignan, I. and Ferrell, O.C. (2001) Corporate citizenship as a marketing instrument: Concepts, evidence and research directions. *European Journal of Marketing*, **35**(3/4):457–484.

Marcoux A.M. (2000). Balancing act. In: DesJardins, J.R., McCall, J.J. (eds) *Contemporary Issues in Business Ethics* (4th edn). US:Wadsworth, pp. 92–98.

Matten, D. and Moon, J. (2008) 'Implicit' and 'Explicit' CSR: A conceptual framework for a comparative understanding of corporate social responsibility. *Academy of Management Review,* **33**(2):404–424.

Meyer, J.W. and Rowan, B. (1977) Institutionalized organizations: formal structure as myth and ceremony. *American Journal of Sociology*, **83**:340–363.

Mitchell, R., Agle, B. and Wood, D. (1997) Towards a theory of stakeholder identification: Defining the principle of who and what really counts. *Academy of Management Review*, **22**(4), 853–886.

Moir, L. (2001) What do we mean by corporate social responsibility? *Corporate Governance: International Journal of Business in Society*, **1**(2):16–22.

Morgan, G. (2001) Transnational communities and business systems. *Global Networks*, **1**(2): 113–130.

Munro, R. (1996) Alignment and identity work: The study of accounts and accountability. In Munro, R., and Mouritsen, J. (eds) *Accountability – Power, Ethos & the Technologies of Managing*. London: Thompson Business Press.

Neu, D., Warsame, H. and Pedwell, K. (1998) Managing public impressions: Environmental disclosures in annual reports. *Accounting, Organisations, and Society* **23**(3):265–282.

Olen, J. and Barry, V. (1992) *Applying Ethics* (4th edn) CA: Wadsworth publishing company.

Oliver, C. (1991) Strategic responses to institutional processes. *Academy of Management Review*, **16**:145–179.

Owen, D.L., Swift, T.A., Humphrey, C. and Bowerman, M.C. (2000) The new social audits: Accountability, managerial capture or the agenda of social champions? *European Accounting Review*, **9**(1):81–98.

Pfeffer, J. (1981) *Power in Organizations*. Boston, MA: Pitman.

Pfeffer, J. and Salancik, G. (1978) *The External Control of Organizations*. New York: Harper.

Phillips, R. (2003) *Stakeholder Theory and Organizational Ethics*. San Francicso, CA: Barrett-Koehler.

Phillips, R., Freeman, E. and Wicks, A.C. (2003) What stakeholder theory is not. *Business Ethics Quarterly*, **13**(4):479–502.

Phillips, R.A. (1997) Stakeholder theory and a principle of fairness. *Business Ethics Quarterly*, **7**(1):51–66.

Powell, W.W. and DiMaggio, P.J. (1991) *The New Institutionalism of Organizational Analysis*. Chicago: University of Chicago Press.

Pratt, C. (1991) Multinational corporate social policy for ethical responsibility in sub-Saharan Africa. *Journal of Business Ethics*, **10**:527–541.

Ritchie J. and Lewis, J. (2003) *Qualitative Research Practice: A guide for Social Science Students and Researchers*. London: Sage.

Roberts, J. and Scapens, R. (1985) Accounting systems and systems of accountability – Understanding accounting practices in their organisational contexts. *Accounting, Oganizations and Society*, **10**(4):443–456.

Roberts, P.W. and Dowling, G.R. (2002) Corporate reputation and sustained superior financial performance. *Strategic Management Journal*, **23**(12):1077–1093.

Robertson, C.J., Crittenden, W.F., Brady, M.K. and Hoffman, J.J. (2002) Situational ethics across borders: A multicultural examination. *Journal of Business Ethics*, **38**(4):327–338.

Schlegelmilch, B.B. and Robertson, D.C. (1995) The influence of country and industry on ethical perceptions of senior executives in the U.S. and Europe. *Journal of International Business Studies*, **26**(4):859–881.

Scott, W.R. (1987) The adolescence of institutional theory. *Administrative Science Quarterly*, **32**(4):493–511.

Scott, W.R. (1995) *Institutions and Organizations*. Thousand Oaks, CA: Sage Publications.

Scott, W.R. and Meyer, J.W. (1983) The organization of societal sectors. In: Meyer, J. W. and Richard Scott, W. (eds) *Organizational Environments: Ritual and Rationality*. Beverly Hills, CA: Sage, pp. 129–153.

Sharma, S. (2000) Managerial interpretations and organizational context as predictors of corporate choice environmental strategy. *Academy of Management Journal*, **43**(4):681–697.

Sims, R.R. and Brinkman, J. (2002) Leaders as moral role models: The case of John Gutfreund at Salomon brothers. *Journal of Business Ethics*, **35**(4):327–339.

Sivanathan, N. and Fekken, G.C. (2002) Emotional intelligence, moral reasoning and transformational leadership. *Leadership & Organization Development Journal*, **23**(4):198–204.

Slinger, G. (1999) Spanning the gap – The theoretical principles that connect stakeholder policies to business performance. *Corporate Governance: An International Review*, **7**(2):136–151.

Starik, M. (1994) What is a stakeholder? Essay by Mark Starik. pp. 89–95 of the Toronto Conference. *Reflections on Stakeholder Theory, Business & Society*, **33**:82–131.

Starik, M. (1995) Should trees have managerial standing? Toward stakeholder status for nonhuman nature. *Journal of Business Ethics*, **14**:207–218.

Suchman, M.C. (1995) Managing legitimacy: Strategic and institutional approaches. *The Academy of Management Review*, **20**(3):571–610.

Swift T. (2001). Trust, reputation and corporate accountability to stakeholders. *Business Ethics: A European Review*, **10**(1):16–26.

Thelen, S. and Zhuplev, A. (2001) Comparing attitudes toward *ethical dilemmas* in small business: Russia versus the United States. *Journal of East-West Business*, **7**(4):29–54.

Unerman, J. (2003) Enhancing organizational global hegemony with narrative accounting disclosures: An early example. *Accounting Forum*, **27**(4):425–448.

Visser, W. (2007) *Meaning in the Life and Work of Corporate Sustainability Managers*, PhD dissertation (Business and Management), University of Nottingham Business School, UK.

Walley, L. and Stubbs, M. (1999) Greenjacking – A tactic for the toolbag of environmental champions? Reflections on an SME success story. *Eco-Management and Auditing*, **6**(1):26–33.

Wartick, S.L. and Cochran, P.L. (1985) The evolution of the corporate social performance model. *Academy of Management Review*, **10**:758–769.

Weber, M. (1947) *The Theory of Social and Economic Organization*. Glencoe, Ill: The Free Press.

Whitley, R. (1998) Internationalization and varieties of capitalism: the limited effects of cross-national coordination of economic activities on the nature of business systems. *Review of International Political Economy*, **5**(3):445–481.

Whitley, R.D. (1992) *European Business Systems: Firms and Markets in their National Contexts*. London: Sage.

Williams, P.F. (1987) The legitimate concern with fairness. *Accounting, Organizations and Society*, **12**(2):169–189.

Wolfe R.A. and Putler, D.S. (2002) How tight are the ties that bind stakeholders groups? *Organization Science* **13**(1):64–80.

Zairi, M. and Peters, J. (2002). The impact of social responsibility on business performance. *Managerial Auditing Journal*, **17**(4):174–178.

3 A Historical Overview of Stakeholder Management

Frank Harris

3.1 Introduction

The progressive development of different forms of contract is explained against the background of historical formal reviews of construction industry performance. The principles and arrangements for managing the growing trend of more active stakeholder influence on the activities of the construction industry in general and firms and projects in particular are highlighted in this context.

Achieving quality of the built environment may be observed as sustainable and sensitive land use; inspired, sympathetic and aesthetically appealing architecture; affordable functionally well-designed and constructed domestic, commercial and industrial buildings, facilities and infrastructure; high standards of in-built utility/health and safety/energy efficiency/maintainability/environment-friendly features.

The more successful projects are usually achieved by enlightened and informed clients/developers/promoters/patrons acting intelligently in concert with imaginative and able project leaders or advisers, reinforced by competent designers and contractors, all having common purpose aimed to ensure client satisfaction for the finished product, i.e. fulfilment of conception, function, cost, time, quality, utility and after-sales service objectives. Hence the importance of selecting the appropriate contractual arrangements for the stakeholders engaged to undertake the necessary responsibilities and functions, namely funding and finance; aesthetic aspect, form, function and utility; design, engineering and cost planning; construction and finally successful operation and maintenance. All of which need to be carried out within the normal bounds of official building controls, regulations and corporate social responsibilities (CSR). Best results impacting on the project and stakeholders involved normally exhibit the following key attributes:

- extent of design complete before construction commences;
- complexity of the project;
- experience of the contractor's project manager (PM);
- contractor's past performance;
- supervision of quality and progress by the client's project leader.

To this end, Figure 3.1 identifies the development phases conventionally present in the construction process. The explanation of these phases identifies the key stakeholders involved in their attainment.

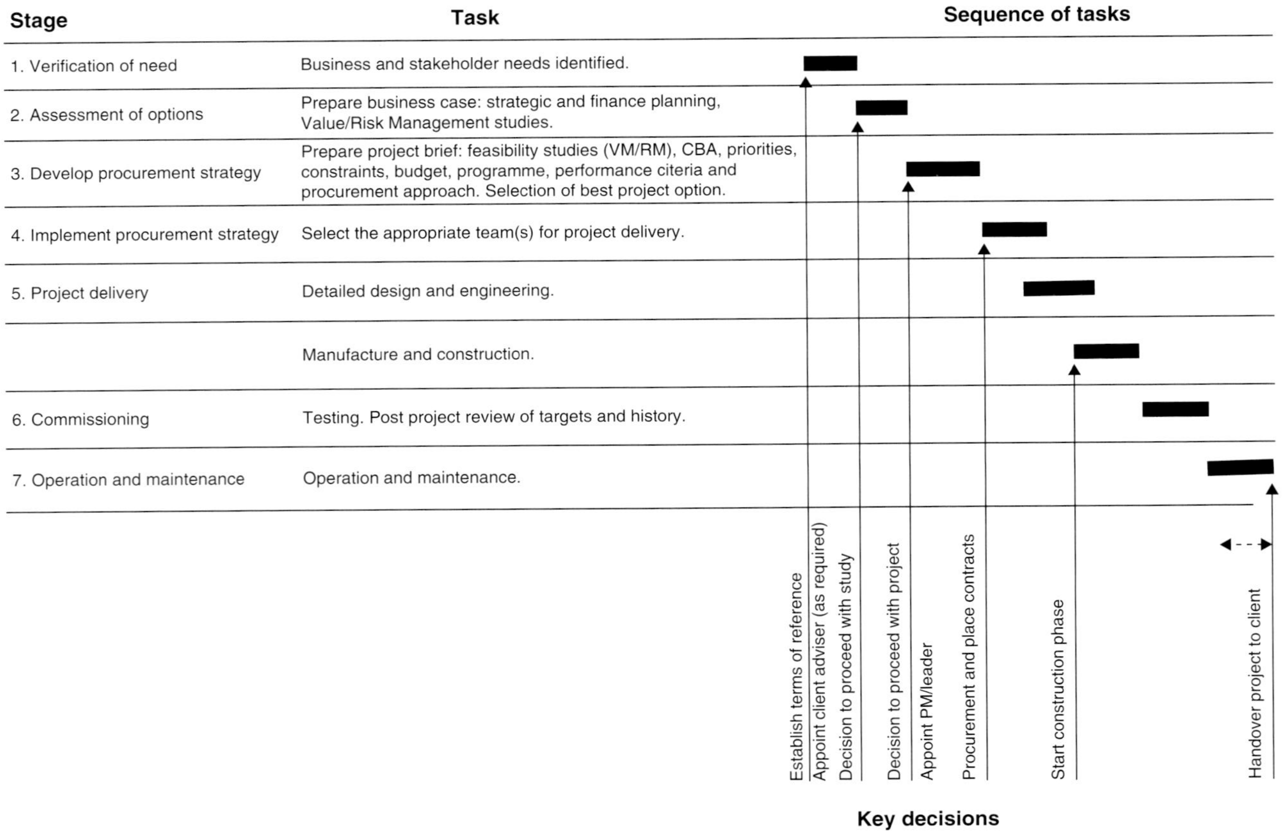

Stage	Task	Sequence of tasks
1. Verification of need	Business and stakeholder needs identified.	
2. Assessment of options	Prepare business case: strategic and finance planning, Value/Risk Management studies.	
3. Develop procurement strategy	Prepare project brief: feasibility studies (VM/RM), CBA, priorities, constraints, budget, programme, performance criteria and procurement approach. Selection of best project option.	
4. Implement procurement strategy	Select the appropriate team(s) for project delivery.	
5. Project delivery	Detailed design and engineering.	
	Manufacture and construction.	
6. Commissioning	Testing. Post project review of targets and history.	
7. Operation and maintenance	Operation and maintenance.	

Key decisions:
- Establish terms of reference
- Appoint client adviser (as required)
- Decision to proceed with study
- Decision to proceed with project
- Appoint PM/leader
- Procurement and place contracts
- Start construction phase
- Handover project to client

Figure 3.1 Development phases in the conventional construction process.
(*Source*: Harris and McCaffer, 2006; Used with permission from Wiley-Blackwell)

3.2   Verification of need

3.2.1   Stage 1: Verification of need

At this initial stage, the *terms of reference* are established, assisted by an external adviser as required in the case of an inexperienced sponsor. Aspects to be investigated include determining the need and objectives for the project, identifying who will act as the project sponsor in the client organisation; identification of other stakeholders; prioritising the constraints regarding time, cost, quality, finance, legal issues, impact on the business, etc. including for publicly funded projects such as local government; and fundamental issues concerning demonstration of 'Best Value' procurement. Subsequently a decision to carry out *further study* on the potential project options can be implemented.

3.2.2   Stage 2: Assessment of options

The potential project options are evaluated for feasibility and value appraised with techniques such as Value Management (VM), Risk Management (RM) and Cost Benefit Analysis (CBA). A firm business case is subsequently prepared giving for each option a physical outline as appropriate, base estimate, investment and financial appraisal, risks/life cycle costs, time plan, and technical, legal and regulatory issues together with a provisional budget. Thereafter the *decision to go ahead* with the project can be made based upon the options information and a *PM/leader is appointed* to act for the client, which in the case of an experienced client in building procurement might be an in-house function, otherwise an external specialist professional organisation (stakeholder) is sought.

3.2.3   Stage 3: Develop procurement strategies

The PM/leader in conjunction with the client reviews the business case, and takes on the major task of preparing the *project brief* covering details such as dealing with statutory regulations and approvals, execution plan, performance criteria, constraints, budget, control and reporting procedures. In particular VM/RM techniques are further applied to assist in the evaluations, thereafter the decision to proceed with the project can be reconfirmed, the procurement strategy articulated and the perceived best approach selected.

3.2.4   Stage 4: Implement the procurement strategy

The most appropriate contract category (separated, management, integrated or discretionary contract), contractual arrangement or type and form of agreement are chosen, which may have already become evident in the business plan and project brief. The best team/parties to deliver the business solution for the client/sponsor is now *procured* and contracted. In the case of public sector projects, the European Union Procurement Directives require projects exceeding specified threshold values to be advertised across member states in the *Official Journal of the European Union*.

3.2.5   Stage 5: Project delivery

Essentially execution of the project in terms of *design and/or construction* is now the task of the selected project team or parties. However, the PM continues to act for the client/sponsor, and has specific responsibility for briefing the selected stakeholders on all the aspects of the project, inducting new members as and when appropriate, implementing agreed performance, control and modification measures, advising and making decisions relating to the project objectives, requirements, agreeing payments, changes, contractual and legal issues, etc.

3.2.6 Stage 6: Commissioning

The PM produces a historical review of the project, assesses the asset's performance against the set targets, including any necessary physical testing and minor adjustments to ensure that the completed asset has achieved the objectives and functions satisfactorily; finally accepting handover on behalf of the client.

3.2.7 Stage 7: Operation and maintenance

Some projects, if procured through Private Finance Initiative (PFI), DBFO, etc., require the facility to be both operated and maintained profitably by the constructor under a business agreement; hence, the impact of life cycle costs need to be carefully considered, particularly materials and technology obsolescence over a long period.

3.3   Historical context

The 8–10% of GDP typically generated by the construction industry and the relatively low level of productivity growth in this domain has constantly exercised Government attention, manifested through the long series of joint government and industry initiatives directed towards improving performance. The reports often produced in this regard mostly concern perceived inefficiencies, practices of the industry's major stakeholders, and business and contractual arrangements. The main features of the most recent and pressing reports are catalogued in Table 3.1.

3.3.1 Chronology

Early reports – largely concentrated on the placing and management of contracts; problems before the construction industry; communications, interdependence and uncertainty; large industrial sites; the public client and the construction industries; and faster building.

Latham 1 – dealt with mistrust between the client's professional advisers, contractors and subcontractors, particularly delayed payments, disharmony, poor quality and client dissatisfaction of project outcomes.

Latham 2 – laid stress on the potential for productivity improvements through better procurement practice, tendering arrangements, conditions of contract application and most notably teamwork.

Egan 1 – went further by raising key issues relating to inefficiency, unpredictability and customer dissatisfaction. The key recommendations being eliminating waste and increasing value for the customer, partnering through supply chains, integration of processes, benchmarking and quantified outturn measures, standardised processes and offsite fabrication. The report lead to M4I (Movement for Innovation) which attempted to drive the identified potential improvements by raising awareness towards value for money, reliability and respect for people through the demonstration and dissemination of best practice and innovation. Important targets included annual 10% increases in productivity, 20% reduction of reportable accidents and 10% reduction of project completion times.

Egan 2 – stressed the importance of collaboration between members of the supply team. The concurrent engineering concept inherent in 'world-class' manufacturing was also given emphasis, whereby design and build are brought together seamlessly. Key performance indicators (KPI) for all aspects of the project, the partners in the process and stakeholders in general formed an important element.

Table 3.1   Overview of some reports on improving construction practice

(Associated) Author	Title/Subject matter	Date
Early reports – Simon:Emmerson: Banwell:National Economic Development Office:Tavistock	Various reports on placing of contracts	1944:1962:1964:1967:1966
Latham 1	Trust and money	1993
Latham 2	Constructing the team	1994
Levene Efficiency Scrutiny	Construction procurement by Government	1995
CSCS	Construction Skills Certification Scheme	1995
MCG	Major Contractors Group	1996
Egan1	Rethinking Construction (Movement for Innovation)	1998
National Audit Office	Modernising construction	2001
Egan 2	Accelerating Change Construction	2002
National Audit Office	PFI Construction Performance	2003
National Audit Office	Improving public services through better construction	2005
Conditions of Contract	New forms of contract	2006
CDM	Construction Design and Management regulations	2007, 1994, 1974
National Audit Office	Building for the future: sustainable construction and refurbishment on the Government estate	2007
Department of Communities and Local Government (DCLG)	Developing the Local Government Services Market: Working Paper on Local Government Asset Management and Construction Services	2007

National Audit Office – like Latham and Egan adversarial relationships were given prominence and examined the benefits of partnering in the construction process, most notably
modernising construction procurement by government departments and agencies. The
report recommended that government adopt PFI, Public Private Partnership (PPP), Prime
Contracting or Design and Build as the best way forward in securing more successful outcomes for public sector construction work, reiterated in the 2003 report on PFI Construction
Performance. The progress 2005 report, based on three major case studies, further identified
six key areas for further improvement, namely effective construction programme, developing
and supporting capable clients, basing design and decision making on 'whole life value', using
appropriate procurement and contracting strategies, working collaboratively through fully
integrated teams, evaluating performance and embedding project learning. Most recently the
2007 study reported on sustainable construction and refurbishment of the Government estate.

Levene – made a number of recommendations to improve the procurement and management of construction projects, including better communication within the construction industry to reduce conflict; adoption of a more commercial approach; negotiation of deals justified
on value for money grounds; and increased training of civil servants on procurement and RM.

Major Contractors Group (MCG) and Construction Skills Certification Scheme (CSCS) – emphasised the need to properly assess the competencies of construction personnel and introduced

the CSCS, which requires possession of the appropriate CSCS card to work on the construction site of participating contractors. All personnel, including the design and other professional consultants, are embraced, and passing the CSCS Operative Health and Safety Test also became mandatory requirement.

The scheme stakeholders were CSCS Limited, controlled by a management board drawn from the Construction Confederation, Federation of Master Builders, GMB Trade Union, National Specialist Contractors Council, Transport and General Workers Union and Union of Construction Allied Trades and Technicians. Observer status extended to the Department for Education and Employment, the Department for the Environment, Transport and the Regions, the Health and Safety Executive and the Confederation of Construction Clients.

The CITB – ConstructionSkills administered the scheme and verified documented formal evidence of competency. A Certificate of Achievement was also available for individual companies depending on the proportion of their onsite registered cardholder personnel.

The MCG has enthusiastically adopted the CSCS, which through the interests of major contractors is able to effectively lobby government and other decision makers drawing on the expert practitioners of its members.

Construction Design and Management (CDM) – Health and Safety compliance in the United Kingdom is governed by EU legislation largely through the 'Health & Safety at Work Act 1974' and Construction (Design and Management) regulations CDM (2007). Under these revised regulations, the client has become the statutory Health and Safety duty holder, but can be discharged through an appointed independent and competent Health and Safety coordinator for the project.

Department of Communities and Local Government (DCLG) – this recent working paper proposes greater strategic leadership and use of more integrated and innovative solutions for asset management and construction services.

3.3.2 *Progressive contracts development and stakeholder interactions*

Judged purely on the basis of these reports, the client, project leader, designers, contractors and supply chain all tied in by formal contracts, and acting in concert with the regulatory and standards bodies, have clearly been regarded as the dominant primary stakeholders in the process by getting the project up, running and commissioned. These formal contracts place certain expectations on the different stakeholders; and so the relevant responsibilities of these stakeholders are broadly varied in the different types of contract. Figure 3.2 shows different types of formal contracts which have evolved over time.

The current approximate proportions of total contracts embraced by these new measures are shown in Table 3.2.

The evolution of these formal contracts and their associated demands on stakeholders are briefly examined below.

Traditional separated contract

Traditionally, the separated type of contract formed the popular option for clients, who discharged the development/design risks to architectural or engineering consultants on the basis of fee-based contracts, and more rigorous fixed price contracts for the actual production deals with the main contractor and/or subcontractors.

In this approach, ideally, the design would be complete before tendering; hence, any later variations caused by changes in the quantity of work or site conditions were issued as written instructions by the architect or engineer. In some forms of contract, the role of a quantity surveyor is defined. However, this approach demonstrated a tendency to contractual disputes around variation requests, interface management, time delays, quality performance and cost over budget.

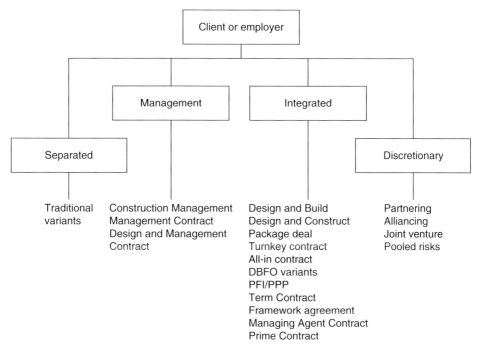

Figure 3.2   Types of formal contracts.
(*Source*: Harris and McCaffer, 2006; used with permission from Wiley-Blackwell)

Table 3.2   Usage of different types of contracts

Type of contract	Proportion	
	By value	By number
Separated	43%	83%
Integrated	43%	15%
M&CM	12%	1%
Discretionary	2%	1%

Management oriented contracts (1980s–)

Over recent years, clients of major building and civil engineering work in particular too often experienced poor construction performance, especially when designs were incomplete before contractors were appointed, exacerbated further in not securing timely planning approvals, with intricate projects being especially affected. Thus for innovative highly complex projects requiring progressive development of the design details as construction proceeds, relatively inexperienced clients occasionally adopted this revised form of contract, described as a management contract, construction management contract, design and management contract.

In the management oriented contract, the construction manager or managing contractor joins the project management team at the earliest possible time prior to construction on equal terms to other consultants. Responsibilities cover preparing the overall construction programme and works packages, steering these through the design stage, recommending/appointing the works (sub) contractors and securing their smooth integration. The combined design and supervisory functions of the traditional architect/engineer are removed, being executed by the appointed team. Coordination and cooperation is largely achieved by requiring the contracted designers and works (sub) contractors to provide detailed programmes with procurement and key dates clearly defined together with explicit statements on working methods and equipment resources. Managing all these demands considerable programming expertise and effort necessitating regular progress meetings and budgets for each work package to be carefully monitored. Indeed successfully bringing together all the project interfaces and ensuring good fit of others' designs, technical proposals and site work is a major challenge.

Even so the traditional choices of contractual arrangement still generally applied, with contracts formed typically between client and designer, client and contractor, and contractor and subcontractors. The main difference to the tradition arrangement being that the main contractor has few, if any, obligations to execute actual construction work. A fee is charged for coordinating the subcontractors and advising the design team.

The construction management contract in contrast is devised for a particular project, with the client having a direct contract with each individual works contractor.

Contrastingly, quality management is largely executed by works (sub) contractors themselves, with independent site inspection and testing executed as necessary, All encouraged by only inviting for tender those works (sub) contractors with an ISO 9000 or similar accredited quality system and a proven record of producing good quality work. Similarly for safety performance in meeting specified requirements laid down by the managing firm embraced by the CDM regulations.

Labour management required the managing firm to lay down a labour relations policy with clear terms and conditions of employment for all works contractors including disputes procedures agreed with unions both at local and national levels. The managing organisation should also have a firm policy on the engagement of self-employed labour; similarly for welfare and site facilities.

Integrated contracts (1990s–)

Problems of project management and contractor relations nevertheless continued unabated, consequently, the design and build contract focused attention towards placing the project under the control of a single party leader in an attempt to provide better *value for money* in reduced whole life cost and improved quality. This approach stresses the merits of concurrent engineering so dominant of 'world-class' manufacturing, which for the construction context implies integrated design and construction, particularly the key elements of product development, project implementation, partnering in the supply chain and increased usage of manufactured components, i.e. 'lean construction'. Ideally the method is best suited to routine construction procured from contractors marketing well-developed standard products, i.e. those which can be readily evaluated by a well-informed client as to what is to be provided, its specification, method of payment and means of monitoring performance, cost and quality.

Regular projects such as office blocks, high-rise buildings, schools, hospitals, housing, factory units, prefabricated modular system industrial assemblies, car parks, etc. represent typical examples.

Significantly greater value creation is predominantly sought through an established, well-managed supply chain or network, rigorous measurement of performance against KPI targets and the relentless drive to achieving sustained safety and quality improvements.

Inevitably the client has had to relinquish some control over the design, although **early contractor involvement** and most recently partial contractor involvement *post the feasibility stage* using tendered fees with subsequent as-built sharing of savings on target cost estimates is being tried, partly to overcome this difficulty – on condition that a thorough procurement process (see Holt above) is well established in the client organisation.

Notwithstanding, aspects of design detail may still be incomplete when the contract is awarded, which may subsequently require costly time and effort (to be absorbed by the contractor) in agreeing responsibilities, duties and site coordination with the chain of designers, specialists, suppliers and subcontactors, to achieve the project specification and quality expected by the client for the contracted price.

Significantly, the contractual arrangement is a single point responsibility and accountability of the contractor to the client for both the design and construction facets is secured either by competitive negotiation or by tendering.

The most novel forms of the integrated contracts are the PFI and PPP schemes initiated by central government, one of its agencies or local government. Normally the engaged contractor is required to finance, design, construct, maintain, and where appropriate operate the facility over a stipulated concessionary period of ownership of perhaps 30 years or more, and be paid a rental or lease sum by the client for the services provided. Significantly participants may sell their stakes in the asset at market values during the concession.

This phase of public sector experimentation has also seen the emergence of framework agreements, managing agents for highways maintenance and prime contracts for the MOD.

Discretionary contracts (2000s–)

Large specialised projects such as power stations, airports, oil refineries and similar complex utilities have still proved difficult to manage satisfactorily from conception through to commissioning and handover. The reasons are many, including securing project approval and construction progress in a climate of pressing concerns around environmental and social responsibility issues. In essence, however, poor construction performance has continued to relate mainly to unique single project procurement and the lack of stability in relationships between stakeholders and especially made more difficult by the inability of clients to adequately define their needs at the outset.

The new element of discretion is attempting to address and improve working relationships between the parties by encouraging teamwork and reducing potential adversarial attitudes manifested in the options of partnering, alliances, joint ventures and pooled risks contracts.

The formulation of contractual conditions reflecting the intentions of the parties concerned present a challenge from conventional practices, especially in expressions of goodwill and cooperation which if tested in court should a dispute arise could prove legally binding. However, third parties not involved in the partnering agreements such as suppliers and subcontractors continue to pose potential points for arrangements to unravel into conventional contractual elements if disputes occur. Indeed, unscrupulous contractors have the scope to hide from the client transactions detail with suppliers unless vigilance is strongly upheld.

Universal 'value for money' is yet to be proven with this kind of arrangement and awaits further trials and reported results from users of the system before firm conclusions can be truly finalised.

3.3.3   Associated conditions of contract

Some of the latest forms of contract were introduced to embrace the recommendations in the wake of the formal reports described earlier. Conjointly, (revised) conditions of contract have evolved to underpin these contract forms. Notable amongst these are:

- Joint Contracts Tribunal (JCT 2005)
- EPC (Engineer Procure Contract)
- BOT (Build Operate Transfer)

- Government Prime Contracting Contract
- JCT (i) Design and Build Contract 2005; (ii) Major Project Construction Contract 2005; (iii) Framework Agreement 2005; (iv) Adjudication Agreement
- GC/Works/1 Design and Build Contract
- ICE Design and Construct Contract
- NEC3 (i) Engineering and Construction Contract; (ii) Professional Services Contract; (iii) Term Service Contract; (iv) Adjudicator's Contract, see Fig (8.10); (v) Framework Contract.
- Project Partnering Contract 2000 (PPC2000) developed by the ACA.
- Public Sector Partnering Contract (PSPC) endorsed by the Federation of Property Societies.
- BeCollaborative form developed by Collaboration for the Built Environment.
- ECC Partnering Agreement.
- JCT 2007 form.
- JCT(2008) Pre-construction Services Agreement
- NEC option X12
- FIDIC conditions of contract for: Design, Build, Operate (DBO) projects - 1st edition

The standard form of construction contract will often identify the responsibilities of client and contractor. In addition, some functions of other stakeholders are mentioned in these documents, e.g. quantity surveyor, subcontractor, PM and architect.

3.3.4   Stakeholder considerations

While the changing contractual circumstances of the industry are aimed towards effectively engaging the traditional players, many more stakeholders have emerged, as social responsibility and environmental groups came to the fore. Consequently, even more fresh, practicable, robust and measurable management processes have needed to be devised and audited to try and ensure that no stakeholder is overlooked and all are satisfied as far as possible, depending on the personal or specific interest involved, with any actual or potential impact of construction work on the perceived concerns embraced. Thus in the quest for improved communications, project clients and their project leaders find it useful to categorise stakeholders according to one of the following headings, listing out the key stakeholder motivations and interests.

3.3.5   The stakeholders

Stakeholders may be internal members of the project team or coalition, while others are externally affected by the project, either as a threat or benefit. Thus, stakeholder interests may run in many directions, with each stakeholder regarded as contributing something and/or receiving something from the project in the case of a specific construction contract, or from each other when acting as separate parties in a business situation. Hence the differing stakes may conflict and need to be well managed.

In the case of a construction project, a stakeholder is an individual or organisation holding an interest or a share. Many stakeholders may be involved in a project. Figure 3.3 shows the grouping of these stakeholders who are also identified and classified in Table 3.3. The following paragraphs explain the influences of these stakeholders on construction projects.

3.3.6   Primary players

The client whether a firm, public body or private individual acting as the promoter or developer is commonly the *first level primary player* or stakeholder in any construction project in

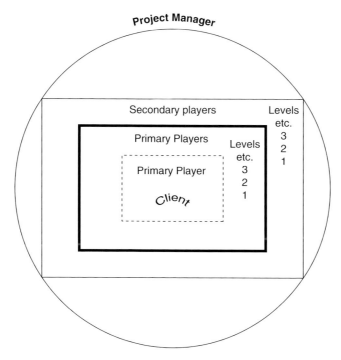

Figure 3.3   Grouping of stakeholders.

having strategic and operational responsibilities, interfaces with the markets, is concerned with competitive position and determines the need for the facility. Authority is usually demonstrated by setting the needs and requirements for the project, i.e. scope, extent, character, design and overall budget including any land issues, has ultimate control of progress, quality, funding and interim payments to the project delivery team, and finally signs off the project when commissioned to full satisfaction. Particularly important for the client will be the development of solid relationships and confidence building with the key *second level* player(s) responsible for financing the scheme, whether government investment, shareholding stakes, bank loans, bonds, private equity funds or even the derivatives market.

At the *second level*, the project leader or manager has the prime responsibility for organising the project team and will be selected from construction management firms, general contractors, architects, engineers, owners themselves or specialty contractors. The task is to guide the many separate phases and parties involved towards producing a successfully completed building or facility. Particularly interpreting the plans and specifications; preparing cost estimates and programme schedules to meet the client's requirements; determining and implementing the best development and construction practices, means and methods to meet time, cost and quality set by the client, including overseeing and managing all the design and construction work. Notably the critical or complex architectural or design interfaces, and the key (sub) contractor possession, handover points and interim payments need to be identified and properly managed.

Also at level 2, design professionals such as architects, engineers and specialist consultants deal with end user needs and requirements, aesthetics, form, function, utility, safety and maintenance issues of the proposed building, including the facility programme, construction plans, construction details and specifications.

Table 3.3   A mapping of stakeholders

Primary players			Secondary players		
First level	Second level	Third level	First level	Second level	Third level
• Clients • Promoters • Patrons • Customers	• Project leader/manager • Financial institutions and lenders • Designers and professional consultants • Contractors • Health and Safety coordinator	• Subcontractors • Business Process Outsourced suppliers • Equipment and maintenance suppliers • ICT vendors • Directors, managers and employees • Supply chain vendors and associates • Insurance organisations • Joint venture partners and alliances • Logistics and transport vendors • Overseas agents and partners • Security vendors	• Building inspectorate • Unions • CSCS • Planning, Regulatory and Standards authorities • HSE inspectorate • Police • Local government and agencies	• Investors–shareholders • Government departments • HSE • Education and training providers • Qualifications authorities and professional societies • Internal R&D including 'suggestion box' input • Individuals, groups or organisations concerned with environmental, social responsibility and human rights issues	• Private organisations interests • External R&D innovators and entrepreneurs • Public body interests • Visitors • General public and lobby groups • Competitors • Local communities and citizens • Trade associations and manufacturers • General users of the facility

Again at this level, the general or principal contractor, who may not actually perform any of the actual construction work, manages all the functions relating to estimating, scheduling, purchasing, equipment provision, supervision any of its own field staff such as site management, field engineers, foremen, lead workers and sub/speciality contractors.

Subcontractors typically work independently under a contract or agreement with the general contractor and in turn may engage others in the supply chain. Vital services are provided but at the expense of added risk regarding scheduling, interface management, safety and quality. Recently the imposition of responsibility of the client for Health and Safety has raised the stakeholder value of the Health and Safety coordinator contracted to execute the legislation.

3.3.7 Secondary players

Secondary players exercise influence on the construction process beyond the direct control of the primary players. Contractual relationships with primary players are generally not involved but nevertheless exercise influence. For example, local government, public agencies and private individuals concerned with environmental issues, noise and pollution. While others such as labour unions, training organisations and professional societies may be more interested in worker remuneration, occupational welfare, health and safety, employment conditions and qualifications. Indeed even employees themselves can cause considerable disruption in a shortage market, especially through taking poaching opportunities. Shareholders are another separate group who by having invested in the firm, or indeed indirectly in the project itself, expect rewards in terms of dividends, share appreciation, capital repayments etc. and may even demand board level representation, e.g. pension company investors. Potentially influential R&D innovators and entrepreneurs able to offer fresh thinking may prove invaluable in maintaining comparative advantage and business model improvements. Contacts such as business partners, customers, competitors, consultants, trade shows, conferences, employees, internal sales and service units, academia and employees are all likely candidates with ideas to exploit. Not least is the general public or third party user of the facility who may be concerned with an unpredictable variety of issues both during development and the construction phases.

3.4 Stakeholder management

Customarily stakeholders involved with a construction project have largely been regarded as the primary players with secondary interests handled only as they arose, most typically observed in the traditional form of contract. However, the greater concerns in society for environmental and social issues, together with the evolving types of contract, require more proactive attention be given to managing the issues raised by stakeholders, both perceived and those randomly occurring during the course of the works.

The objective of stakeholder management is therefore to manage relationships in order to motivate stakeholders to behave in ways that support the firm's objectives, with commitment demonstrated visibly at the executive level, and audited integrally within the corporate governance process. Necessarily the approach to date has tried to embrace matters either arbitrarily in relation to the impact on profitability alone, but lately more readily as part of CSR policy. Here a multi-fiduciary emphasis offers some response to shareholder demands at the expense of normal business results. Ultimately stakeholder values may be fully *synthesised* and incorporated into the firm's business model by embracing a moral, political, technological and economic non-obligatory tone. While the latter degree of commitment is currently hardly evident, societal trends are heading in this direction.

3.4.1 Management process

Practical stakeholder management requires identification of the following key considerations:

Who are the stakeholders and their interests?
What opportunities do these interests offer the project or firm?
What challenges or threats are thereby presented?
What level of responsibility is appropriate in meeting stakeholder requirements?
What are the necessary strategies demanded, e.g. direct dealings, aggressive attitudes or accommodating or a combination of different courses of action.

Clearly, therefore, the management process requires a good understanding of the current stakeholder circumstances, the influential factors and crucially obtaining of feedback information, so that the best approach may be tailored to achieve maximum benefit. The project's sponsor should appoint a 'champion' best placed to address the relevant stakeholder issues and proactively promote the stakeholder programme. Practically, the position is likely to be undertaken by the project leader/manager, broadly the responsibilities being typically carried out bimonthly along the following lines:

- Establish the project stakeholder 'champion' to drive relationships forward and manage the process.
- Identify the risks from stakeholders.
- Incorporate into the process appropriate individuals responsible for each necessary stakeholder/player level depending on the needs of the project and/or degree of stakeholder interest.
- Involve stakeholders at the appropriate stage in the process and identify new stakeholders as required.
- Assess the current level of support provided to each stakeholder.
- Continuously evaluate changed interests or influences in the stakeholder programme, and identify any significant potential events, risks or impacts from the stakeholders.
- Address the issues raised and devise remedies, contingency and continuity plans.
- Encourage teamwork and problem solving with the stakeholders.
- Reiterate the benefits and outcomes of the programme to remind stakeholders of the end goals, particularly during challenging times.
- Identify other supportive personnel to help engage commitment.
- Set up a review mechanism for feedback, ideas and knowledge exchange.
- Champion carries out a regular review of the effectiveness of the programme and amends as necessary.
- Primary players conduct similar reviews for internal monitoring and control of effectiveness.
- Champion periodically reports results and issues to the sponsor for audit and corporate governance, CSR evaluation and overall impact on the sponsor's business.
- The individual players investigate impacts for internal consumption and management, as appropriate for their areas of responsibility.

The whole process can be prepared as a spreadsheet document to constitute the task register for each manager, project, department, company, etc. depending upon the organisational level carrying out the evaluation.

The proposed actions are installed and the anticipated outturn costs of stakeholder compliance including those that arose unforeseen are evaluated and recorded. Feedback information is subsequently redirected to help inform and adjust contingency/continuity plans and estimates as necessary.

Formalisation of the procedure using a suitably designed coding system should help evaluation and subsequent identification of 'best practice' measures of control to better inform the enterprise on stakeholder management and thereby rationally guide and manage the relevant issues.

3.4.2 *Corporate social responsibility*

Finally the firm's CSR policies concerning governance, ethics and the environment also need to be fully embedded into the stakeholder management process and ideally directed through the stakeholder champion. The objective being to forestall PR disasters by using Relationship Marketing (RM) techniques to develop codes of practice and conduct aimed at offering greater transparency in commercial undertakings. By achieving common CSR rules, risks may be effectively spread and opinion shaped, such that CSR decisions across the company's operations become so completely ingrained in the stakeholder management process that the notion of 'doing well by doing good' becomes part of the company's competitive advantage.

3.5 Conclusions

Carefully developed positive and mutually supportive stakeholder relationships that encourage trust and stimulate collaboration by reducing conflict and suspicion can help avoid time delays and increased costs, whose reduction would be beneficial to the normal business of profit generation. Moreover the gaining of reputation for ethical and socially responsible behaviour may offer the opportunity to steal a competitive edge. But to take effect, there must be an awareness by the enterprise, its management and workforce of the diverse and multiple issues likely to upset stakeholders. Tension will nevertheless arise even between different stakeholder interests, which need to be recognised by a responsible management implementing appropriate procedures and the auditing of performance and review.

Reference

Harris, F.C. and McCaffer, R. with Edum-Fotwe, F. (2006) *Modern Construction Management* (6th ed.). Oxford: Wiley-Blackwell.

4 Uptake, Applications and Best Practices in Stakeholder Management

Michael Thompson

4.1 Introduction

For many years, the construction industry, particularly in the United Kingdom but elsewhere also, has considered itself to be so different to other industry sectors that the applications of best practice in the other sectors could not possibly be relevant. This reflects attitude of mind at the time.

At the same time, the industry during the 1960s and 1970s and beyond was becoming more and more confrontational, as competition increased and with it, any margins relating to profitability reduced. The rules of the game had unwittingly been set up by client organisations and their agents, often mindful of their responsibilities when spending public money but not always mindful of value for money or of the risks associated with cost reduction. A confrontational approach meant that not all stakeholders were being kept informed of what was planned, increasing the risk of changes (e.g. in design or construction, or even operation) at a later stage.

As the competition increased, so did successful tenderers rely on there being flaws in the contracts that they had won. The first stage, on award, became to scrutinise the contract in detail, determining where there were discrepancies between what had been planned and what was actually there, seeking out omissions and errors and from the start, developing claims that would enable the contractor to make a reasonable profit, and more.

This as often as not lead to overspend on projects, sometimes by substantial amounts, reducing the industry's ability to undertake all of the work that it had to do, and helping peripheral elements of the industry to prosper where they had not necessarily prospered before, while at the same time restricting the ability of the architects and designers to develop the very best products which fitted not only what the clients wanted but what they needed.

Whereas the quality of the finished product was often excellent, commercially, the project would be a disaster, either for the client or for the contractor, or for both. Client budgets were frequently exceeded, sometimes by as much as 50%.

Confrontation had taken over from common sense, and few were benefitting. Not only that, but in the European and world arena, the construction industry was not seen as competitive, putting it at risk, just as had been the situation for the automotive and other manufacturing industries before it, to more efficient organisations from overseas.

Something had to be done to improve the situation that the industry had found itself in, within the United Kingdom.

4.2   Recognising the need

Speaking at an event several years later, Latham (2007) summarised the problem in 1993 as follows:

- there was client dissatisfaction
- there was poor performance by all
- there were adversarial relationships
- there was a 'Claims culture'
- there were non-existent profit margins
- there was heavy lobbying by specialist contractors

4.2.1   Constructing the Team

'Constructing the Team' (Department of Environment, 1994) was born as a joint government/ industry initiative with the aims of reducing conflict and litigation and improving productivity and competitiveness. Sir Michael Latham was its principal author.

The main recommendations for clients to come out of the Latham Report included:

- procurement guidance for clients
- checklist on briefing
- quality/price mechanism to choose consultants
- the setting up of a clients' forum
- the wishes of clients to be paramount

The main recommendations for Industry included:

- adopting a target of 30% real cost reduction by the year 2000;
- improving tendering arrangements/registration (with government);
- drawing up a joint code of practice for selecting subcontractors and main contractors;
- implementing previous reports on training and on the education of professionals;
- improving the public image;
- producing a coordinated Equal Opportunities Action Plan.

There were also recommendations for contracts including the development of standard contract documentation based on a set of 13 principles. The focus of the principles was a mix of cultural and process orientation. Hence, the concepts of partnering and collaborative working within the UK Construction Industry were born, and were put into action in a very short space of time.

This is possibly an apocryphal story but it illustrates the point. The newly introduced 'Partnering' was being applied to a road project in the West Midlands in 1996. The Project Manager was being interviewed by the local radio station.

'Tell me', said the Interviewer, 'What is so special about this project?'

'It is really great!', said the Project Manager, 'We are all working together for the benefit of the project, without confrontation between the various stakeholders involved.'

'Oh' said the Interviewer, 'Isn't that the way you always do it?'

Common sense was beginning to take over from confrontation.

Example 1: Wastewater Treatment Works in the South of England (1995–1998)

The Wastewater Treatment Works was probably the first partnering project in the UK Water Industry, formed out of a partnership between four major organisations, the client, the project manager, the designers and the contractor. This was the first integrated project team as suggested later by Sir John Egan, set up in July 1995, where amongst the core team, there were no boundaries and everyone was working more for the project than for their own organisations. The same could not be said of the supply chain, although in the upper tiers, suppliers were aware that there was something special about the project.

The challenge for the team was to work closely with stakeholders and to design, construct and commission within 21 months a new Wastewater Treatment Plant, virtually from scratch, on a so-called 'green field site' which was in fact the site of two large borrow pits, filled with water. It was recognised that probably the most likely way to complete the project within budget and within time was by adopting a partnering approach. 'Get the culture in place and everything else, if successful, will fall into place'.

July 1995 turned out to be a hot summer, and when tempers should have frayed in the heat, the team worked together well, involving such stakeholders as were relevant to the development of the design. Using the tried and tested value methodology within a partnering environment, it was possible to develop a well judged and practical design which provided the client with what he needed while at the same time recognising both construction and operation and maintenance requirements to achieve best value throughout the life of the works.

Because of the critical path of the construction programme, it was necessary for the construction to commence before the design had been fully completed. This required strong collaboration between the designers, the client operations team, the constructors and the various supply chain members, and it worked well.

The net result was an operating works with no surprises that was handed over to the client on time and within budget.

One unusual aspect of the management was the use of a single, common filing system, available to all stakeholders.

In retrospect, there were things that the team could have done better, but it was agreed by all at the time that this project, worth around £12 million (late 1990s), was a success in terms of the good, strong collaborative working that took place between the various partners.

One lesson learnt was to do with the change of personnel associated with the project. This occurred during the natural change over from essentially civil engineering personnel to mechanical and electrical engineering personnel, the latter not fully buying in to the culture of the team, developed during the early stages of the project. This occurred again shortly after the commissioning of the works when the original client operations staff were replaced by staff who had not been involved in the development of the project. They started to introduce what they considered would be 'nice to have' rather than what was 'needed to have'.

4.3   So who are the stakeholders?

During the 1960s and 1970s and perhaps later, the term 'stakeholder' would not have been given a lot of consideration if any at all. Equally the term 'risk' was also something that was not given a lot of consideration, although every budget or cost estimate would have contained something called a 'contingency', often estimated in percentage terms against the capital spend for a project.

Accepting that the term 'stakeholder' means 'any party who affects, or can be affected by the actions of another party', there is a clear link between 'stakeholder' and 'risk'.

Figure 4.1 illustrates various stakeholders associated with a particular project. This particular diagram related to a road scheme, but it is equally applicable in principle to other

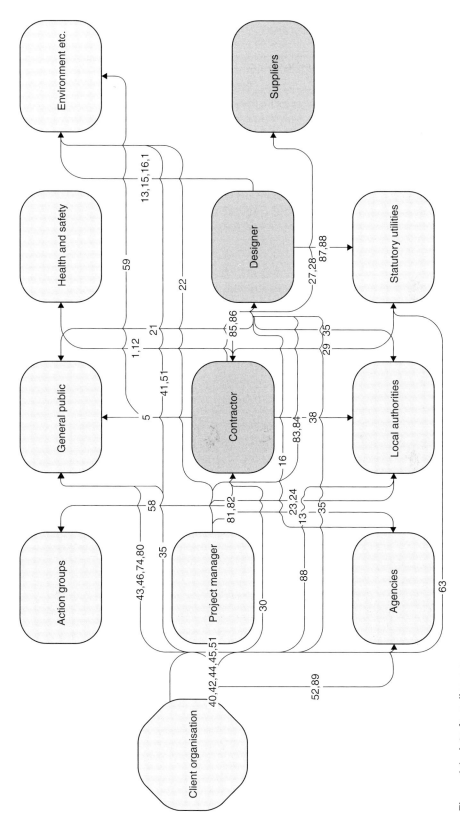

Figure 4.1   Interface diagram.

types of scheme. Only the names and the relationships will change. The names are unimportant, except in as much as they represent different organisations, and could equally represent departments within larger organisations where there are interfaces which could affect or be affected by another party. Each of those interfaces (in this diagram, numbered to reflect the identity number of a particular risk) represents a risk or potential uncertainty that should be addressed by the project team.

Recognising one of the main recommendations of the Latham Report is that the client's wishes are paramount, the client must be the starting point for understanding the priorities and interests of project stakeholders. However, there is a distinction between what the client wants and what he actually needs. It is important that the latter is established before proceeding very far with a project, ideally in a collaborative environment involving all relevant stakeholders, using something like the following value methodology (BSI, 2000):

- know what the problem is that has to be solved,
- establish what data is available to enable decisions to be made,
- what data is missing that would need to be obtained,
- create ideas that might enable the problem to be solved,
- judge which of those ideas might enable a relevant solution and which might not,
- developing proposals that will be acceptable to stakeholders,
- obtain buy in to the proposals from all stakeholders.

The project illustrated above (Example 1) used the value methodology to good effect in the first weeks that the project team came together, reducing risks by, amongst other things:

- Providing a good audit trail that showed how the project had been created, showing what ideas had been considered and why they had been accepted or rejected, thereby reducing the chances of fundamental changes to the design at a later stage (all stakeholders).
- Providing an installation that would meet the client and end users' needs – because the end users had been involved in the decision-making processes (end user in particular).
- Deciding a construction process which would minimise the risk of overrun in terms of time and cost (client manager).
- Involving local authority planners early so that there would be no surprises at the time of application, having already observed the requirement for maintaining appropriate sight lines in an area considered to be of beauty (local authorities at rural district and parish levels).
- Identifying environmental risks associated with the chosen land filling process and establishing that they were minimal (environment agency and others).
- Keeping the local public informed of activities on the project which could affect them, particularly to do with night working (local public).
- Working closely with different suppliers to be more certain that the assembled whole functioned as required (suppliers and end users).

In contrast to the Pennington project and towards the end of the last century, there was the 'Millennium Dome' project in London. Here was a situation of change, as one government departed and another came on board, one government minister replaced another. Everyone knew that we needed something notable to celebrate the coming of the new Millennium. In developing the Dome, little consideration had been given to how it might be used in the longer term; its use during the Millennium year being paramount. There were uncertainties. It could have been argued that if during the early stages of the design of the Dome, time had been spent considering its longer-term use, there would have been less risk of it standing empty for a long period following the Millennium year. Did we spend £800 million of public money wisely? The jury is out on that one.

4.4   Rethinking construction

The Latham Report: Constructing the Team was a major landmark in 1994 in the addressing of the problems within the construction industry, as already described in this chapter. Another major landmark was to appear in 1998. This was the report, 'Rethinking Construction', published by the Construction Task Force on behalf of the deputy prime minister. Sir John Egan chaired the Task Force and his name became attached to this report. The document formulated proposals for improvement of the construction industry by studying the experience that had been gained at the cutting edge of construction and in other industries that had transformed themselves in recent years.

The report makes it clear that the Construction Task Force is not asking the construction industry to do the things that it does better. It is asking it to join with government and major clients to do it entirely differently. What was proposed was a radical change wherein five key drivers of change were identified:

- committed leadership
- a focus on the customer
- integrated processes and teams
- a quality driven agenda
- commitment to people

Not surprisingly, several of these drivers focus on stakeholders.

It was also identified that ambition targets and effective measurement of performance are essential to deliver improvement. A series of targets for annual improvement were proposed, based on the Task Force's own experience and evidence obtained from projects both in the United Kingdom and overseas.

The targets were set as follows (forming the original basis of key performance indicators); changes being year on year:

1. Capital cost (all costs excluding land and finance) – reduce by 10% annually.
2. Construction time (time from client approval to practical completion) – reduce by 10%.
3. Predictability (number of projects completed on time and within budget) – increase by 20%.
4. Defects (reduction in number of defects on handover) – reduce by 20%.
5. Accidents (reduction in the number of reportable accidents) – reduce by 20%.
6. Productivity (increase in value added per head) – increase by 10%.
7. Turnover and profits (turnover and profits of construction firms) – increase by 10%.

The report recommended that integration should be focused around the four key elements of:

- product development
- project implementation
- partnering the supply chain
- production of components

It was strongly suggested that the Best Practice Programme run by the deputy prime minister should be turned into a knowledge centre for construction which would give the whole industry and all of its clients' access to information and learning from demonstration projects.

4.5 Accelerating Change

The next landmark report was 'Accelerating Change' in 2002, prepared by the Strategic Forum for Construction, again chaired by Sir John Egan. This report built on and reaffirmed the principles that had been set out in 'Rethinking Construction'. It acknowledged that change was already on the way, although it also recognised that whereas some of the Forum's proposals seemed controversial to some, they were common sense to others.

The vision of 'Accelerating Change' was for the UK Construction Industry to realise maximum value for all clients, end users and other stakeholders and to exceed their expectations through the consistent delivery of world-class products and services. This was to be achieved by setting targets so that by the end of 2004, 20% of construction projects by value should be undertaken by integrated teams and supply chains, and 20% of client activity by value should embrace the principles of the Clients' Charter. By the end of 2007, both these figures should rise to 50%. This objective warrants stakeholders working together in a much closer manner.

In addition, the Forum was determined to reverse the long-term decline in the industry's ability to attract and retain quality workforce by developing and implementing strategies which will enable the industry to recruit and retain qualified people, so that this would result in a 50% increase in suitable applications to built environment higher and further education courses by 2007.

A 'toolkit' was to be developed by April 2003 to help clients and others to assemble integrated teams, mobilise their value streams and promote effective team working skills. This was successfully undertaken, and is available to all on the Internet, although it is probably one of the best kept secrets within the industry.

4.5.1 Changing a long established culture

The advent of the work of Sir Michael Latham and Sir John Egan and others around the mid- to late 1990s and early 2000s produced for some the expectation that at long last the UK Construction Industry was changing for the better. Partnering workshops at the start of projects were becoming the norm, although the majority were organised by the contractors, not the client organisations, suggesting that the contractors had recognised the need more than the clients. Many had expectations that the industry was changing for the better and that it would be a new and different industry by 2007. It was therefore a disappointment when it became obvious that this was not the case, and that it would take very much longer. Change was slow in coming. The Egan targets announced in 'Accelerating Change' stated that by the end of 2004, some 20% of all projects by value should be constructed by integrated project team and integrated supply chains, and 50% by the end of 2007. Whereas some progress has been made within the industry, these targets were not met within the timescale defined.

Part of the learning process depends on understanding that when things go wrong, it is necessary to persevere until they are put right, not revert to old-fashioned ways. The hope that people within the industry, whichever stakeholder they might represent, would recognise the need for change, was not realised fully and it is suspected that there will be several years yet before the Egan targets are met.

Example 2: Bridge Strengthening (2002–2006)

The bridge in question (in the North of England) was in need of strengthening to cater for the increasing loads that were being imposed on it. The client decided on adopting a tendering process based mainly on quality and invoking an Engineering and Construction Contract which recognised partnering principles and which enabled Early Contractor Involvement (ECI). For the client, this project went well and although it took longer to complete than had been intended, not helped by the bankruptcy of one of the major suppliers, it was completed within the budget. The contractor had been asked to take on various risks identified during Stage 1 of the ECI process, which was accepted; although the allowance for those risks later turned out to be insufficient for what they actually cost the contractor and their subcontractors.

The project was undertaken by an integrated project team in line with the recommendations of Sir John Egan and for the most part, this worked well. Offices were shared by the client, their agents, the contractor and suppliers alike but there was little or no evidence of an integrated supply chain.

By way of an example of learning, during the course of the work, there were perceptions by the project team that there were issues to be addressed, particularly between the inspectors on the job and those being inspected. During a workshop designed to tease out such issues, it was discovered that what had been suspected and what were actually the issues, were different, highlighting the need for better knowledge between the various tiers of the project team. As a result, a more formal communication structure was put into place to improve the communication between the various tiers.

This example illustrates the significance of communication between stakeholders in any endeavour. Imagine the power of such a process to reduce risk and consequential cost, by communicating down in to the lower tiers of the supply chain. Figure 4.1 demonstrates the numerous interfaces inherent in most (construction) projects. A breakdown in communication at each of these interfaces can impact on a project negatively and their compounding effect can be very consequential.

Some stakeholders actively resisted change in the magnitude advocated by Latham/Egan and reverted to the old practices, often against the wishes of their own clients. Paradoxically, the legal profession, initially suggesting that partnering would not work, embraced its principles and have consistently over the years encouraged others to do the same. In partnering, the stakeholders operate via mutual objectives and joint problem–resolution. This augers well for stakeholder management as potential conflicts are identified upfront and resolved communally. This minimises conflicts and ensures more satisfied stakeholders at the end of the day.

Speaking to a Best Practice Club in the South of England during the early 2000s, the question was put to the 40 or so people attending, mainly from the Construction Industry, and representing clients, contractors, designers and SMEs: 'How many of you have experienced any form of partnering during the course of your work?'. There was a show of hands, representing about 75% of those present. 'How many of you have experienced successful partnering?' Only one hand showed.

Whether or not this was a representative sample of people working within the Construction Industry, the success rate was very low and suggested that whereas the majority had been working in some sort of collaborative working environment, they had apparently not understood the reasons why they were doing it. If they had, when the partnership was showing signs of difficulties, its members would have made every effort to make it a success, increasing the overall success rate significantly. Some 10 years after the culture had been introduced to the industry, there was still a very long way to go. Hence there is some scope for developing stakeholder management.

4.6   Conclusion

It is said that it takes at least 30 years for changes to fully take place and for what is new today to become the norm tomorrow. At the time of writing of this chapter, it is 14 years since Sir Michael Latham wrote 'Constructing the Team' and 10 years since Sir John Egan chaired 'Rethinking Construction'. Whereas it has not been as rapid as might have been hoped, progress has been made and the industry has to be congratulated for that. Whereas there is still evidence of major projects which have not adopted the practices recommended by Latham and Egan, e.g. Wembley Stadium, there are many that have, e.g. Emirates Stadium.

Most importantly since 'Rethinking Construction' (Egan, 1998), the Constructing Excellence Demonstration Programme has proven to be an excellent vehicle for the capture and dissemination of best practice knowledge in the built environment sector (www. constructionexcellence.org.uk). Some 10 years later, over 600 projects had been recruited, and some 1300 organisations UK-wide had been involved, representing project value approaching £10 billion. To continue this type of success rate, we have to convert such practices from the 'new' to the 'norm' so that it becomes practice that is automatic. Fundamental to this is the continued and more effective involvement of all relevant stakeholders through good, effective communication, so that risks are reduced and value is enhanced, to the benefit of not only the industry, but of the United Kingdom as a whole.

Good effective communication between stakeholders leading to trust and common focus is the key to the future success.

References

BSI EN 12973 (2000) *Value Management*. London: HMSO.

Constructing Excellence Demonstration Programme (www.constructionexcellence.org.uk).

Department of Environment, Transport and the Regions/Sir Michael Latham (1994) 'Constructing the Team: Final Report of Government/Industry Review of Procurement and Contractual Arrangements in the UK Construction Industry'. London: HMSO.

Egan, J. (1998) *Rethinking Construction*: The report of the Construction Task Force to the Deputy Prime Minister, John Prescott, on the scope for improving the quality and efficiency of UK construction. London: Department of the Environment, Transport and the Regions.

Latham, Sir Michael (2007) 'Industry change since 1994'. Speech delivered at The Building Services Summit 2007, Wembley, 24 October 2007.

5 The Contextual Approach to Stakeholder Management in Finland

Adekunle Sabitu Oyegoke

5.1 Introduction

Construction projects have direct and indirect impacts on different project interest groups. These different interest groups are referred to as stakeholders. The aim of this chapter is to examine the management of stakeholders in a typical Finnish project. The chapter explores the Finnish project environment and stakeholders' management principle employed in Finnish practice along different project phases. The study could be used in understanding some key organisation principles and could serve as subsequent transferable learning opportunities for a wider stakeholder management.

5.2 Understanding project needs and stakes

The demand for a facility is a derived demand that is driven by the interaction of macroeconomic and microeconomic variables, and local supply and demand conditions. There are two main principal needs in a construction project, namely the physical and the financial requirements. The physical requirements underline the real need for the project, while the financial requirements deal with construction cash flow, profitability of the scheme, viability of the investment and income generation. In achieving the project needs and requirements and stakeholders' expectations, different players and parties are involved in the demand and supply chain as well as in auxiliary activities that enhance the achievement of the project's objectives. For instance, construction project can be wholly owned and financed by private individuals/entities or by public entities. Because of the restriction/implication placed on public funding, integrated funding approaches have emerged.

There are a number of dimensions to the number and type of stakeholders in a project. According to Mitchell et al. (1997), a stakeholder should posses one or all of the following attributes: power to influence; legitimacy in relationship and the urgency of the stakeholder's claim. There are some basic issues that are intrinsic to construction projects which result in the involvement of stakeholders such as ownership, usage/user, finances, supply chain and sustainability among others.

Figure 5.1 presents two major drivers of stakes in a project: ownership and interest. A good example of ownership stake is shareholders who have legal right and title in their stake. Conversely, a community has moral right and claim to protect the interest of the public for fairness, justice and equity.

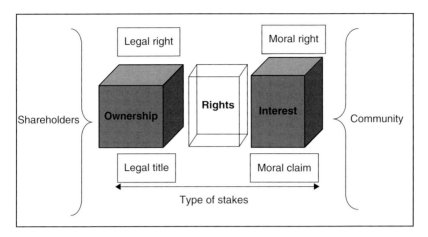

Figure 5.1   Major drivers of stakes in a project.

5.3   The project environment in Finland

According to Barrie and Paulson (1992), the building industry is characterised as a custom-oriented, incentive dependent, and predicated on human factors, which consequently leads to the industry being fragmented and sometimes divisive. The Finnish building practice is characterised by industrial prefabrication, energy efficiency and ability to build in sub-zero temperature. At industry level, there are many stakeholders who are construction decision-makers or those who influence decision-making process. The share of building construction in the Finnish construction industry increased in 2004 to just under 5%, while civil engineering increased to 4.5% of the gross domestic product (GDP).

At firm level, the stakeholders' environment focuses more on production and managerial views of the firm. The former tends to focus on product/project delivery between the supply chain and the users/owner of the project, while the latter in addition focuses on the firm management of the shareholders and employees. This group comprises of those who have stakes in the industry in a form of legal right and title as well as moral right and claim. According to Statistics Finland, there are approximately 33 000 licensed contractors that include skilled subcontractors and very few large firms that also engage in international construction business. In 2006, the construction industry employed about 162 000 people; 80 000–100 000 of them work as employees in the private sector. The construction sector is strongly concentrated in the growth areas, especially in the southern part. Ageing is a factor that will reshape the nature of the industry. The current workforce is getting older and retiring from work; a substantial number of workers will disappear from the construction sector unless the number of newcomers grows. Another issue of concern for stakeholder management is the declining number of skilled domestic labour and the growth in the number of the foreign workers.

At project level, construction projects offer a challenging combination of product-development and value-chain production management. This is the stage where different components/disciplines are interchangeable and interlinked according to the prevailing conditions and work environment. There are different types of projects in Finland. More than 14 billion cubic metre of residential space was completed in 2006, followed by 4.36 billion cubic metre of industrial buildings and 3.81 cubic metre of agricultural buildings. In the dwelling sector, 33 683 houses were completed in 2006, comprising of 15 991 detached houses, 11 943 blocks of flats and 5 426 attached houses (Statistics Finland, 2007).

The foregoing scene setting shows many players in the demand and supply sides of the construction business. On the demand side is the owner or group of investors who have equity stakes in projects, while on the supply side are consultants, contractors, specialist subcontractors and suppliers, among others. At the task/trade level, the industry is typified by its temporary and multi-organisational nature that results in the involvement of tasks/ trades organisations in the execution of the projects. It should be borne in mind that the project environment is affected by economic, social and political factors which may mar the outcome of a project. Oyegoke (2006) asserted that the volatility in the Finnish construction industry can be attributed to the restructuring of asset portfolios by the investor and owner-occupiers. In recent time, the prime position of the banks has been relegated due to the active involvement of other players. The gaps created by the banks have been filled with a surge in investments from the pension insurance companies and construction companies acting as developers especially in the housing market.

5.4   The project stakeholders in Finland

According to the Project Management Institute (2004), project stakeholders are individuals and organisations that are either actively involved in a project or whose interest may be affected as a result of project execution or project completion. According to Carroll and Buchholtz (2006), there are three dimensions to the categorisation of stakeholders: core, strategic and environmental stakeholders. The core stakeholders are essential for the survival of the organisation, while the strategic stakeholders are vital to the organisation and the particular set of threats and opportunities it faces at a particular point in time. They postulate further that environmental stakeholders are all others in the organisation's environment that are not core or strategic. Wheeler and Sillanpää's (1997) two-fold categorisation denotes primary and secondary and social and non-social stakeholders. The primary social stakeholders have a direct stake in the organisation and its success and therefore are influential. The secondary social stakeholders are extremely influential and are more representational of public or special interest.

Blockley and Godfrey (2000) differentiate between project stakeholders and players. They emphasise stakeholders as the people and organisations that have interests in a project. Their functional role and participation depend on their demand and supply functions and their involvement in making buying or purchasing decisions. There are two categories of stakeholders: internal stakeholders, those who are actively involved in project execution, and external stakeholders, who are those affected by the project. However, the interaction of the supply and demand chain within and between the financial and other supporters, community and the natural environment delineate project stakeholders as shown in Figure 5.2.

The project stakeholders in Finland are similar to other practices. It primarily consists of stakeholders as listed and discussed below:

- construction clients
- project users
- supply chain members
- financial supporters
- community/public

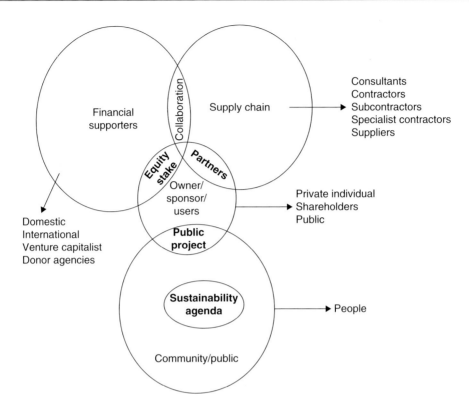

Figure 5.2   Interaction of project environment and stakeholders.

5.4.1   Construction clients and users

A construction client is usually centrally located employing the services of contractors and consultants. A client also sources funds for the project and will have direct link with the financial institutions. The project client is the most active stakeholder in the network, being centrally located and serving as a main connector or hub within the stakeholder's network. In Finland, an interest group known as RAKLI (The Finnish Association of Building Owners and Construction Clients) represents the most prominent real estate owners, investors and service providers on issues concerning legislation, taxation and common policies. With over 400 private and public real estate companies in its memberships in 2002, it represents over 50% of the total extent of commercial facilities, and nearly 90% of the income-producing commercial real estate volume in Finland (RAKLI, 2006).

5.4.2   Users

The project client or another party may sometimes be the user of the project. This is a person or an entity that will put the facility into use. End user's requirements are vital factors in an owner investment strategy, because they provide revenue streams to pay for the project, user's satisfaction is one of the determinants of project success and the users are the ones who put the project into effective use. Finnish users can be broadly categorised into four types based on the type of the property: residential, industrial, agricultural and government (public) property. According to the construction statistics, in 2006, the amount of space completed in all building category amounted to 37.92 billion cubic metres. Residential property users in the same year amounted to more than 38%, followed by industrial users with 11%, agricultural users with 10% and others with over 40%.

5.4.3   Supply chain members

Supply chain members denote member of delivery network often comprise different organisations, linked upstream and downstream to deliver the project based on agreed quality standards. The consultants act as a connector between the client and the contractor. The contractor acts as a leader in constructing unit. On the consulting side are a number of consultants in architectural, engineering and cost management disciplines. The type of project often dictates the level of their involvement. On the contracting side are a number of firms with varying contracting capacities: contractors, subcontractors and suppliers.

5.4.4   Financial supporters

In Finnish practice, there are a number of big players in the financial sector in addition to the services provided by the banking operators. Funds may be raised through joint venture partners (private equity capital) who provide the developer with equity funding during the development period. Another player is the long-term equity investors (private equity with long-term commitment) who may or may not be involved during the construction period. A financial arrangement that encourages equity holding by the players in the form of direct equity responsibility or as a guarantee-trust for a third party has been tested in Finland. Risk allocation to the players in the form of equity stake holdings increases their commitment and technical input, contributes to overall project quality and leads ultimately to a reduction in project costs. There are series of financial players in Finnish construction environment. Sometimes these financial players play the role of a sponsor and/or project owner. At the forefront of these companies are monetary institutions, financial institutions, and security and insurance institutions.

5.4.5   Community/public

The public has vested interest in a project either privately or publicly owned. There are a number of interest groups working to protect the interest of the public, e.g. a local community group, social activists and Greenpeace. Their work is focused on achieving sustainability in social, economical and environmental terms. These groups have been successful in influencing government policies and business environmental policy. The Finnish stakeholder external environment is relatively stable due to the fact that the general public is highly enlightened, the level of awareness in Finland is very high, a balanced objective between achieving the project goals and societal criteria, maximum community involvement prompting a high public confidence and a high level of acceptance or rejection rate of a project among the public.

5.4.6   Approach to stakeholder management

The stakeholder management process can be carried out in many ways: (i) by evaluating the needs and expectations of stakeholders in relation to the project's main objectives and (ii) by designing a broader project management process that enables active interaction between the stakeholders from project inception to completion. A number of tools can be used to manage stakeholder's requirements in a project. This often depends on the type of the client and the source of funding. For instance, a project can be financed through development aid by donor community, which comprises international agencies (e.g. World Bank), donor government (Ministry for Foreign Affairs of Finland), etc. The donor project often entails project cycle management, procurement strategies, value and technology assessments and analyses, environmental evaluations, stakeholder analysis, pre-construction, construction, and facility management planning. A good example of a donor project sponsored by Finland is a health sector programme that was aimed at improving the implementation of the provincial health strategic plan in Manica Province – Mozambique from 2003 to 2007. Cleland (1986) postulated that stakeholder management is vital in order to determine the stakeholder's reactions to project decisions, the influence of their reaction and possibility of interaction with each other. The involvement of stakeholders in project implementation enable to satisfy the notion of value for money, management of risks and responsibilities, project cycle management, environmental assessments and sustainability, openness, fairness and effective competition, among others.

5.5   Stakeholder management in Finland

There is no exclusively single method of managing stakeholders in Finland. The process entails identification, understanding, analysing, educating and managing stakeholders through six key management principles.

5.5.1   Management model

1. Early identification of the stakeholders
2. Identify potential conflict areas
3. Educate stakeholders on potential risks and harms and how they are mitigated and potential gains
4. Engage with the stakeholders – communication line
5. Involve other entities, especially government and show that due process is followed
6. Managing the process

All of these six principles are applied in every phase of a project. All the stakeholders are identified at the onset and further assessment exercises are carried out throughout the project. This is to identify the stakeholders or interest group which might develop in the course of carrying out the project. At the same time, potential conflict areas are identified and special remedial actions are planned and undertaken. The identification phase is a participatory planning process where people, organisations, institutions, communities who could influence and contribute to the planning and management processes are identified. This is carried out by a basic understanding of the social and institutional context associated with the project and its environment. The stakeholders are enlightened through information outlets on the potential gains, risks and harms associated with the project. An effective line of communication is required in order to engage with the stakeholders. The engagement (e.g. through series of meetings) will allow the stakeholders to be conversant with the progress made in the project, be acquainted with the steps taken to mitigate against risks and harms, and raise their awareness on the strict adherence to rules, regulations and code of practice connected with the project.

5.5.2   Application of Finnish stakeholder's management model in practice

In this section, the project phases are oversimplified into five phases: identification, programming, appraisal, implementation and facility management.

Identification phase

Early identification of the stakeholders is often commenced as the project is been conceived in the project identification phase. The *identification phase* is one where the project needs are prioritised in relation to benefits and costs. It involves pre-feasibility studies and may also be used to identify, investigate, and select or prioritise ideas. The selected idea or project is further tested on whether or not to develop it further taking into consideration costs, benefits, environmental and other stakeholder's parameters. In addition, social and ethical concerns are brought forth for consideration. The challenge to the management team is to identify the nature and legitimacy of a group's stakes. The management approach is to formulate relatively quick responses for planned and unplanned fallouts from the project, e.g. protest. The first thing is to identify and gain adequate knowledge about those that will be affected internally and externally with the outcome of the project. The opportunities and challenges each of the identified stakeholder posses are analysed. The management team prioritises and determines salient stakeholders as well as the degree of attention that will be given to them. The knowledge gained is used to predict and manage the stakeholders' behaviours and actions.

A good example from a statutory point of view in Finland is the earlier interaction between a new developer and her neighbours. The law stipulates that the consent of the neighbours should be sought before a development is carried out. The consent should not unreasonably be denied.

Another example is a nuclear plant project which was the first in Europe in the last 10 years (Power Technology, 2007). Due to negative public perceptions to nuclear technology, the government energy policy of 1997 stressed availability, security, diversity, price and the need to meet international environmental commitments. This was the beginning of educating the public about the need and security of energy supply and arousing public interest and debate on health and safety issues. There are few selling points to the public at the identification phase: economic criteria of lowest cost kWh, energy security, economic stability as a result of lowest sensitivity to increment in fuel prices, and environmental grounds as a result of substantial reduction in greenhouse gases.

Other selling points are the provision of radioactive waste disposal, approval of new generation of Pressurised Water Reactor (EPR – Evolutionary Power Reactor) and good safety records in the existing plants. Even with all these positive attributes, there was a strong division in public opinions and intense debate in the parliament. In the parliament for instance, 107 parliamentarians voted in support while 92 voted against. The story was different in 1993 when a similar proposal was rejected by the parliament due to the global political climate surrounding nuclear plants.

Programming phase

The *programming phase* entails overall guidelines and principles to be followed in executing the project. The management of the internal stakeholders (supply chain members) is agreed upon through the form of procurement method. The procurement method spells out the boundary between participating firms and allocation of risks and responsibilities. Procurement management is a tool that is designed to manage project activities and its stakeholders. It also defines project organisation setups and inter-firm coordination, collaboration and interactions in order to achieve sustainable project objectives. The investors in Finland have positive experiences regarding price and quality with construction management (CM) models in a range of projects by size and complexity. The management approach has been extended to outsourcing maintenance, especially in the building sector, and similar models have also been introduced to the infrastructure sector (railway-maintenance management).

This method relies on inter-organisational management principles in socially integrating the supply chain members. In addition, a public owner (investor) is able to attract long-term financing with a substantial savings gained through the competitive subcontracting of all project and maintenance work packages. Potential conflict areas among the project internal and external stakeholders (e.g. a fear of toxic pollution) are identified and adequate enlightenment is carried out through educating the stakeholders.

Appraisal phase

The *appraisal phase* involves the assessment of key project factors regarding finance, implementation, quality, from the stakeholder's point of view, especially from those that are direct or indirect beneficiaries. This is a 'make or break' point where the viability of the project is determined. The shareholders will like to know if the project is feasible and whether adequate return on their investment could be made. The project team will like to know if the execution/implementation process is realistic and if the project can be delivered within the estimated cost, time and schedule. Intensive enlightenment campaign (e.g. via the Internet) and the external stakeholder analysis are carried out. The stakeholder analysis entails identification of their characteristics, interest and expectations, sensitivity, potentials and deficiencies and wider implication to the success of the project. The reaction of the external stakeholders is now measured and the decision to 'continue or abandon' the project is finally made. Openness and adequate communication is very important.

Before an investment decision was made, environmental impact assessments (EIA) were carried out for new nuclear power units in Finland before seeking government approval. The worries in nuclear plant projects are the fuel and enrichment supplies, security of the plant and disposal of waste. The power companies carry out waste management responsibility until its final disposal. Designs were sought by the company which were submitted to the national regulatory authority for review. Afterward tenders were submitted by the three selected vendors for four designs. One design was then chosen upon which a construction licence was sought.

Implementation phase

Implementation involves the bringing together of different groups of professionals in the project supply chain. This is the phase where the actual construction work is carried out and the designing organisations, management firms, contracting firms, financial institutions and other players interact. The work of McElroy and Mills (2000) shows the influence of stakeholders' values on project management and the ways how the stakeholders could be managed respectively. Walker et al. (2001) examine the client's role on power and influence over design and procurement. They found out that at the high end of the scale, the client or client representative has the major input and influence on determining the working relationship, i.e. defining the procurement route. The contractor takes the majority of cost risk and cannot make design decisions nor has substantial power to effect or affect design decisions in the traditional and cost reimbursable procurement approaches. Conversely, the contractor holds most influence over design detail and even the design strategy in BOOT-type (Build Own Operate Transfer) projects and design and construction (Walker et al., 2001).

The model that is gaining prominence now in Finland is Specialist Task Organisation (STO) approach that allows for upper and lower management levels through integrated management systems. The approach enables for integrated product development and fragmented project implementation by varying numbers of specialist contractors. The external stakeholders are monitored, effective communication line is open and adequate information about the project is available through web-based Internet modalities. During the implementation stage, more interest groups either collectively or individually will arise due to the activities on site (presence of workforce and physical structure). The identification process of newer stakeholders and potential conflict areas should be carried out. Adequate management through educative and communication strategies is needed in order to show that due process is followed.

5.5.3 Facility management

There is a need for *facilities management* arising out of the executed project. At this stage, the shape of the stakeholder will depend on the form of agreement in implementation stage. On one hand, the contacting agreement may terminate at the end of the project, resulting in a new facility management contract with another firm. On the other hand, contracting agreement will extend beyond implementation stage and cover facility management stage. Also, at this stage, there is a very active involvement of the final users, project owner and the community.

   This phase is also crucial and should be incorporated in overall stakeholder management at the onset of the project. A good example is how nuclear waste will be managed or how the facility will be run and maintained. For instance, the decommissioning of the nuclear reactor is the responsibility of the power companies. The regulation, supervision for waste disposal and overall safety are the responsibilities of the Ministry of Trade and Industry assisted by many advisory committees. Following due process in facility management is very important. Transportation of nuclear wastes is a vital issue to anti-nuclear campaigners. The Finnish Amendment Act of 1994 states that nuclear wastes should be handled wholly in the country and a deep geological disposal should be provided. In the interim, a facility is designed to hold used fuel or wastes for 50 years. In order to avert any fear about nuclear waste disposal, the locals have an absolute right of veto on the siting of the facility.

5.6 Conclusions

Stakeholder management is becoming increasingly important in achieving social, environmental and economic objectives in a project. Project stakes could be through legal right or legal title, or through moral right or moral claim. In Finland, six management processes have been identified which are applied in every phase of a project. This list is not exclusive, the uniqueness, complexity and the project environment will determine the nature of stakeholder planning and management process that is required. Assessing the stakeholder rights and responsibilities is important in stakeholder management process where rights and legitimacy of the stakeholders are determined.

References

Barrie, D.S. and Paulson, B.C. (1992) *Professional Construction Management*. New York: McGraw-Hill.
Blockley, D. and Godfrey, P. (2000) *Doing It Differently: Systems for Rethinking Construction*. London: Thomas Telford.
Carroll, A.B. and Buchholtz, A.K. (2006) *Business & Society: Ethics and Stakeholder Management* (6th edn). Thomson South-Western: Mason.
Cleland, D.I. (1986) Project stakeholder management. *Project Management Journal*, **17**(4):36–45.

McElroy, B. and Mills, C. (2000) Managing stakeholders. In: Turner, R.J. and Simister, S.J. (editors). *Gower Handbook of Project Management* (3rd ed.); Surrey – UK: Gower Publishing Ltd.

Mitchell, R.K., Bradely, R.A. and Wood, D.J. (1997) Toward a theory of stakeholder identification and salience: defining the principle of who and what counts. *Academy of Management Review*, **22**(4):853–886.

Oyegoke, A.S. (2006) Construction industry overview in the UK, US, Japan and Finland: a comparative analysis. *Journal of Construction Research*, **7**(1&2):13–31.

Power Technology (2007) Olkiluoto 3 1600MW Nuclear Power Plant, Finland, (Online) http://www.power-technology.com/projects/Olkiluoto/ (Accessed in 2007).

Project Management Institute (2004) *A Guide to the Project Management Body of Knowledge, PMBOK* (3rd edn). Newtown Square, PA: Project Management Institute Inc.

Rakli (2006) *The Finnish Property Market 2006*. Helsinki: KTI.

Statistics Finland (2007) *Construction Statistics*. Available (Online) from http://www.stat.fi/ (visited March 2007).

Wheeler, D. and Sillanpää, M. (1997) *The Stakeholder Corporation*. London: Pitman.

Walker, H.T.D.; Hampson, D.K. and Peters, R. (2001) *Relationship-Based Procurement Strategies for the 21st Century*. Brisbane - Australia: AUSInfo.

6 Risk and Construction Stakeholder Management

Mei-yung Leung and Paul Olomolaiye

6.1 Introduction: the essence of stakeholder management in construction

To manage stakeholders in construction successfully, the answers to two fundamental questions should be clarified from the outset: Who are the stakeholders and what is stakeholder management? Although numerous studies have been devoted to examining the stakeholder concept, no single definition of a stakeholder has been universally accepted. A stakeholder can be an individual, a group or an organization. Most studies indicate that there are two categories of definitions of the concept of stakeholder: narrow definitions and broad definitions (see Table 6.1).

There are many narrow definitions of the term stakeholder in the literature (Clarkson, 1994, 1995; Madsen and Ulhoi, 2001; Cleland and Ireland, 2002; Bourne and Walker, 2005; Olander, 2007). Stakeholders need to have some kinds of stake, right or ownership in the organization (Post and Je, 2002); bear some risks in the investment of capital, human resources or something of value in a firm (Clarkson, 1994) or have a stake in the project outcomes (Olander, 2007). Such a view is useful for identifying those parties that have direct economic relationships with the organization. However, it excludes those parties that do not have ownership or a stake in the organization but are able to exert influence on the implementation of a project using non-economic methods. For instance, local residents may support or oppose the construction by the government of an incinerator adjacent to a residential estate or residential building, although they may not have any direct stake in the project. Therefore, the definition of stakeholder should not simply be based on economic factors.

The term stakeholder can also be defined broadly to include those who only have an interest in a particular issue (Savage *et al.*, 1991; PMI, 1996, 2004; Scheffran, 2006); those who actually affect or are affected by the achievement of organizational objectives (Freeman, 1984); those who have a vested interest in the success of a project and the environment within which the project operates (Olander, 2007) or those who have a contractual, financial or ethical interest in the decisions and actions of the organization (Rotarius and Liberman, 2000). However, such definitions are open to the criticism that there is little value in the stakeholder concept if everyone is a stakeholder (Mitchell *et al.*, 1997; Sternberg, 1997; Phillips, 2003).

Construction projects normally involve such diverse stakeholders as clients, end-users, customers, consultants, contractors, financial institutes, green groups, governmental agencies and local communities. In order to gain an overall view of construction stakeholders, it is helpful to classify them into different categories. Construction stakeholders can be categorized into two groups according to their legal or contractual relationship with a project: **internal** (or primary) stakeholders and **external** (or secondary) stakeholders (Madsen and Ulhoi, 2001; Cleland and Ireland, 2002; Winch, 2002; Olander, 2003). **Internal** stakeholders refer to those who are members of the project coalition, provide finance or have a legal or contractual relationship with

Table 6.1   Definitions of the concept of stakeholder

Category	Definition	Author
	In a broad sense, stakeholders can be considered as those individuals or groups that **have an interest or concern** in a particular issue.	Scheffran, 2006
	Project stakeholders are individuals and organizations that are actively **involved** in a project or whose **interests** may be affected as a result of project execution or project completion.	PMI, 2004
	Those individuals, groups or organizations having a contractual, financial or ethical **interest** in the decisions and actions of the organization.	Rotarius and Liberman, 2000
Broader view	Stakeholders include those individuals, groups and other organizations who have an **interest in the actions of an organization** and who have the ability to **influence** it.	Savage *et al.*, 1991
	An individual, individuals, team or teams **affected** by a project.	Juliano, 1995
	Any group or individual who can **affect** or is **affected** by the achievement of an organization's objectives.	Freeman, 1984
	Individuals and organizations who are actively **involved in** a project, or whose **interests may be positively or negatively affected** as a result of project execution or successful project completion.	PMI, 1996
	Stakeholders supply a project with critical resources, bear additional risk or have the power to affect the outcome of a project.	Post and Preston, 2002
	Stakeholders are individuals or groups with **a legal, economic, moral and/or self-perceived opportunity to claim ownership, rights or interest** in a firm and its past, present or future activities – or in parts thereof.	Madsen and Ulhoi, 2001
	Stakeholders are individuals or groups who have an **interest or some aspect of rights or ownership** in a project, can contribute in the form of knowledge or support or can impact or be impacted by a project.	Bourne and Walker, 2005
	Stakeholders are persons or groups that have, or **claim, ownership rights or interests in a project** and its activities: past, present or future.	Clarkson, 1995; Cleland and Ireland, 2002; Preble, 2005
Narrower view	Those bearing some form of risk as a result of **having invested some form of capital, human or financial, or something of value**, in a firm. These stakeholders are those without whose participation the corporation cannot survive.	Clarkson, 1994
	Project stakeholders are individuals or a group of people who **have a vested interest** in the success of a project and the environment within which the project operates.	McElroy and Mills, 2000; Olander, 2007
	The fundamental idea of the stakeholder is that he or she or it **has a stake in an organization**. Stakeholders are those that contribute voluntarily or involuntarily to the organization's wealth-creating capacity and activities. They are, therefore, its potential beneficiaries and/or risk bearers.	Post and Je, 2002

Category	Definition	Author
	Stakeholders **have a stake in the outcomes of a project**. It could be an interest, a right or ownership. An interest is a circumstance in which 'a person or group will be affected by a decision; it has an interest in that decision'. A right is either a 'legal right when a person or group has a legal claim to be treated in a certain way or to have a particular right protected' or a 'moral right'. Ownership is a circumstance 'when a person or group has a legal title to an asset or a property'.	Olander, 2007
	Stakeholder theory should be concerned with who **has input in decision-making** as well as who **benefits from the outcomes of such a decision**.	Phillips, 2003

the project. **External** stakeholders are those who influence or are influenced by the project, but are not normally engaged in transactions with the project and may not be essential to the survival of the project. Parties such as owners, consultants, suppliers, customers, users, contractors and financial institutes are usually internal stakeholders, while the public community, local residents, local or national authorities, interest group may vary according to the project. Whether a stakeholder is classified in the internal or external group depends on his or her or its specific situation in each project. For example, a government department (e.g. a highways department) may be an external stakeholder in a project for the development of an estate and, simultaneously, be a client of other projects for the construction of bridges, roads or highways.

Each participant has his or her own background, and will contribute towards the success of the construction project. Since participants' inputs are often interdependent, conflicts may arise in some circumstances. For instance, a developer (as an internal stakeholder) expects to fully utilize a site area and cut down all the trees on it, while a green group (an external stakeholder) will emphasize the need to protect the natural environment. Stakeholder management requires stakeholders to simulate as many risks and conflicts as possible, identifying project goal specificity and ensuring goal commitment among stakeholders in the implementation process (Leung and Liu, 2003). It is essential to effective stakeholder management to ensure that stakeholders work in a team throughout. However, one cannot expect stakeholders to attain the necessary synergy automatically, given that each stakeholder has his or her or its own interests in a particular project. Therefore, the identification of key stakeholders and their objectives are important to achieve project success.

6.2   Stakeholders in construction: their objectives

Although different kinds of stakeholders are involved in each particular project, stakeholders in construction can also be classified into five main groups: clients, consultants, contractors, external public parties and external private parties. Clients, consultants and contractors can be grouped together as internal stakeholders, while the remaining parties are considered external stakeholders. Table 6.2 lists some key stakeholders in construction projects under these five main headings.

The **Client** is the initiator of all construction projects. His or her or its requirements are often crucial to the project success, as the client finances the project, determines the objectives and scope of the project, specifies the functions that the project outcome should satisfy

Table 6.2   Example of construction project stakeholders

Categories	Individuals/groups	Objectives and roles
Internal stakeholders		
I1 – Clients	Private clients	Ensure the project will support the organization's strategy
		Ensure the organization's resources will be used economically and effectively
		Learn skills, earn wages, work on the frontline
		Link between the client and consultants, ensure the project is completed successfully in terms of quality, time and cost
		Provide financial support; maximize return with minimized risk
		Purchase the construction product
	Public clients	Serve public interest based on the organization's strategic objectives
		Consume what is delivered in order to satisfy functional and basic needs
		Allocates funds to the project
		Ensures that public funds will be used properly
		Link between the client and the consultants, ensure the project completed successfully in terms of quality, time and cost
I2 – Project professionals (in-house/ out-of-house)	Architect	Develops the design of the project; produces drawings and specification; ensures that a project is implemented within cost and time, and according to quality control
	Quantity surveyor	Advises client on financial and budgetary matters; assists in preparing tender documents; examines and reports upon tenders; monitors costs during construction and seeks to understand valuation and measurement; assesses the legitimacy of claims from contractors and prepares final accounts
	Structural engineer	Designs all structural calculations and elements; designs building structure; ensures statutory compliance

Table 6.2   Example of construction project stakeholders   (*Continued*)

Categories	Individuals/groups	Objectives and roles
	Building service engineer	Designs electrical and mechanical building service systems, such as HAVC, fire, water, electronics, lift, etc.
	Other consultants	Assistance in developing the brief; advice on special studies and surveys for design development; collaboration with the design team to develop design and cost control; advice on developing drawings, specifications and other tender documents; prepare design drawings; monitor work on site with regard to quality, cost and time; attending commissioning and acceptance testing and completion of relevant work; assist in valuations and the settlement of accounts
I3 – Contractors/suppliers	Main contractor	Carries out and completes the work designed by consultants to meet time, cost and quality objectives; supervises and manages operations on site; sometimes assists in design; coordinates and supervises all sub-contract work, materials and suppliers
	Sub-contractors	Carry out work assigned by main contractors
	Labourers	Finish tasks assigned, earn living, learn skills
	Suppliers	Supply, install and commission the hardware that constitutes the finished building (e.g. materials suppliers, equipment suppliers and manufacturers)
External stakeholders		
E1 – External public parties	Government authorities	Ensure that the project abides by laws and regulations; may be indifferent to any project so long as it complies with codes (e.g. planning department, electrical and mechanical services department, transport department, highways department, etc.)
	Consultation bodies such as district board	Ensure the local communities' requirements will be reflected in the project
	Town planning board	Ensures the project will be in line with district planning
	Labour union/employers' association	Influences the conduct of its members (privilege protection function)

(Continued)

Table 6.2   Example of construction project stakeholders   (*Continued*)

Categories	Individuals/groups	Objectives and roles
	General public	Participate in and contribute to the government process of a society as a whole
	Media	Influence project decisions (influence company reputations)
	Institutional forces/ nationalized industries	Influence professional institutions upon the activities of their members through rules of conduct, education, conditions of engagement and fee scales
E2 – External private parties	Local residents/community	May fear a fall in amenity, therefore against the project
	Local landowners	Own land; ensure that their interests will not be hurt by the project
	Archaeologists	Concerned about the loss of important historical artefacts
	Environmentalists/ conservationists	Wish to protect the environment from destruction or pollution
	Competitors	Seek to gain competitive advantage
	Tourists	Enjoy the scene
	Others	Their connection to the project is not immediately clear, but their operation and support may be vital to the project success

and largely determines the implementation methods of the project. However, in practice, a client's basic requirements are not often sufficiently clarified or understood. Sometimes, a client's needs may even vary according to the dynamics of organizational structures and strategies, as well as to the changing organizational environment. Hence, most project failures are attributable to the vague and uncertain requirements of clients.

Clients can be divided into two types, according to the ownership of a construction project: private clients (either a corporation or an individual) and public clients (government departments). **Private clients** normally focus on the financial aspect and the economic returns of a project, while public clients need to consider the interests of the public. The objectives of an *individual private client* are relatively simple, as an individual client only needs to consider his or her preferences, required functions and needs.

A *corporate private client* may consist of several interested groups. It is useful to break them down into different sub-groups and analyze their objectives and requirements separately. Senior managers are primarily concerned with aligning the project mission (e.g. the project nature, the type of project, the rate of return, etc.) with the corporate strategy. Therefore, they have to understand the actual market direction and the corporate resources in order to make the correct primary decisions for the corporation.

The **public clients** include publicly owned organizations that have the authority to raise public finance from the initial stage to the commission stage of a construction project (Walker, 1989). They consist of different sub-groups such as the environment bureau (e.g. environmental protection department), the transport and housing bureau (e.g. highways department and housing department), the development bureau (e.g. architectural services department, building department, electrical and mechanical service department and civil engineering and development department), the education bureau (e.g. secretariat of the university grants committee, student financial assistance agency) and so on. As the funding is raised from the government, this type of client is generally more concerned about public interest than individual preferences.

In both corporate private client and public client organizations, it is common to employ *multi-disciplinary in-house professionals* for various construction projects. In practice, the objectives of the professionals are normally adjusted according to the wishes of the client, although they also need to contribute their expertise to the project. For instance, community value and image (public interest) are emphasized by the architect in the case of a public client, while full utilization of the plot-ratio (the economic factor) is important for the architect of a corporate private client.

On the other hand, *end-users* normally focus on the functions of the project product. The end-users in a building product could include operators, visitors, residents and customers, while the building product could involve safety, a green environment, hygiene, a sea or panoramic view and facilities. The level of importance for each criterion may be different in the actual environment, since each of the parties has their particular objectives for the project. Therefore, construction stakeholders should take into consideration the needs of every user, and incorporate their opinions in the design stage.

Construction projects typically involve the use of various consultants. Appropriate and capable project **consultants** are indeed fundamental to the success of a project. A *project manager* is the key person in the project team, and his or her capability and working style largely determines the way the other consultants operate, and to a great extent determines the fate of a project. At the pre-design phase, the project manager is mainly responsible for the indication of the major direction of the project and the preparation of the feasibility study, while he or she has to discuss the overall scheme and detailed designs with clients and different consultants. The project manager needs to plan, direct, organize and supervise the overall construction process as well as develop staff at the construction phases, while specific *consultants*, such as architects, geotechnical engineers, structural engineers, building service engineers and surveyors, mainly contribute their professional knowledge to the project throughout the design, tendering and construction stages (RIBA, 1991).

Apart from the clients and project consultants, the **contractors**, **sub-contractors** and **suppliers** are the other key internal stakeholders during the construction phase. Both the main contractor and the sub-contractors have to make sure that the work is completed within the time and cost, and according to the quality specified in the contract documents, while the suppliers (e.g. materials suppliers, equipment suppliers and manufacturers) need to provide reliable material and equipment for carrying out the work on site.

Conventional project management focuses mainly on the management of internal stakeholders. However, it is also crucial for the success of a project to establish a good relationship with the external stakeholders, as a building product often influences a whole environment and is affected by the expectations of society and the requirements of government authorities. The external stakeholders can be divided into two categories: external public parties and external private parties. Each party has its own interest in the project and exerts a different kind of influence on it. They constitute a critical environment in which a construction project operates.

External public parties are public organizations involved in construction projects, including government authorities, labour unions, trade associations and nationalized industries. The influences of these parties on a project are diverse. Some of the public agencies of *government authorities*, such as planning departments and building departments, have a legitimate authority within the project as construction projects have to be designed and built according to the building regulations and have to be approved by government authorities. Associations such as *labour unions* and trade associations are not directly involved in a project, but they often protect the interests of their members through their guidelines or through regular meetings with employers or developers. Therefore, it would be risky to ignore their existence when managing a construction project.

As the implementation of a construction project (e.g. the development of a new road) always involves permanent changes in the environment, *local people* will have their own opinions on the project. Landowners, for example, will be interested in any resulting increase in the land price, residents may be concerned about a possible increase in noise pollution and local communities will be interested in any possible enhancement of tourism development. Key **external private parties** may not have a legitimate authority to influence a project, but they will still play an important role among the stakeholders in construction projects. In order to ensure sustainable development in society, the consultation of external private parties should not be ignored.

Construction projects include multi-stakeholders from the development stage through the construction process to the operation stage. As each party has different objectives and there may be some conflicts between the objectives and the actual situation, it is necessary to identify the risks involved in construction projects and the stakeholders associated with them.

6.3   Stakeholders in construction: the associated risks

According to the RIBA, a construction development project viewed as a whole consists of twelve phases within five main stages: (1) the pre-design stage: the inception, the feasibility and the outline proposals; (2) the scheme design stage; (3) the detailed design stage; (4) the tendering stage: production information, bills of quantity and tender action; (5) the construction stage: project planning and operations on site and (5) the completion and feedback stage. The stakeholders often bear different risks at various stages (see Table 6.3). The identification of risks is essential if the stakeholders are to manage the project.

A project is initiated to satisfy a **client's** specific purposes. However, *clients' requirements* are *not* always fully *clarified* in the project briefing stage. Sometimes, *conflicts* between those with differing professional knowledge may be involved, or conflicts regarding governmental regulations or the concerns of various interest groups may arise. Clients need to accept a *dynamic environment* and adjust their expectations throughout the project according to the actual situation. Any *uncertain information or any change in a client's requirements* can directly induce a revision of the project scope, an adjustment of the project budget and schedule or other significant changes. This then affects the implementation of the construction project in the subsequent stages; for example, there will be *design changes* in the sketch and detailed design stages, the selection of *contractors or sub-contractors in the tendering stage may become irrelevant* and the *variation orders* may have to be issued in the construction stage.

Thus, the client has to seek support from project consultants from the initial stage of a construction project. The capabilities of project **consultants** are thus critical to project success. However, the incompetence of consultants (e.g. project managers, architects, engineers, surveyors and so on.) and ineffective teamwork may also cause *inaccurate investigation, incorrect assumptions, inappropriate designs, design discrepancy* and *inaccurate cost estimations*, and subsequently *delay the overall management and governmental approval processes. Incomplete Tender Documents* may result in the contract being awarded to *inappropriate or unqualified tenderers* at the tendering stage.

Construction work is implemented by **contractors, sub-contractors** and **suppliers** based on the Main Contract Documents and the Sub-contract Documents. Their capabilities and

Table 6.3   Major construction project risks and associated key stakeholders

Phases	Risk events	Internal stakeholders			External stakeholders	
		Clients	Consultants	Contractors	Public	Private
Pre-design stage	Conflicts of requirements of interest groups	✓	✓		✓	✓
	Vague client's requirements	✓				
	Errors in cost estimation		✓			
	Market environment changes				✓	✓
	Incorrect assumptions		✓		✓	✓
Sketch design stage	Delay of management approval	✓	✓			
	Inaccurate cost data		✓			
	Change of client requirements	✓	✓		✓	✓
	Change of market environment		✓		✓	
Detailed design stage	Conflicts between different drawings		✓			
	Delay of approval from government department		✓		✓	
Tendering stage	Unqualified tenderers		✓			
	Incompleteness of tender documents		✓			
Construction stage	Site congestion	✓				✓
	Delay on site possession	✓		✓		
	Variation order	✓				
	Weather and acts of God			✓		
	Testing and sampling		✓	✓	✓	
	Errors/omissions/discrepancies in bills of quantities		✓	✓		
	Defective design		✓		✓	✓
	Financial failure	✓		✓		
	Delay of management approval	✓	✓	✓		

(Continued)

Table 6.3 **Major construction project risks and associated key stakeholders** (*Continued*)

Phases	Risk events	Internal stakeholders			External stakeholders	
		Clients	Consultants	Contractors	Public	Private
	Default by main contractors	✓		✓		
	Default by sub-contractors			✓		
	Default by suppliers			✓		
	Inaccurate physical conditions investigation	✓	✓	✓		
	Disputes	✓		✓		
	Damage and injury to person and property	✓	✓	✓	✓	
	Union strike					✓
	Resource shortage			✓		✓
	Ambiguous contract	✓	✓			
	Changing laws and regulations				✓	
	Delay of certificate and payment	✓	✓	✓		
	Protest from local parties/unions			✓	✓	✓
	Omissions and discrepancies in design information	✓	✓			
Completion and operation stage	Delays of approvals from government authorities	✓	✓	✓	✓	
	Defective work		✓	✓		
	Design not matching with end-users' requirements		✓	✓		
	Changing end-users' requirements	✓				

experiences play particularly important roles in the construction stage. The work may *bring damage or injury to persons or property* during the construction period, if contractors or subcontractors make mistakes or carry out an unsafe procedure. Any *error or discrepancy* in bills of quantities or *lack of contract clarity* may lead to *disputes* between clients and contractors, as contractors or sub-contractors cannot handle the deliverables within the specific time and limited budget for the required quality. In the end, such an error or discrepancy may lead to *defective work* during the construction process. Moreover, *shortages of resources* (labour, materials or financial support) and *late payment* can also affect the cash flow of construction companies and delay the overall construction process. This may eventually *terminate the contractual relationships* between the client, the contractor, sub-contractors and suppliers. A reliable supply of materials and equipment is also important to a construction project, since *insufficient or defective materials or equipment* may lead to delays or pauses in the project, or even a complete halt to it.

External public parties may introduce risks to the project as well. Any changes in laws, regulations or guidelines issued by government agencies may also increase the risks in a project. For example, labour registration policy in Hong Kong now requires all labour on site to be registered with the government; and housing development policy in mainland China now requires all developers to reserve at least 70% of the total number of units in a development for low-income customers. The internal stakeholders in construction projects, including clients and contractors, may be heavily fined if they do not carry out the work according to the revised regulations. In Hong Kong, the formwork labour union held a *strike* for a period of 36 days in 2007. All construction projects, especially those that were at the stage of the construction of structural elements, were seriously affected, and their estimated times of completion were then postponed.

External private parties have no contractual relationship with the construction client and no authority over construction projects, but *failure to recognize them or their concerns* in the construction project may also create risks. As construction projects always involve permanent changes to the immediate environment, local residents and other interest groups, such as green groups and archaeologists, will thus be very sensitive to a project. If their interests are infringed, they can create a surprisingly powerful opposition to a project. In recent years, a large developer in Hong Kong planned to erect a 54-storey luxury residential tower in the Mid-Levels district on Hong Kong Island. The proposed development prompted an outcry from local residents due to the possible air pollution and traffic congestion issues arising from the project. Only two solutions were open to the developer: to solve the conflict using legal procedures or to terminate it altogether.

In sum, both the internal stakeholders (clients, consultants and contractors) and the external stakeholders (public and private external parties) can bring various risks to construction projects. Ignorance of these risks may well cause substantial problems for a project. Successful project management needs to bear this in mind and to ensure that all the key stakeholders and associated risks have been identified and appropriate strategies have been developed to engage with them. It should also be noted that the stakeholders are not isolated from each other, and the interplay between the stakeholders in construction have significant impacts on construction projects.

6.4   The complex interplay of stakeholders in construction

The stakeholders in a construction project are by no means isolated from one another. They are closely interrelated through formal or informal ties during the implementation of a project. The internal stakeholders are linked via legal contracts (e.g. a consultancy agreement between a client and a consultant, and the main contract between a client and a contractor) or direct interest (e.g. customers in a shopping centre and the client), while the external stakeholders will also affect or be affected by the construction product through indirect ties. Although the external stakeholders are not decision-makers in the project, they still have a link to the project and can influence the project in a number of ways. The relationships of project stakeholders are depicted in Figure 6.1.

Clients, consultants and *contractors* generally have official communication channels, as they are closely interrelated through legal contract ties.

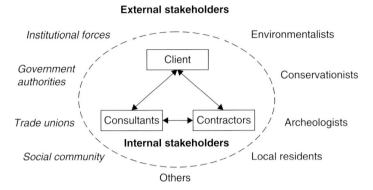

Figure 6.1   Interplay between some key stakeholders.

In the **initial stage**, the client normally employs consultants to prepare a marketing report in order to understand the existing economic situation and the *customers'* expectations of the site. The client will also seek comments from *general practice surveyors* regarding the purchase of land from the *government*, after describing the client's requirements to the government. If necessary, a *private client* may seek advice from *financial institutions* on a development project. *Public clients* may not have any financial difficulties, but, in recent years, they have also placed an emphasis on the need for the support of the *community* and *local residents*. Although clients try their best to describe their requirements clearly, it is still difficult or impossible to write down every requirement accurately at this preliminary stage. A client's requirements often change and this further affects the overall development after the consideration of all the limitations on the proposed project in the later stages.

In the **pre-design stage** of a construction project, clients normally employ a group of consultants as internal stakeholders to further investigate the project. In order to ensure that a project will be constructed and operated in a functionally, technically and financially feasible manner, a group of *internal stakeholders* such as the *project manager*, the *architect*, the *engineers* and the *quantity surveyor* have to work together on a feasibility study. Since the information provided is not clear at this stage, it is difficult for designers to assess the technical problems and surveyors to estimate the budget. As each consultant concentrates on their particular area (e.g. the architect concentrates on aesthetic value, the structural engineers on structural support and the quantity surveyor on cost saving), there may also be some conflicts among consultants at both the pre-design and the design stages. The client needs to make decisions about the type of building, the scope of the project work and the investment in the development, on the basis of his or her provisional requirements, town planning considerations, site conditions, design considerations, financial conditions, construction costs and so on. Therefore, clients may also need to interact further with *financial institutions*. Factors such as interest rate and credit policy will affect the performance of the project investment, and consequently influence the client's decisions.

During the **design stage**, clients and project *consultants* have to work hand-in-hand to ensure that the design is satisfactory. In the scheme design phase, clients normally decide on one or two proposal(s) based on the recommendations contained in the feasibility report (RIBA, 1991). The project manager will then cooperate with other consultants to study the approved proposals in detail and make a preliminary design for the project based on the client's requirements. During this stage, constant and frequent communication is essential. All the technical issues and the conflicts between different types of drawing have to be resolved by the consultants. Any delay in approvals by a client may cause the whole project to be delayed, as all the drawings will have to be submitted to the *building control department* and *other relevant departments* for formal approval. The project manager needs to give frequent reports to the client in order to provide a clear picture of the project's progress, and to avoid design discrepancies or conflicts between the client's requirements and the professional opinion, and ensure an accurate and effective cost estimation for the project.

During the **tendering stage**, there are close interactions between *clients, consultants* and *tenderers*. Once they are in possession of all the specifications for the designs (architectural, structural engineering and building services engineering), quantity surveyors need to prepare Tender Documents for a particular project. During this period, cooperation among all consultants is necessary. The architect should carefully check the Tender Documents with quantity surveyors to ensure their correctness, while the surveyors have to clarify any design discrepancies with the architect and the engineers. Detailed design and accurate Tender Documents definitely help the client to select the most appropriate tenderer as the contractor for the project. All the consultants need to clarify any query raised by tenderers. They should assess the Tender Documents they receive and recommend to their client the most appropriate tenderers as contractors. In order to ensure they understand all the details of each Tender Documents, the client and the consultants normally have pre-tender meetings or interviews with interested tenderers at this stage. Once the project is awarded to the selected contractor, Contract Documents have to be prepared. Although the tendering period is not long compared with the other stages in a project, a lengthy interplay between clients, consultants and a number of tenderers (including the potential contractors) is normally involved.

In the **construction stage**, a contractor has to carry out the work on site, working to the Contract Documents. A *construction team*, including the *main contractor*, the *sub-contractors* and *suppliers*, should be formed at this stage. From the commercial point of view, a contractor has to minimize expenditure and maximize income. Therefore, project managers and other consultants, on behalf of the client, need to monitor contractors' work frequently, in order to ensure that the project is on the right track. If there is any problem (e.g. design error, design discrepancy or design changes) in the construction process, the *consultants* should coordinate any variation and cost adjustment with the contractors. Contractors also need to cooperate with the various *labour unions* and suppliers to ensure that the work is carried out and the materials and equipment are supplied as specified. Quantity surveyors will normally issue interim certificates confirming the progress of the contractors, and the client will make corresponding payments to the contractors. To secure a healthy cash flow in a construction company, it is common for a contractor to claim an amount for each interim payment that is higher than required for the actual work done. Hence, in order to maintain the channel of communication, normal progress on the project should be constantly reported to the client. If any major change is necessary, approval from the client is required, and the *client*, the consultants, the contractors and all other relevant parties should sit down together to examine all the possible alternatives. If the issue cannot be resolved, the question of termination may be raised by any party during the construction process. In the end, new contractors, sub-contractors or suppliers might be required.

In the **completion and operation stage**, the project has to be handed over to the *client*. The *contractor* will have to deal with *government agencies* to obtain certain permissions (e.g. an occupation permit for a housing development project). The *project manager* must coordinate his or her work with *other designers, surveyors* and contractors to ensure that all the terms of the contract are fulfilled. If there are any defects in the building product, in principle, the contractor has the responsibility to rectify them within the defects liability period. Where defects appear after the defects liability period or the causes of defects are unclear, the client, the consultants and the contractors need to thoroughly investigate them in order to establish who is responsible. Property and facility management parties are responsible for maintaining the building during the operation stage. The developer and contractors will collect feedback from the end-users on the performance of the project in order to improve future projects.

The interactions between construction stakeholders are rather complex and they vary with the nature and scale of construction projects. Wise project managers need to constantly review the interplay of stakeholders throughout the whole life cycle of construction projects.

6.5   Assessing the risks posed by different stakeholders

A construction project will have multiple stakeholders (multi-stakeholders). Each stakeholder plays a different role in the planning, development, construction and operation of a building product. However, each one can also create certain risks in construction projects. They influence each other throughout the project from its initial stage to the operation stage. In order to assess the risks imposed by construction stakeholders, a systematic risk assessment process should be followed (also see Table 6.4).

1. Risk identification
2. Risk impact analysis
3. Risk occurrence probability estimation
4. Risk prioritization

Risks can be introduced by various stakeholders (both internal stakeholders and external stakeholders). Stakeholder **risk identification** can be broken down into two steps: (1) to identify as many project risks as possible; and (2) to determine the stakeholder associated with a particular risk. There are several techniques for identifying risks and the stakeholders associated with them. *Brainstorming* is a powerful tool for tackling project risks, as a comprehensive list of risks can then be addressed in the risk assessment processes that follow (Olander, 2003). This technique enables construction stakeholders to hear what the other construction stakeholders see as risks, and promotes the inspiration to identify additional

Table 6.4   Risk assessment process and the relevant techniques

Risk assessment process	Risk assessment techniques	Purpose
1. Risk identification	Brainstorming	→ indicating a comprehensive list of risks
	Checklist	→ listing foreseeable risks
2. Risk impact analysis	Power–Interest Matrix	→ identifying key stakeholders for the risks
	Discussion and interview	→ finding out hidden key stakeholders
		→ understanding the impacts of stakeholders
	Delphi technique	→ obtaining consensus on main risks
3. Risk occurrence probability estimation	Interview and discussion	→ collecting qualitative data from relevant stakeholders
	Monte Carlo analysis	→ obtaining estimated probabilities of cost overrun and delayed schedule from relevant stakeholders
4. Risk prioritization	Impact–Probability Matrix	→ obtaining risk score from stakeholders for planning risk response

project risks associated with stakeholders (Smith *et al.*, 1999). Since the selection of members of a brainstorming session is crucial, it is suggested that all key stakeholders are included in the session to stimulate the identification of as many risks as possible. In order to have a clear concept of project risk, the *checklist* is another good tool for listing foreseeable risks in the project by studying other similar projects. As the stakeholders gain more experience and knowledge of the project, the checklist can be correspondingly expanded (Harrison and Lock, 2004). Once the risks have been identified, they can be prioritized according to their impact and probability of occurrence.

The purpose of the **risk impact analysis** is to assess the severity of a risk in a project. The impact of stakeholders consists of two dimensions: power and interest. *Power* is the ability to influence others and to get things done (Lovell, 1993; Pinto, 1998). The more powerful a stakeholder is, the more damage his or her or its influence might cause to the project. The more *interest* a stakeholder has in the project, the more active the stakeholder may be in project activities and the more uncertainties this will bring to the project. One of the frameworks to assess the impact of stakeholders is the Power–Interest Matrix (Mendelow, 1981; Johnson and Scholes, 1997; Olander, 2003). The stakeholders in a project can be classified into four categories according to the degree of power and interest they have (see Figure 6.2): (1) those who have great power in the project and a high level of interest; (2) those who have great power but a low level of interest; (3) those who have little power but a high level of interest and (4) those who have little power and a low level of interest.

Category 1: *Key player* stakeholders (high levels of power and interest): These are the most important stakeholders; as key players, they exercise a great deal of power to influence project implementation and completion (Olander, 2003). The internal stakeholders such as the representatives of the client, the consultants, the contractors and the suppliers can be placed in this category, as authority is delegated to them by the client to implement the project. If they are incompetent in their work, flaws in design or construction tasks can result.

Category 2: *Keep-satisfied* stakeholders (high level of power but low level of interest): Though their interest in the project is low, they can exercise great power over the project. For instance, governmental agencies have full authority to approve detailed drawings. Any revised regulation issued by the government agencies may cause additional work for the internal stakeholders [e.g. as previously noted, according to the labour registration policy issued by the government in 2007, all labour on site must now be registered (BLIS, 2007)].

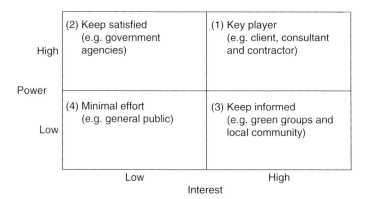

Figure 6.2   Stakeholder risk impact by Power–Interest Matrixes.
(*Source*: Adapted from Johnson and Scholes, 1997; Olander, 2007)

Category 3: *Keep-informed* stakeholders (high level of interest but low level of power): Although these parties, such as local residents or environmental groups, have limited power over project decision-making, they can still influence a project indirectly or undermine the image of the client's company. Therefore, client and client representatives need to consider their expectations and keep them informed of progress and key decisions.

Category 4: *Minimal-effort* stakeholders (low levels of interest and power): Such stakeholders will have less risk impact on the project than others. If such risks develop to a certain extent, coping strategies might become necessary.

The impact of each stakeholder is different in each particular project. In general, the stakeholders in category 1 can have a high level of risk impact on a construction project (e.g. the internal stakeholders such as the client, the consultants and the contractors), while those in category 4 can have only a low level of risk impact on a project (e.g. the general public). The stakeholders in categories 2 and 3 can have a medium level of risk impact on a project (e.g. government departments and green groups). Once the key stakeholders are identified in a construction project, there are several techniques, such as discussion, interview and the Delphi technique, that can help us to analyze the impact of the stakeholders and their influence on the project risks.

Discussion and interview with key stakeholders (Field and Keller, 1998) are useful methods of understanding the impact of construction stakeholders. Interviewing project personnel from each discipline and staff within the organization can not only lead to the acquisition of the knowledge of their requirements and the associated risks, but can also uncover other hidden key stakeholders who are not easily detectable in the development and construction processes.

The *Delphi technique* is a useful technique to reach consensus among experts on project risk impact. Sometimes, risk information about stakeholders cannot be obtained from stakeholders directly, so experts in this area are needed to give information on the project stakeholder risks. Project risk experts are identified and invited to participate in this exercise anonymously. A facilitator uses a questionnaire to solicit ideas about the important project stakeholder risks. Their response is submitted and is then circulated to the experts for further comments. Consensus on the main risks can be reached in a few rounds of this process (Olander, 2003).

Once the stakeholders have achieved a consensus on the main risks, a risk event occurrence probability should be estimated. **Risk probability analysis** identifies the likelihood that the stakeholder response will actually occur. The probability of a risk event occurring can be estimated by factors that cause risk, which is determined by the unique nature of each project. A number of techniques, such as interview and discussion, Monte Carlo analysis or sensitivity analysis, can be used to analyze the probability of occurrences qualitatively or quantitatively.

Monte Carlo analysis is a sophisticated technique facilitated by computer. It can be applied to evaluate the probabilities of a risk event occurrence. If project data are inputted, the computer can calculate the statistical data of completion time for the final activity in the networks (Harrison and Lock, 2004). With this calculation, the client can obtain statistical risk event materialization data such as estimated probabilities of cost overrun and delays to the schedule.

Risk severity or risk impact can be measured on a scale from high (10) to low (1); risk probability can also be rated on a scale from high (10) to low (1). Risk event occurrence can generally be categorized into four main priority groups: great danger, probable catastrophe, challenge and minor risk (see Figure 6.3).

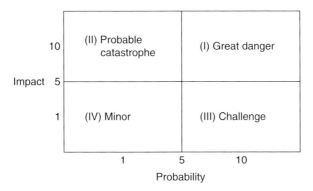

Figure 6.3   Risk Priority Matrix based on power and probability.
(*Source*: Adapted from Winch, 2003)

Category I: *Great-danger risks* (high impact and high probability) bring probable disaster to a
    project, and require the most attention. When a majority of the risks belong to this type of
    risk, the project might be in great danger. It may be better to modify the plan or cancel the
    project (Smith *et al.*, 1999).
Category II: *Probable-catastrophe risks* (high impact but low probability) will cause great
    damage to a project if events occur. Such risks also need to be closely watched by project
    managers. Formal and constant communication with the stakeholders will be helpful in
    keeping these risks under control. Modification or contingency plans are necessary to
    ensure that the project can still survive if these risk events occur.
Category III: *Challenge risks* (low damage impact but high occurrence probability) pose great
    challenges to project managers who have to cope with them. In order to reduce the prob-
    ability of this type of risk event occurring, some actions to change certain features of the
    project should be taken. The stakeholders may need to balance the cost of the actions and
    the gains obtained from them in order to reduce the risks.
Category IV: *Minor risks* (low impact and low probability) pose the least danger to the project.
    Sometimes, it may be economical simply to let a minor-risk event go ahead. However, key
    stakeholders still need to keep an eye on these risk events as they may evolve as time goes
    by. If they evolve to a critical point, action may become necessary.

Once risks have been identified and assessed with regard to the impact and probability of
an occurrence, they can also be scientifically **prioritized**. The risk score is calculated by mul-
tiplying *risk severity* and *risk probability*. Hence, risk can be prioritized from the highest to the
lowest. Such information can then be considered as an input into a risk response plan in the
subsequent step.

Table 6.5 illustrates an example of three risks posed by various stakeholders in a development project for a private school. The dynamic nature of society, an inappropriate design
and injury to a person are three common risks in construction projects. The event occurrence
probability for each risk changes depending on the particular project, while the impact and
severity of a risk event differs for different stakeholders. In the case shown in Table 6.5, inappropriate design is prioritized as the first serious risk influencing the most stakeholders.
Therefore, action appropriate to the stakeholder should be taken to manage the risks among
stakeholders in the projects.

Table 6.5   Examples of risk assessment for various stakeholders in the development of an international private school project

Risk	Risk probability	Pro. deficit	Risk severity/impact	Stakeholder	Stakeholders' impact deficit (power–interest)	Risk score	Prioritize for international stakeholder
Market changes	Birth rate in each year	6	Insufficient students	Client Consultant General public	5 2 1	48	2
Inappropriate design	A set of standard drawings for other similar projects	2	Change drawings Errors in cost estimation/over budget Delay management process Delay construction period	Client Consultant Contractor Government General public Green group	8 8 4 3 2 1	52	1
Injury to person	No safety training and security guideline	2	Accidents on site Affect the subsequent projects	Client Consultant Contractor Government Labour union	4 3 8 1 6	44	3

Note: 'Pro. deficit' – Probability of deficit.

6.6   Managing stakeholders using risk management principles

Risk management requires cooperation between the stakeholders. By adopting a risk management framework, construction stakeholders can manage risks using the following steps:

1. Identify project risks and relevant stakeholders.
2. Assess risks systematically.
3. Prioritize stakeholders' risks.
4. Respond to risks.

As each action involves expenditure, the stakeholders need to balance the cost and the benefits when considering action plans. The strategies that stakeholders can adopt to deal with risks include the following:

- Avoid risks.
- Mitigate risks.
- Transfer risks to other stakeholders.
- Share risks with other stakeholders.
- Contingency plans.
- Accept stakeholder risks.

Avoiding risk caused by stakeholders means completely removing the risk from the work to be done. A method of this kind always involves redefining the project scope and redesigning the project plans. It can only eliminate specific risk events and is only applied on some occasions. When stakeholders are dealing with risks they are unfamiliar with and fail to control – risks that will, if they materialize, lead to substantial damage to the project – it is appropriate to use this method.

Risk mitigation involves reducing the impact of a risk event caused by stakeholders or reducing the probability of its occurrence (Schwalbe, 2000). In most cases, risk mitigation can be a cost-effective method for engaging with stakeholders. Other stakeholders predict their actions and the consequent impact on the project, and then take the necessary action such as the establishment of formal communication mechanisms and the extension of the coordination period for each task, in order to reduce risk event probability and impact.

Transferring risk means transferring the impact to other parties if the risk event occurs. A construction company can claim compensation for any loss due to fire or theft from an insurance company if a risk event occurs. A client can try to transfer the costs of calculation errors to contractors through the Drawings and Specification contract, or the design faults, calculation errors or management mistakes through the Design and Build contract or the Construction Management contract.

However, this strategy may introduce other additional risks to the project. For example, the insurer may refuse to pay for the loss, the financial compensation may not be adequate in compensating for the time and resources lost or the compensation may not be sufficient to maintain a project on track if the risk event occurs. Therefore, transferring the risk does not always help the parties to a project in the long term (Field and Keller, 1998). The strategy of transferring risk should therefore be used tactically in combination with other methods.

Sharing risk is a useful method for handling a complex project. If there are too many risks for a specific stakeholder, he or she or it should seek more partners to join the project as a joint venture (Smith *et al.*, 1999; Harrison and Lock, 2004). Using this method, the impact of failure could be shared among a greater number of partners. As the method often needs the support of a senior manager and the involvement of multiple parties, the stakeholders in the management process are required to have good coordination and communication capabilities.

The **contingency plan approach** requires the development of alternative plans to cope with the losses in the event of a risk materializing (Field and Keller, 1998). An amount may be allocated in the budget to recover minor cost overruns and to keep the project on track, and the benefits and costs of any alternative plan therefore need to be evaluated so that the optimum solution can be selected. This method usually applies to critical stakeholder risks; that is, those that have special impacts on the project and would probably lead to a crisis if the risk materialized. Key stakeholders need to propose and submit the contingency plan to their senior managers for approval at the outset in order to allow sufficient additional resources to be allocated to managing the risks in the project.

Accepting risk is a method used to deal with the less harmful risk events in construction projects. Where a risk event has a minor impact on the project and has less probability of endangering the entire project, there is little the stakeholders can do to manage it as it is too costly to develop a contingency plan for everything (Field and Keller, 1998; Harrison and Lock, 2004). But since project conditions keep changing during the implementation processes, it is still necessary to monitor the state of such risks regularly. The stakeholders need to take appropriate action to tackle a risk event if it has become critical to the project.

Further to the examples in Table 6.5 (dynamic society, inappropriate design and injury to a person), various risk management methods that stakeholders can use in construction projects are listed in Table 6.6. Risks can be *avoided* or eliminated altogether by cancelling the project

Table 6.6 Risk management by stakeholders

Risk source	Risk event	Stakeholders					Risk management method					
		I1	I2	I3	E1	E2	Avoiding	Mitigating	Transferring	Sharing	Contingency	Accepting
Market changes	– Vague client's requirement.	✓					I1 – Cancel the project.	I1 – Allocate additional resources for project scope change.	I1-3 – Adopt Build–Operate–Transfer procurement.	– Allow provisional item(s)/bill(s) in tender documents.	I2 – Allow contingency sum in tender documents.	I1-3 – Issue variation order for any change during post-contract stage.
	– Change of client's requirements.	✓	✓	✓	✓	✓		I1-2 – Communicate with *clients* formally and constantly.				
	– Changes of project scope and requirements.	✓	✓					I1-3, E1-2 – Consult *all parties* on project scope and requirements in detail during project briefing.				
	– Conflicts of requirements of interest groups.	✓	✓	✓	✓	✓		I1-3, E1-2 – Consider as many stakeholders' requirements as possible.				
Lack of experienced consultants	– Inaccurate physical condition investigation.	✓	✓	✓			I1,I2 – Don't award the contract to *consultants* with insufficient experience.	I1-2, E1-2 – Consult *external parties* such as government departments and environmental organizations during design stage.	I1-3 – Adopt design and build contract for single point responsibility.	I1-3 – Insert a clause into *consultant* agreement for sharing the loss caused by the misconduct of consultants.	I2 – Allow contingency sum in tender documents.	I1-3 – Issue variation order to adjust any error during post-contract stage.

Risk				Mitigation measures		
– Errors in cost estimation.	✓			– Employ an experienced project manager to supervise the project. (I1-I2) – Insert a clause into consultant agreement as a penalty for any error or discrepancy. (I1-3) – Employ other experienced consultants, if necessary. (I1-2)		
– Inappropriate design.	✓	✓		– Request consultants to (1) employ a senior experienced professional for supervising their junior staff; or (2) recruit new experienced staff. (I2)		
– Conflicts between different drawings.	✓					
– Incompleteness of tender documents.	✓					
– Errors/omissions/ discrepancies in bills of quantities.	✓	✓				
Lack of safety knowledge. – Injury to person.	✓	✓		– Don't carry out any risky work. (I1,I3) – Insert a training item in Contract Documents. (I1,I3) – Request contractor to purchase insurance for all accidents and their effects. (I3) – Share responsibilities of all accidents between client and contractor.. (I1, I3) – Prepare a guideline for handling different accidents (e.g. fire, falling objects, death, etc.) (I3) – Record all accidents and report to government without any further action. (I3, E1)		

(Continued)

Table 6.6 Risk management by stakeholders (*Continued*)

| Risk source | Risk event | Stakeholders | | | | | Risk management method | | | | | |
		I1	I2	I3	E1	E2	Avoiding	Mitigating	Transferring	Sharing	Contingency	Accepting
	– Damage property.	✓	✓	✓			– Don't move any heavy machine. *I1,I3*	– Employ a safety officer on site as required by law. *I3*			*I3*	*E1*
	– Pay compensation to injured person.	✓	✓	✓				– Provide safety equipment (e.g. shoes, helmet, etc.) on site. *I3* – Provide some safety training. *I3*				
	– Labour union complaint regarding insufficient compensation.	✓		✓		✓		– Consult external parties such as labour union on actual work processes and occupational safety organization for site safety. *I1,E2*				

Note: I1, I2, I3, E1 and E2 refer to the stakeholders (client, consultant, contractor, external public parties and external private parties, respectively, as noted in Table 6.2.

or simply terminating the contract. However, it is difficult for stakeholders to implement a construction project without being subject to any risk at all. Stakeholders can *mitigate* risks by using alternative methods such as allocating additional resources to allow for potential changes in the scope of a project, by employing an experienced project manager to supervise inexperienced consultants and by providing safety equipment to be used if there are safety problems. Alternatively, clients can *transfer* risks or *share* them with their consultants or contractors. They can also, for example, adopt different contractual procurements, include provisional items and insurance clauses in Bills of Quantities and so on. In some cases, construction stakeholders may prepare *contingency* plans (e.g. contingency sums for variations and guidelines for handling different accidents) for handling the potential risks at the outset or simply *accept* the risk in the real situation.

6.7   Summary

In summary, it is necessary for the stakeholders to manage construction projects throughout the development and construction processes due to the dynamic nature of society, the complex nature of construction work and the interplay of multi-stakeholders. This chapter has classified five main stakeholder groups under two major categories (i.e. client, consultants and contractors as internal stakeholders; and external public parties and external private parties as external stakeholders). The construction project is subject to various risks from its initial stage right through to the operation stage. The stakeholders, firstly, have to identify the critical risks using a systematic risk assessment process: risk identification, risk impact analysis, risk event occurrence probability estimation and risk prioritization. In order to manage the risks well, the stakeholders need to balance the costs and the benefits of different risk management strategies (i.e. avoiding risk, mitigating risk, transferring risk, sharing risk, contingency plans and accepting risk) for coping with the potential critical risks involved in construction projects.

References

BLIS (2007) Construction Workers Registration Ordinance. Available from http://translate.legislation. gov.hk/gb/www.legislation.gov.hk/blisexport.nsf/chome.htm. (Viewed on 11 February 2008).

Bourne, L. and Walker, D.H.T. (2005) Visualising and mapping stakeholder influence. *Management Decision*, **43**(5):649–660.

Clarkson, M.B.E. (1994) A risk based model of stakeholder theory. *Proceedings of the Second Toronto Conference on Stakeholder Theory*, Toronto, pp. 18–19.

Clarkson, M.B.E. (1995) A stakeholder framework for analyzing and evaluating corporate social performance. *Academy of Management Review*, **20**:65–91.

Cleland, D.I. and Ireland, L.R. (2002) *Project Management: Strategic Design and Implementation*. Boston: McGraw-Hill.

Field, M. and Keller, L.S. (1998) *Project Management*. Boston: International Thomson Business Press.

Freeman, R.E. (1984) *Strategic Management: A Stakeholder Approach*. Boston: Pitman.

Harrison, F.L. and Lock, D. (2004) *Advanced Project Management: A Structured Approach*. Aldershot, England: Gower.

Johnson, G. and Scholes, K. (1997) *Exploring Corporate Strategy*. New York: Prentice Hall.

Juliano, W.J. (1995) External communication as an integral part of project planning. *Project Management Network*, February:18–20.

Leung, M.Y. and Liu, A.M.M. (2003) Analysis of value and project goal specificity in value management. *Construction Management and Economics*, **21**(1):11–19.

Lovell, R.J. (1993) Power and the project manager. *International Journal of Project Management*, **11**(2):73–78.

Madsen, H. and Ulhoi, J.P. (2001) Integrating environmental and stakeholder management. *Business Strategy and the Environment*, **10**(2):77–88.

McElroy, B. and Mills, C. (2000) Managing stakeholders. In: Turner, J.R. and Simister, S.J. (eds) *Gower Handbook of Project Management* (3rd edn). Aldershot, England: Gower, pp. 757–775.

Mendelow, A.L. (1981). Environmental Scanning: The impact of the stakeholder concept. *Proceedings of the International Conference on Information Systems*. Cambridge, Mass, pp.407–417.

Mitchell, P.K., Bradley, R.A. and Wood, D.J. (1997) Toward a theory of stakeholder identification and salience: Defining the principle of who and what really counts. *The Academy of Management Review*, **22**(4):853–886.

Olander, S. (2003) *External Stakeholder Management in the Construction Process*, PhD thesis, Department of Building and Construction, Lund Institute of Technology University.

Olander, S. (2007) Stakeholder impact analysis in construction project management. *Construction Management and Economics*, **25**(3):277–287.

Phillips, R. (2003) *Stakeholder Theory and Organizational Ethics*. San Francisco: Berrett-Koehler.

Pinto, J.K. (1998) *Power and Politics in Project Management*. Sylva, NC: Project Management Institute.

PMI (1996) *A Guide to the Project Management Body of Knowledge* (*PMBOK Guide*). Newtown Square, PA: Project Management Institute.

PMI (2004) *A Guide to the Project Management Body of Knowledge*, (*PMBOK Guide*) (3rd edn). Newtown Square, PA: Project Management Institute.

Post, E.L. and Je, P. (2002) *Redefining the Corporation: Stakeholder Management and Organizational Wealth*. Stanford, CA: Stanford University Press.

Post, J.E. and Preston, L.E. (2002) Managing the extended enterprise: The new stakeholder view. *California Management Review*, **45**(1):6–28.

Preble, J.F. (2005) Toward a comprehensive model of stakeholder management. *Business and Society Review*, **110**(4):407–431.

RIBA (1991) *Architect's Handbook of Practice Management*. London: RIBA.

Rotarius, T. and Liberman, A. (2000) Stakeholder management in a hyper-turbulent health care environment. *The Health Care Manager*, **19**(2):1–7.

Savage, G.T., Nix, T.W., Whitehead, C.J. and Blair, J.D. (1991) Strategies for assessing and managing organizational stakeholders. *Academy of Management Executive*, **5**(2):61–75.

Scheffran, J. (2006) Tools for stakeholder assessment and interaction. In: Stoll-Kleemann, S. and Welp, M. (eds) *Stakeholder Dialogues in Natural Resources Management*. Heidelberg: Springer-Verlag.

Schwalbe, K. (2000) *Information Technology Project Management*. Florence, KY: Course Technology Press.

Smith, N.J., Merna, T. and Jobling, P. (1999) *Managing Risk in Construction Projects*. Oxford: Blackwell Science.

Sternberg, E. (1997) The defects of stakeholder theory. *Corporate Governance*, **5**(1):3–10.

Walker, D.H.T. (1989) *Project Management in Construction*. London and Malden, MA: Granada Blackwell Science Inc.

Winch, G.M. (2002) *Managing Construction Projects*. Oxford: Blackwell.

7 Mapping Stakeholders

Lynda Bourne and Patrick Weaver

7.1 Introduction

The objective of every stakeholder mapping process is to develop a useful list of stakeholders. The next step should then be to assess their key characteristics and present this information in a way that helps the project team implement effective stakeholder management initiatives. The key element of an effective mapping process is as far as possible to replace subjectivity with objective measures and to make the assessment process transparent. This transparency will allow the basis of any assessment to be clearly understood by others and will facilitate review and updating as appropriate.

The challenge with stakeholder mapping is that the elements being assessed, such as the level of a stakeholder's support, are driven firstly by the perceptions of the project held by the stakeholder and secondly the perceptions of the stakeholder's attitudes held by the people undertaking the mapping process. These factors are influenced by the 'hardwiring' in each person's mind and because of this influence two people can have completely different 'views' of the same situation.

The brain hardwires everything it possibly can, and defines ways to store data and retrieve it that best suit the history and personality of an individual (Rock, 2006). Because our brain must, consciously or unconsciously, manage all and every stimulation it receives, it will attempt to automate as much as possible. Therefore it is important when presenting information, to consider ways to assist the brain to process the stimulation, and be consciously aware of important and/or new information. Presenting data in graphical or pictorial form will help the audience map connections more readily: the brain processes ideas fastest visually (Rock, 2006:90).

People learn and also retain information by using the mode they are most comfortable with in the first instance, whether visual, auditory or kinaesthetic. Other studies have shown that everybody learns best and also retains information longer when offered the data in more than one mode. For example, people will learn better by listening and seeing, and even better by listening, seeing and doing (Glasser, 1998). Therefore the complex data collected about stakeholders will be most easily understood when presented in several complementary forms; graphical or pictorial views supported by tabulations and/or sorted lists.

This chapter will firstly discuss the importance of mapping stakeholders; this discussion will be followed by a review of mapping methods and techniques used today. The next section will describe a specific mapping method and technique, the *Stakeholder Circle®* that provides guidance to knowing who the right stakeholders are for any time in the project, providing the information needed for developing strategies for targeted communication. A final step of the methodology supports monitoring the effectiveness of the communication.

7.2   A picture tells a thousand words

Researchers or reporters must develop ways to present data they have collected. Organising the data allows researchers to categorise and review the information they have collected to gain valuable insights. Different methods of representation may allow researchers or others to recognise patterns that support comparison or contrast of this data to known data, or may simply allow others to absorb or make sense of the data more easily. In the world of construction projects, the most effective presentation of the data will be graphical or pictorial.

7.2.1   The evolution of stakeholder mapping

Chapter 3 has discussed the history of stakeholder management and the introduction to this chapter has highlighted the importance of presenting complex data to aid understanding. One of the consequences of the emerging understanding of 'stakeholders' through the 1980s and 1990s was the need for business managers and researchers to be able to visualise the 'stakeholder community' surrounding their particular business unit or project. Consequently, the concept of representing data collected about stakeholders as maps – tabular, graphical or pictorial – has been adopted by researchers and consultants from the earliest studies. We suggest that there are three basic approaches used to help visualise, map and understand stakeholders: customer relationship management – CRM, influence and social networks, and techniques for listing and mapping stakeholders.

The approach with the highest profile in general business is the customer relationship management or CRM approach. This approach requires substantial data sets to be gathered about a key segment of the business' stakeholder community (typically customers) followed by the use of data mining techniques that allow trends and opportunities to be identified, graphed and communicated. These reports inform management decision-making and help the business prosper. CRM works effectively in situations where the business is relatively stable and there is a large class of stakeholders interacting with the business in a reasonably consistent way: its focus is to build and maintain a customer-centred enterprise cost-effectively, and generate a good ROI (Bligh and Turk, 2004). CRM has little application to the construction industry due to the individual significance of most stakeholders and the relatively low levels of repeat business from customers. For example most families will buy a new house once or twice in a lifetime but will visit their local supermarket once or twice every week.

A second approach that cannot be ignored is the extensive body of work focusing on influence networks. This research focuses on the importance of relationships through the study of 'influence networks', 'social networks', 'social capital', viewing projects as temporary knowledge organisations – TKOs (Sbarcea and Martins, 2003) and more recently in complex relationships (Weaver, 2007). All of these theories emphasise the critical importance of the relationships between different stakeholders both within and around the project team. The strength and effectiveness of the internal relationships enable the project team to function effectively and allows the team (or the project) to interact and influence its surrounding stakeholder community. The difficulty in using these strands of research lies in building the influence/relationship maps. The work is difficult, time consuming and invasive requiring extensive interviews with the stakeholders. Consequently, whilst an appreciation of these ideas is critical for effective stakeholder management, the opportunities to undertake a detailed analysis of a particular stakeholder community are very limited and typically only occur as part of an academic research assignment.

The need for a practical, usable approach to visualising many different stakeholder communities has led to the development of a range of listing and mapping techniques by academics, consultants and businesses over the years. These approaches trade the richness of data available under the CRM approach for a holistic view of the whole stakeholder community and largely ignore the complex network of relationships considered in CRPR and the other network theories outlined above for a simpler consideration of 'importance' in some form. Obviously the 'importance' of a stakeholder is directly associated with his or her ability to influence the project through their network of relationships. The difference in the analysis is in the way this is assessed. All of the mapping techniques discussed below use a qualitative perception of a stakeholder's importance rather than a quantitative analysis of the influence networks and relationships surrounding the stakeholder to determine an absolute value for that person's importance.

The following list identifies some of the best known and most commonly used methods for stakeholder mapping:

- Mitchell *et al.* (1997) proposed a classification of stakeholders based on power to influence, the legitimacy of each stakeholder's relationship with the organisation and the urgency of the stakeholder's claim on the organisation. The results of this classification may assess the fundamental question of 'which groups are stakeholders deserving or requiring manager's attention, and which are not?' This is salience – 'the degree to which managers give priority to competing stakeholder claims' (Mitchell *et al.*, 1997:854).
- Fletcher *et al.* (2003) defined a process for mapping stakeholder expectations based on value hierarchies and Key Performance Areas (KPA).
- Savage *et al.* (1991) offered a way to classify stakeholders according to potential for threat and potential for cooperation.
- Turner *et al.* (2002) developed a process of identification, assessment of awareness, support, influence leading to strategies for communication and assessing stakeholder satisfaction, and who is aware or ignorant and whether their attitude is supportive or opposing.

Mapping techniques include the following sub-set of results from a Web search of analysis techniques being used by aid agencies, governments or consultant groups:

- Influence–interest grid (Imperial College London, 2007);
- Power–impact grid (Office of Government Commerce, UK 2003);
- Three techniques used by the Australian Department of Sustainable Environment (2007):
  - ○ Influence–importance grid;
  - ○ Venn diagrams;
  - ○ CLIP analysis (Collaboration/conflict, Legitimacy, Influence and Power);
- Power–interest grid (Moorhouse Consulting, 2007);
- Three-dimensional grouping of power, interest and attitude (Murray-Webster and Simon, 2007).

The first step in building any stakeholder map is to develop a categorised list of the members of the stakeholder community. Once the list is reasonably complete, it is then possible to assign priorities in some way, and then to translate the 'highest priority' stakeholders into a table or a picture. The potential list of stakeholders for any project will always exceed both the time available for analysis and the capability of the mapping tool to sensibly display the results. The challenge is to focus on the 'right stakeholders' who are currently important and to use the tool to visualise this critical sub-set of the total community.

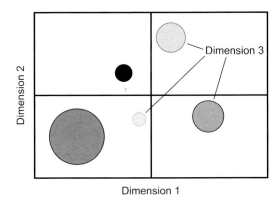

Figure 7.1   Traditional stakeholder mapping.

The most common presentation styles use a matrix to represent two dimensions of interest, with sometimes a third dimension shown by the colour or size of the symbol representing the individual stakeholders. This is summarised in Figure 7.1.

Some of the commonly used dimensions include:

- power (high medium low);
- support (positive, neutral, negative);
- influence (high or low);
- interest (high or low);
- attitude (supportive or obstructive).

Where used, the methods of gathering and recording data for stakeholder maps such as these tend to be subjective, with results usually derived from open questions that allow subjective and inconsistent responses. In many cases, the people preparing the map simply draw symbols in the map based on their personal assessment.

Many of the proprietary methods also indulge in terminology for categories in these maps that may be amusing or interesting expressions of the nature of the categorised stakeholders, but not always acceptable business terminology. Reporting to a senior manager that they have been categorised as 'dangerous', a 'time bomb', 'demanding' or a 'trip wire' is unlikely to be helpful! Using such terminology may add an element of 'fun' for team members, but it involves an extra learning challenge and does not add significantly to the effectiveness of the methodology.

The need for simplicity and flexibility in data gathering and reporting should be reflected in developing guided steps through a series of processes that can be cumulative or can be approached in parts, depending on the needs and maturity of the organisation. Consistency can be achieved through a system of ratings against a consistent set of statements rather than reliance on subjective and variable answers to open questions.

The *Stakeholder Circle*® provides a methodology and a mapping technique to represent data about stakeholders in consistent, staged and guided steps, providing stakeholder data in tables, graphs and pictures. The *Stakeholder Circle*® methodology consists of five steps: identify all stakeholders, prioritise them, display the current members of the stakeholder community, develop an engagement strategy and communication plan and monitor the effectiveness of the communication. Figure 7.2 shows an overview of the five steps of the methodology.

The *Stakeholder* *Circle*® Methodology
Are you managing the right stakeholders?

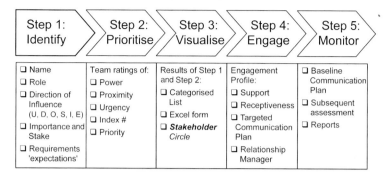

Step 1: Identify	Step 2: Prioritise	Step 3: Visualise	Step 4: Engage	Step 5: Monitor
❑ Name ❑ Role ❑ Direction of Influence (U, D, O, S, I, E) ❑ Importance and Stake ❑ Requirements 'expectations'	Team ratings of: ❑ Power ❑ Proximity ❑ Urgency ❑ Index # ❑ Priority	Results of Step 1 and Step 2: ❑ Categorised List ❑ Excel form ❑ *Stakeholder Circle*	Engagement Profile: ❑ Support ❑ Receptiveness ❑ Targeted Communication Plan ❑ Relationship Manager	❑ Baseline Communication Plan ❑ Subsequent assessment ❑ Reports

Figure 7.2   Summary of the Stakeholder Circle® methodology.

7.3   The Stakeholder Circle® methodology

The *Stakeholder Circle*® is a stakeholder relationship management methodology supported by a visualisation tool that profiles a project's key stakeholder community. It is based on the premise that a project can only exist with the informed consent of its stakeholder community, and that managing the relationships between the community and the project will increase the chances of project success.

The *Stakeholder Circle*® methodology provides a means for the project team to identify and prioritise a project's key stakeholders, and then to develop an appropriate engagement strategy and targeted communication plan[1] to ensure that the needs and expectations of important stakeholders are understood and managed. The visualisation tool charts a project's key stakeholders according to their ability to influence the project's success or failure. Categorisation and mapping of stakeholders holds the key to targeting the *right* stakeholders at the *right* time in the life of the project. Additional mapping of stakeholder support and receptiveness to messages about the project provides the project team with the key to the *right* level of engagement, information and communication.

7.3.1   Gathering data about each stakeholder

The output from each of the steps of the *Stakeholder Circle*® methodology builds information that is essential for designing effective, targeted communication. There are a number of ways to accumulate this information. The first is the use of the *stakeholder-on-a-page*™, a Word document that can either be used in soft copy or hard copy to gather information about each stakeholder. Figure 7.3 shows the template. The second approach is the use of a *Stakeholder Circle*® software tool, either a database that supports all steps of the methodology and stores and presents the data as graphic and tabular reports[2] or a simpler spreadsheet that focuses on key elements of the overall methodology. Figure 7.4 shows the area on the *stakeholder-on-a-page*™ template for collection of information from *step 1* discussed below.

7.3.2   How to identify your stakeholders

In the methodology, *Step 1: identify* consists of three activities:

1. Develop a list of stakeholders with the test of 'which individuals or groups are impacted by the project, or can impact the project'.

Stakeholder Management Pty Ltd

stakeholder-on-a-page

Stakeholder Name _____ | **'stake'**

Directions of Influence: **U__ D__ O__ S__** and **I__E__**   I __ R __

Requires from the work: _____   O __ N __

Importance to the work: _____   I __ C__

Prioritise the Stakeholder

Assessment	Power Rate 1 - 4	Proximity Rate 1 - 4	Value Rate 1 - 5	Action Rate 1 - 5	Index # / Priority #

Build Engagement Profile (see note)

Support / Receptiveness — First assessment

Support / Receptiveness — Next assessment

Support / Receptiveness — Next assessment

Use '**X**' when assessing the current engagement profile of each stakeholder and '√' to indicate the optimal engagement profile

Influence on? _____   **Influenced by?** _____

Communication Plan

Message	Messenger	Format (W/O F/I)	Frequency	Assessment date	Comments

Relationship Manager: 'owns' the relationship _____

Figure 7.3   The *stakeholder-on-a-page* template.

Stakeholder Name _____ | **'stake'**

Directions of Influence: **U__ D__ O__ S__** and **I__E__**   I __ R __

Requires from the work: _____   O __ N __

Importance to the work: _____   I __ C__

Figure 7.4   The identification section from the *stakeholder-on-a-page* template.

2. Identify the two aspects of the relationship between the project and its stakeholders – how is each stakeholder important to the project, and what does he or she expect from success (or failure) of the project. This is *'mutuality'* – key data for understanding and managing stakeholder expectations (and therefore manage their perceptions of success or failure of the project).
3. Begin the categorisation process by documenting each stakeholder's *directions of influence* – *upwards, downwards, outwards, sidewards, internal* and *external*: this data is important for developing targeted communication. These categories will be described later in this chapter.

Develop the stakeholder list

Developing the stakeholder list requires two actions; the first action is to select a team[3] for identification and analysis of the stakeholder community. This team will ideally consist of three to five members, including the project manager, some core team members and someone who understands the power structures and politics of the organisation[4]. The team, which may be considered as a sub-set of the project team, should be formed as early as possible in the project life cycle, where practical, before the planning phase.[5] If possible, membership of this team should remain constant over the entire life of the project. Maintaining consistency within the team will provide some assurance of reduced subjectivity in decisions made about the stakeholder community and its membership throughout the life cycle of the project. An additional benefit to using teams for identification of stakeholders is the sharing of the knowledge that each team member has about certain stakeholders. This process of team decision-making will ensure that every member of that team has learned something more about the project's stakeholders.

Developing the stakeholder list is then simply the collection of the names of those individuals and groups who can impact or are impacted by, the project's work or its outcomes. This is most often achieved through a brainstorming process, where members of the team contribute names which are later categorised according to their importance to the project, what they require from success or failure of the project and their influence on the project. As with any important project data gathering activity, it is essential to circulate this data for review by other individuals who have knowledge of the project and the organisations involved. These reviews will ensure the list is as accurate as possible.

Identify *mutuality*

Each name on the list resulting from the brainstorming exercise must be tested by applying two questions:

- 'How is this stakeholder important? What is their 'stake'?'
- 'What does this stakeholder require from the success or failure of the project?'

The answer to the first question establishes that this person or group is actually a stakeholder and what their potential contribution to the project's success (or failure) may be. Generally, a stakeholder is important to the project because he (or she) is a source of funds, personnel or materials, or can impact the success or failure of the project through either action or inaction. If there is some doubt about whether an individual or group is a stakeholder, it is possible to analyse the definition of stakeholder further into six subcategories: interest, rights, ownership, knowledge, impact or influence, and contribution. Figure 7.5 summarises definitions of each of these subcategories.

The answer to the second question establishes the stakeholder's expectations of, or requirements from, the success or failure of the project. Generally, a stakeholder will have expectations of either personal gain, or expectations of organisational gain, through the success or failure of the project. Personal gain may be enhanced power or reputation or even career or monetary improvement; organisational gain may be enhanced power or reputation for the organisation as a whole or for a department or group within the organisation.

Understanding the stakeholder's stake and expectations is crucial to all subsequent steps in the stakeholder mapping process and to developing targeted communication strategies. It is never appropriate to guess or make assumptions about a stakeholder's expectations; if there is some doubt about the accuracy of information collected, other sources of information should be referenced. The stakeholder could be asked about what he or she requires from the project[6], a survey could be conducted[7] or others could be asked about the expectations of

The 'stake': can be one of the examples below or a combination:	
Interest	**A person or group of persons is affected by a decision related to the work or its outcomes:** • Street closures for a car racing event • Temporary closure of a supermarket for renovation

Rights	**To be treated in a certain way or to have a particular right protected:**
	• **Legal right:**
	   o Occupational Health and Safety
	   o Privacy
	• **Moral right:**
	   o Heritage protection activists
	   o Environmentalists

Ownership	**A circumstance when a person or group of persons has a legal title to an asset or a property:** • Resumption of property for road works • Intellectual property • Shareholders' 'ownership' in an organisation
Knowledge	**Specialist knowledge or organisational knowledge**

Impact or influence	• **Impacted by the work or its outcomes:**
	   o Staff, customers
	   o Shareholders
	• **Impact (or influence) on the work or its outcomes:**
	   o Sponsor
	   o Governments (legislation, regulation)
	   o Public

Contribution	• **Supply of resources** o People, material • **Allocation of funding** • **Advocacy for objectives or work success** • **Buffer** between organisation and work teams or the performance of the work

Figure 7.5   A stakeholder's 'stake'.

this stakeholder[8]. Other important sources of information can be Key Performance Indicators (KPIs), or other documentation that outlines responsibilities and objectives of stakeholders who have organisational management responsibilities.

Directions of influence

The final activity in *step 1: identify*, is to categorise the listed stakeholders according to their *directions of influence* to, or from, the project manager. This categorisation adds another dimension to the data the project uses to effectively manage the relationships with their stakeholders. It is also essential as a contribution to data needed for targeted communication with a project's stakeholders.

There are two elements to consider:

• Is the *direction of influence* of the stakeholder *upwards, downwards, outwards* or *sidewards*?
• Is the stakeholder part of the organisation (*internal*) or outside it (*external*)?

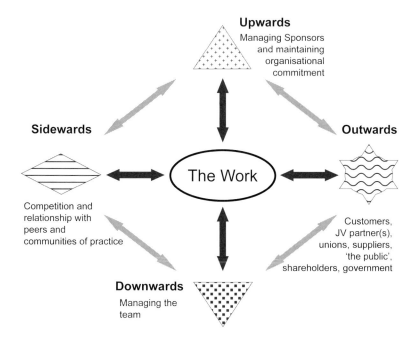

Figure 7.6   The primary directions of influence.

Figure 7.6 shows the first set of *directions of influence* on the project work, and defines the four directions around the project. *Upwards* defines the influence that senior management, especially the sponsor, exert over the project. It is shown in the ***Stakeholder*** *Circle*® colour mapping as orange. *Downwards* denotes team members, whether full-time staff, consultants, contractors or specialists who work with the project manager to achieve project objectives: teams are shown as green. *Outwards* stakeholders are those outside the project and will include individuals and groups such as end-users, Government, regulators, the public, shareholders and lobby groups: *outwards* stakeholders are shown as blue. Finally, *sidewards* stakeholders are peers of the project manager. Peers could be other project managers, industry groups and managers within the organisation who are considered to be at the same level professionally: *sidewards* stakeholders are shown as purple. While adding colour to the depiction of the stakeholder community can add an additional depth of perception, the direction of influence can also be recorded simply as: U for *upwards*, D for *downwards*, O for *outwards* and S for *sidewards*.

Categorisations for *internal* and *external* in the ***Stakeholder*** *Circle*® software will cause the colours denoting the four directions to be light (for *external*) and dark (for *internal*). Otherwise these dimensions will be documented as E for *external* and I for *internal*. Figure 7.7 summarises all *directions of influence*.

The results of these three sets of activities will be a list of stakeholders, categorised according to their direction of influence on the project, with additional information collected about their importance to the project and their expectations of the project. This data is essential for the next step in the stakeholder mapping exercise – *step 2: prioritise*.

7.3.3   How to understand who is important

The results from *step 1: identify* are the starting point for *step 2: prioritise*. For complex projects, the unranked, unrefined, list can be quite large[9]. With large numbers of stakeholders,

Directions of Influence	Stakeholders (areas of interest)
Upwards	Senior management: project sponsor, senior executives, those who represent organisational commitment
Downwards	Team members
Outwards	Outside the team: customers, JV partner(s), unions, suppliers, 'the public', shareholders, government
Sidewards	Peers of the manager or the team
Internal	Stakeholders within the organisation
External	Stakeholders outside the organisation

Figure 7.7   Summary of the stakeholder's *directions of influence*.

project teams will need to understand which of these stakeholders are more important *at this time in the project*. Some project managers and their teams may be able to do this instinctively, but others may not have the necessary experience or understanding. It is also important for long-running complex projects to develop a consistent approach to decisions about who is important at any time in the project life cycle.

Step 2: prioritise in the **Stakeholder Circle**® methodology provides a system for rating and therefore ranking stakeholders according to their relative importance to the project at any time in the project. The ratings are based on three aspects:

- power to kill the project – *power*;
- closeness to the project – *proximity*;
- *urgency* – how important is this project to the stakeholder and how prepared are they to act to achieve their own outcomes (positive or negative). *Urgency* of itself is difficult to rate consistently[10] and should be rated as a combination of *value* and *action*.

The team applies the knowledge they have gained through *step 1: identify*, matching this knowledge to the rating statements, from 1 to 4 for *power* and *proximity* (where 4 is the highest) and from 1 to 5 for the two parts of *urgency*: *value* and *action* (where 5 is the highest rating). Figure 7.8 lists the ratings for *power* and *proximity*, and Figure 7.9 lists the ratings for *value* and *action*.

The index number

An index number is calculated for each stakeholder from the four sets of ratings developed by the team. Calculations are inbuilt in the **Stakeholder Circle**® software; however, for a paper-based use of the methodology, the arithmetic addition of all four ratings will be sufficient. This emphasis on ratings for *urgency* will ensure visibility of stakeholders who may not be considered as important to the project (Mitchell *et al.*, 1997; Bourne, 2005)[11]. After the index number is calculated, the list can be sorted, with the stakeholder with the highest index number being rated as the most important, the second highest next most important and so on. Figure 7.10 shows the section of the *stakeholder-on-a-page*™ that collects the data from *step 2: prioritise*. Once the index number has been obtained, it is then possible to sort the pages into order from highest number to lowest, thus showing which stakeholders have more relative importance than others.

Ratings for *power* and *proximity*	
Power:	4. **High capacity to formally instruct change**: can have the work stopped
	3. **Some capacity to formally instruct change**: must be consulted or has to approve
	2. **Significant informal capacity to cause change**: capacity to cause change
	1. **Relatively low levels of *power***: cannot generally cause much change
Proximity:	4. **Directly involved in the work**: team members working most of the time
	3. **Routinely involved in the work**: part time team members, external suppliers and active sponsors
	2. **Detached from the work** but has regular contact with, or input to, the work processes
	1. **Relatively remote from the work**: does not have direct involvement with processes: clients and most senior managers

Figure 7.8   Ratings for *power* and *proximity*.

Ratings for *Urgency*	
Value: How much 'stake' does the person have in the work or its outcomes?	5. **Very high:** has great personal stake in the work's outcome (success/cancellation)
	4. **High:** sees work's outcome as being important (benefit or threat) to self or organisation
	3. **Medium:** has some direct stake in the outcome of the work
	2. **Low:** is aware of work and has an indirect stake in work's outcome
	1. **Very low:** has very limited or no stake in work's outcome
Action: A measure of the likelihood that the stakeholder will take action, <u>positive or negative</u>, to influence the work or its outcomes	5. **Very high:** self-activated, will go to almost any length to influence the work
	4. **High:** is likely to make a significant effort to influence the work
	3. **Medium:** may be prepared to make an effort to influence the work
	2. **Low:** has the potential to attempt to influence the work
	1. **Very low:** is unlikely to attempt to influence the work

Figure 7.9   Ratings for *urgency* – *value* and *action*.

Prioritise the stakeholder

Assessment	Power Rate 1–4	Proximity Rate 1–4	Value Rate 1–5	Action Rate 1–5	Index #/ Priority #

Figure 7.10   The prioritisation section from the *stakeholder-on-a-page* template.

Name	Project role	Direction of influence	Current priority	Power	Proximity	Urgency	Index
G. Brown	Sponsor	U, I	1	4	3	4	11
F. Green	Designer team member (contract)	D, E	2	2	4	4	10
P. Jones	Architect	O, E	12	2	2	2	6
M. Smith	Builder CEO	U, E	13	2	1	2	5

Figure 7.11   Sections of a typical ranked list of stakeholders sorted by priority.

7.3.4   Revealing project stakeholders

Having identified and prioritised the project stakeholders, it is essential to show this stakeholder community in ways that can highlight who at that time in the project have been rated as the most important stakeholders. The stakeholder community can be mapped as a ranked list, or a table summarising the data gathered as the result of use of the *stakeholder-on-a-page*™ (see Figure 7.11), or a power/impact or influence/impact grid as described earlier in this chapter (Figure 7.1). However, from the perspective of maximum impact a graphical representation is most effective.

The **Stakeholder** *Circle*® tool develops a multidimensional 'map' of the project's stakeholder community. Key elements of the **Stakeholder** *Circle*® are: concentric circle lines that indicate distance of stakeholders from the project or project delivery entity; the size of the block, its relative area, indicates the scale and scope of influence and the radial depth can indicate the degree of power (Bourne, 2005; Bourne and Walker, 2005). This depiction of the stakeholder community represents the project's key stakeholders as assessed by the project team. In the **Stakeholder** *Circle*® shown in Figure 7.12, the most important stakeholder has been assessed as the sponsor: this stakeholder appears at the 12 o'clock position; followed by the project team as the second most important and the CEO as third most important.

Patterns and colours of stakeholder entities (where used) indicate their influence on the project – for example, orange indicates an *upwards* direction – these stakeholders are senior managers within the performing organisation that are necessary for ongoing organisational commitment to the project; green indicates a *downwards* direction – these stakeholders are members of the project team; purple indicates a *sidewards* direction – peers of the project manager essential as collaborators or competitors and blue indicates *outwards* – these stakeholders represent those outside the project such as end-users, Government, 'the public', shareholders. The final colour coding is dark hues and patterns for stakeholders *internal* to the organisation and light hues and patterns for those *external* to the organisation. The **Stakeholder** *Circle*® visualisation of the project community will be different for each project and for each phase of the project – the relationships that visualisation shows will reflect the project's unique stakeholder community. For more information on using the **Stakeholder** *Circle*® visualisation tool for project analysis see Bourne (2005) and Walker *et al.* (2008).

7.3.5   How to gauge support

Step 4: engage is centred on identifying engagement approaches tailored to the expectations and needs of the individuals or groups identified and categorised in the previous three *steps* of the methodology. Developing stakeholder engagement profiles constitutes the final step in

Asset Management Project

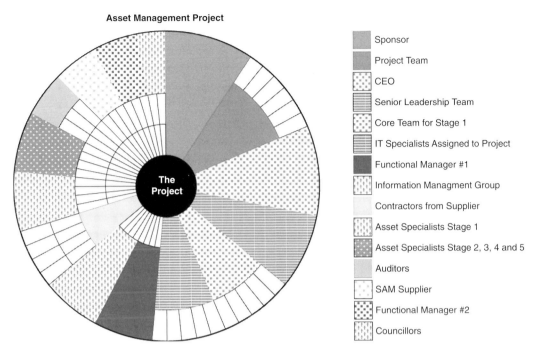

Sponsor

Project Team

CEO

Senior Leadership Team

Core Team for Stage 1

IT Specialists Assigned to Project

Functional Manager #1

Information Managment Group

Contractors from Supplier

Asset Specialists Stage 1

Asset Specialists Stage 2, 3, 4 and 5

Auditors

SAM Supplier

Functional Manager #2

Councillors

Figure 7.12   The *Stakeholder Circle*® software map of the stakeholder community.

data collection, leading to targeted communication plans for effective stakeholder management. This is done by:

- assessing the actual *attitude* – level of *support* and *receptiveness* of stakeholders;
- describing the target *attitude* – level of *support* and *receptiveness* of stakeholders.

The first step of this analysis involves identifying the current level of *support* of the stakeholder(s) at five levels: from 'committed' (5), through 'neutral' (3), to 'antagonistic' (1). The second step is to rate the *receptiveness* of each stakeholder to messages about the project: on a scale of 5, from where 'direct personal contact is encouraged' (5), through 'ambivalent' (3), to 'completely uninterested' (1). Figure 7.13 summarises these rating levels. This information is recorded in a 5 by 5 matrix as shown in Figure 7.14: the *stakeholder-on-a-page*™ template allows for up to three assessments of stakeholder engagement over time.

These steps are repeated to identify the optimal engagement position (target *attitude*) for project success: a realistically achievable level of *support* and *receptiveness* to messages that would best meet the mutual needs of the project and the stakeholder. If an important stakeholder is both actively opposed and will not receive messages about the project, he or she will need to have a different engagement approach from stakeholder(s) who are highly supportive and encourage personal delivery of messages.

Figure 7.15 shows the results of assessments for three different stakeholders. Stakeholder 1 has been assessed as being 'ambivalent about the project, neither supportive nor unsupportive', and 'not really interested in receiving messages' about the project, these results are shown by 'X' in the appropriate boxes in the matrix. However, the team has rated this stakeholder as being important to project success and that the engagement profile SHOULD BE 'passive support' and 'will agree to receive information about the project; this is shown with a '√' symbol.

Ratings for *support* and *receptiveness*	
Support: for the project or its outcomes.	5. **Active support:** provides positive support and advocacy for the work
	4. **Passive support:** supportive, but not actively supportive
	3. **Neutral:** is neither opposed or supportive
	2. **Passive opposition:** will make negative statements about the work, but not do anything to affect its success or failure
	1. **Active opposition:** is outspoken about opposition to the work, and may even act to promote failure or affect success
Receptiveness: to messages or messengers about the project or its outcomes.	5. **High:** eager to receive information
	4. **Medium:** will agree to receive information
	3. **Ambivalent:** may agree to receive information
	2. **Not interested:** not prepared to receive information
	1. **Completely uninterested:** emphatically refuses to receive information

Figure 7.13   Ratings for *support* and *receptiveness*.

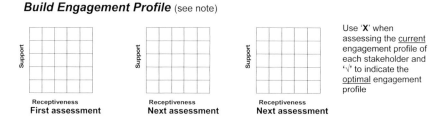

Build Engagement Profile (see note)

First assessment   Next assessment   Next assessment

Use 'X' when assessing the current engagement profile of each stakeholder and '√' to indicate the optimal engagement profile

Figure 7.14   The engagement section from the *stakeholder-on-a-page* template.

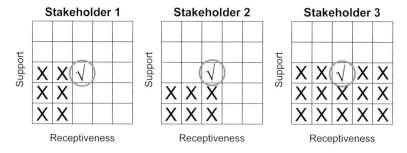

Stakeholder 1   Stakeholder 2   Stakeholder 3

Figure 7.15   Stakeholder *attitude* (baseline).

Stakeholder 2 has been assessed as 'passive unsupportive' and 'ambivalent; may agree to receive information', whereas the engagement profile SHOULD BE 'ambivalent: neither supportive nor non-supportive'. This stakeholder may be someone who is rated as 'not very important' to the project at this time, but nevertheless may cause harm through spreading negative views about the project. In both cases, the gap between the current engagement profile and the optimal profile indicates the level of effort required in developing communication strategies for stakeholders, to encourage their support and interest.

Stakeholder 3 has been assessed as being 'ambivalent about the project, neither supportive nor unsupportive', but 'eager to receive information' about the project, whereas the engagement profile SHOULD BE for receptiveness to be 'ambivalent: neither supportive nor non-supportive'. This is a situation where the current profile is quite different from the optimal profile and will require careful handling from the team.

Based on each stakeholder's engagement strategy, a communication plan will be developed, consisting of: specific messages or message forms (reports); how messages will be delivered; by whom; whether formal or informal, written or oral; at what frequency. The frequency and regularity of delivery of these messages will vary with the level of *support* and *receptiveness* of the stakeholder, the gap between current assessment and optimal assessment, as well as the stage of the project.

One final step to prepare for communication

The final step before developing the communication plan is to categorise each stakeholder into one of three groups:

1. The current engagement position – *attitude* is **equal to** the optimal position.
2. The current engagement position – *attitude* is **greater than** the optimal position.
3. The current engagement position – *attitude* is **less than** the optimal position.

In the first instance where the current *attitude* is **equal to** the optimal position, communication can be maintained at 'business as usual': the defined level and frequency of regular reports, meetings and presentations can be safely maintained. For the situation where the current attitude is **greater than** the optimal position, two possible approaches need to be considered, depending on the results of the engagement matrix. In Figure 7.15, Stakeholder 3 is rated as being well above the level of *receptiveness* to messages necessary for success of the project, but at the appropriate level of *support* of the project to ensure success. The decision the team have to make regarding Stakeholder 3 is whether to reduce the level of information flowing to this stakeholder (and risk a reduction in support from this stakeholder) or to maintain the current level of communication. The decision can only be made in the light of the knowledge the team has gained during the preceding steps of the stakeholder analysis.

The third category where the current *attitude* is **less than** the optimal position, the team needs to focus their efforts on 'heroic' communication if the stakeholder is important; Stakeholder 1 (Figure 7.15) is in this category. This type of communication is generally needed for only a small percentage of stakeholders, but any effort expended on increasing the levels of *support* and *receptiveness* to the optimal position will significantly benefit the project. Generally in this case, a number of different communication approaches needs to be used, from regular reports and meetings, through special presentations and possibly even using the influence of other important but supportive stakeholders to deliver the project information. Multiple complex communication activities must be coordinated by a relationship manager who could be the project manager or a supportive senior stakeholder. Figure 7.16 shows the final portion of the *stakeholder-on-a-page*™, used to record the communication plan for each stakeholder based on all the information and decisions described in this section. Communication strategies will be described in more detail in Chapter 10.

7.3.6   Feedback mechanism

Regular Stakeholder Review meetings, similar to Risk Review meetings will maintain the currency of the project's stakeholder community, or provide information about changes in

Communication Plan

Message	Messenger	Format (W/O F/I)	Frequency	Assessment date	Comments

Relationship Manager: 'owns' the relationship _____

Figure 7.16   The communication section from the *stakeholder-on-a-page* template.

that community that will cause the project's stakeholders to be re-assessed, re-prioritised and re-developed as a new **Stakeholder** *Circle®* (community).

The re-assessment of the engagement matrix of each project stakeholder is an essential part of the project review processes, whether through regular team meetings, specific reviews or in response to other unplanned events around the project. The results of the reviews will provide the necessary trend analysis for the team to know whether or not the communication strategies and activities are being effective. This process will be discussed in detail in the description of *step 5: monitor*.

Monitoring the effectiveness of communication

Once the communication plan has been developed, the strategy relating to the 'who, what, when and how' of delivering the tailored messages planned for the important stakeholders must be converted into action. Monitoring the effectiveness of this communication effort, and providing essential data for corrective actions if required, is the final step of the **Stakeholder** *Circle®* methodology, *step 5: monitor*.

Each time the stakeholder community is re-assessed and the engagement profile updated, any changes in the gap between the current profile and the optimal profile must be considered. This movement (or lack of movement) provides an indicator of the current communication plan's effectiveness in influencing the *attitudes* of key stakeholders. Where the communication is being effective, the current plan should be maintained, where it is not working, the communication plan should be changed.

If there has been a worsening of the gap between the current profile and the optimal profile, this is a strong indicator that the communication strategy developed for this stakeholder is not having the desired effect; it should provide the evidence needed to try a different approach. If there has been an improvement in the gap between the current profile and the optimal profile, this may indicate that the communication strategy is working and encourage its continuation. However, during the review, it is essential to consider the project's overall environment to ensure that any identified changes have been caused by the project's communication efforts rather than by external circumstances.

In the examples, Stakeholder 1 was first assessed as 'passively supportive' but 'uninterested in receiving project messages' (shown previously in Figure 7.15), and an engagement strategy and communication plan was developed to improve the engagement profile. On re-assessment, the level of support has not changed; it remains optimal suggesting the information being communicated is appropriate. On the other hand, Figure 7.17 shows that the level of *receptiveness* has been improved beyond the optimal, suggesting the method of communication is effective. The next decision that the team must make is whether to maintain the

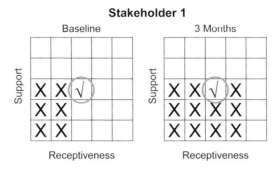

Figure 7.17   Reviewing stakeholder *attitude* after 3 months – Stakeholder 1.

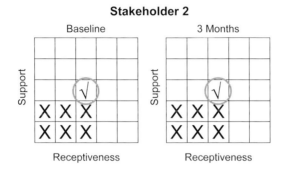

Figure 7.18   Reviewing stakeholder *attitude* after 3 months – Stakeholder 2.

current level of communication to Stakeholder 1 and assess at the next review, or whether to re-plan communications and redirect effort elsewhere.

Stakeholder 2 baseline was assessed to be passively unsupportive but at the level of receptiveness necessary for success of the project. As shown in Figure 7.18, there has been no change. The team will now have evidence that the current communication plan is not being effective: they will have to try a different approach.

Figure 7.19 shows the effects of comparing Stakeholder 1's engagement profile over time. After 3 months, the comparison of the new profile with the baseline showed that the stakeholder's level of receptiveness had exceeded the optimal position. This could mean that the stakeholder had been so influenced by the communication effort of the project team in the first 3 months that he or she required much more information about the project than the project team thought necessary. The project team may have decided to communicate at a lower level as a result of the 3-month assessment, whilst the stakeholder's expectations were of receiving a higher level of attention from the project team in the form of additional communication activities. The consequences of reducing the amount of information to the stakeholder, shown at the 6-month assessment may have been caused by the stakeholder feeling neglected, and losing all interest in the project. This may be the interpretation of the engagement profile at the 6-month review (Figure 7.19). However, it is important to investigate all possible reasons for any such result: the 6-month review may indicate that the stakeholder now regards another project as more important or even that he or she has lost power in the organisation and is now no longer interested in any project.

There are no simple answers: the changes in the profile for each key stakeholder are a strong indicator of the effectiveness of the communications strategy but need to be considered

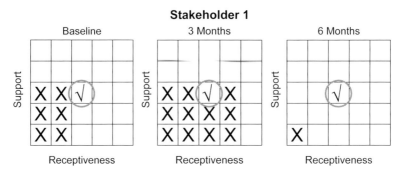

Figure 7.19 Reviewing stakeholder *attitude* after 6 months – Stakeholder 1.

along with changes in the relative importance of the stakeholder and all other pertinent factors. This requires regular maintenance of the whole data set to ensure optimal results from the communication effort.

Managing the stakeholder community

One essential aspect of managing project stakeholders is to recognise that the stakeholder community is not static. Individuals and groups that are essential to project success in one part of the project may not be essential in other parts of the project. For example, stakeholders who are important to the success of the design phase may not be important to the success of the project once it is in build phase; the stakeholder community will change membership as the project moves through its life cycle.

Similarly, the stakeholder community will change when there are changes to the structure or business direction of the performing organisation. Individual stakeholders may lose power within the organisation, or may leave, others may join the organisation. These changes will affect both the membership of the stakeholder community and the relative importance of members of the community over time.[12]

Even when the organisation remains stable, a stakeholder's interests in, and support for, the project may vary due to changes in the focus of the stakeholder. For example, if the stakeholder perceives that the project is not delivering to expectations, he or she may decide the project is no longer worthy of his or her support. Alternatively, another project may become more important to the stakeholder, or senior management may redefine the duties and responsibilities of the stakeholder requiring him or her to focus elsewhere.

Consequently, the process of identifying, prioritising and planning the engagement of project stakeholders cannot be a once-only event. To maintain the usefulness of the stakeholder information, the assessment process may have to be repeated in whole or in part many times. An essential part of the *Stakeholder Circle*® methodology is the repetition of the processes at appropriate intervals and the reappraisal of the stakeholder community, particularly focusing on trends and changes.

Reviews may be triggered proactively, or reactively. A proactive approach would be to include reporting on all aspects of the stakeholder engagement activities as a regular item on the project meeting agenda. Team members should be encouraged to report any information gathered during communication with stakeholders. This may be in the form of rumours about personnel in the performing organisation or other organisations, or pieces of information that may together with other small pieces of information provide some forewarning of changes to funding, resourcing, sponsorship or the importance to the organisation of the deliverables of the project. The issues raised or news collected as a result of the feedback on stakeholder communication may trigger a review of the current community and re-assessment of the importance, attitude and support of members of that community.

Another example of proactive stakeholder management is to factor the principal communication points and regular reviews of the stakeholder community, such as when the project moves into a new phase, into the project schedule. Responsibilities should be allocated so that the communication and review activities will be reported regularly as part of the project status information and the reviews undertaken at the planned times. This should not be too onerous, the review will probably only affect a small number of stakeholders and their relative importance in the community.

The reactive approach to re-mapping and managing the stakeholder community will be to undertake a review only in response to major changes to the organisation or when problems occur.

Regardless of the trigger, the results of the re-assessment will be a redefined stakeholder community and updated engagement profiles. Once enough information has been collected to form a picture of the project environment, the next step would be to refer this information to supportive stakeholders for interpretation. By operating within the political context of the organisation in this way, the project team can be prepared for adverse events and be best place to exploit opportunities.

However, given that the membership of the stakeholder community will change, it is important that the team develops ways to maintain a current view of the stakeholder community, so that they always have a view of who are the *right* stakeholders for any time in the project's life cycle.

7.4   Implementing the methodology

The readiness (or maturity) of the organisations involved in working on the delivery of project outcomes will influence the support that is provided to the project for effective implementation of the **Stakeholder** *Circle*® methodology. The appropriate parts of the methodology to use, and the path to a successful implementation of the full methodology can be gauged by evaluating the organisation's current state against the Stakeholder Relationship Maturity Model – SRMM® (Bourne, 2008). The level of support from different organisations within the overall project delivery team will also be influenced by the procurement strategy adopted for the project. Chapter 12 discusses different forms of procurement and the impact of these different forms on stakeholders; see also Walker *et al.*, 2008 for the influence of stakeholders in supply chain management. The challenge facing the project team is to make the assessment as inclusive as possible.

The probability of a successful stakeholder engagement is enhanced when all of the groups involved in the work of the project recognise the benefits of collaboration in stakeholder identification, mapping and engagement processes. If the collaborative approach is encouraged, representatives from all areas – client, designer, contractors and suppliers will participate in the stakeholder engagement team, as described earlier in this chapter. When the team is constituted from a sub-set of those involved, the richness and completeness of the data gathered will be reduced, possibly reducing the effectiveness of the stakeholder engagement activities with a consequential increase in the possibility of the project failing.

However, even without the full cooperation of all parties involved in the delivery of the project outcomes, the application of the methodology by the project team will provide valuable information and insights. Using the methodology will help the team identify the relationships that need to be nurtured and the stakeholder groups that must be engaged for success of the entire project.

7.5   Conclusion

Any decisions that the project team (people) make about the project's stakeholders (also people) are of their very nature subjective; and because the people involved are ever changing, any attempt at a single, objective analysis or reporting is bound to fail. Successful stakeholder mapping requires a transparent, dynamic process that builds understanding as the project progresses. Project teams must find ways to not only understand who their stakeholders are at any time in the project, but also what their expectations are, and finally, find ways to measure the effectiveness of the team's communication efforts.

Effective mapping systems need simplicity and flexibility in both data gathering and reporting about project stakeholders. These requirements should be reflected in guided steps through a series of processes that can be cumulative or can be approached in parts, depending on the needs and maturity of the organisation. The **Stakeholder** *Circle*® methodology and visualisation tools provide an effective way to achieve this through a five *step* process designed to provide a cumulative collection and mapping of data about a project's stakeholders and through trend analysis monitor the effectiveness of the team's communication efforts.

There are three parts to the **Stakeholder** *Circle*® methodology and visualisation tool that cumulatively add to its effectiveness. The methodology supports the identification and prioritisation of all the project's stakeholders, producing a manageable number of the key stakeholders of that project. The second part of the methodology is the supporting tools, which makes the task of allocating relative importance of stakeholders both time and effort efficient. The final part of the methodology is the processes for developing an engagement strategy and associated communications plan to support understanding of the expectations and perceptions of the stakeholders, and how they can be managed and met.

Organisations that are prepared to invest in an appropriate system will benefit from the increased awareness by the project team members of the importance of project relationship management and the provision of tools to achieve a better understanding of how to achieve it. The ROI can be substantial; Chapter 13 discusses the 'gains and pains' of stakeholder management in construction projects.

Endnotes

1. The data gathered through the steps of the methodology support the development of communications design to address the information needs of the stakeholder and the support needs of the project.
2. For more information on the **Stakeholder** *Circle*® go to www.stakeholder-management.com.
3. Data collected by the authors through a classroom exercise comparing the efficiency of team and individuals in decision-making included in their *Successful Stakeholder Management* workshops, shows that of approximately 500 participants, only 6 individuals scored better than their team.
4. The sponsor of the project would be a valuable team member for this exercise.
5. Ideally, the selection of these team members should be the responsibility of the Sponsor or a senior manager representing the performing organisation. Selection and management of this team as a joint activity of the client PM and prime contractor's PM is also effective.

6. This action could have positive or negative results: positive if the stakeholder is pleased to be actually consulted – and this may lead to a higher commitment from the stakeholder. A negative result may occur when the stakeholder does not give a completely honest answer – he or she may just want to give the team a 'good news' answer, or may just want to be non-confrontational. It is always a useful policy to attempt to get answers to these questions from at least two sources.

7. An expectations survey could legitimately be conducted as the starting point to a customer satisfaction survey planned as part of project closing activities.

8. Supportive stakeholders can be a good source of information about other stakeholders.

9. In working with organisations using the **Stakeholder** Circle® methodology and software for mapping and managing stakeholder relationships, the authors have assisted in projects that have over 300 stakeholders (both individuals and groups) identified in the first step.

10. During the 12 months research in development of this methodology, it became evident that the concept of 'urgency' was too multidimensional for consistency. Once the concept was devolved into two parts – *value* and *action* – the ratings appeared to be applied consistently.

11. By weighting *urgency* more highly than *power* or *proximity*, the methodology helps team members identify less obvious, or less outspoken, stakeholders thus ensuring that 'surprises' are minimised. Generally, those stakeholders with power in the project environment will be relatively easy to identify, but those with high levels of *urgency* may not be.

12. When describing membership of the stakeholder community, it is important to recognise stakeholders as being either individuals, groups or organisations. An individual can be an important stakeholder by virtue of being a key representative of a group or organisation: that same group or organisation is not necessarily at the same level of importance as its individual representatives.

References

Bligh, P. and Turk, D. (2004) *CRM Unplugged: Releasing CRM's Strategic Value*. New York: John Wiley & Sons.

Bourne, L. (2005) *Project Relationship Management and the Stakeholder Circle*, Doctor of Project Management, Graduate School of Business, Melbourne, RMIT University.

Bourne, L. (2008) SRMM®: Stakeholder relationship management maturity. *Proceedings of PMI Global Congress – EMEA*, St Julian's Malta, PMI.

Bourne, L. and Walker, D.H.T. (2005) Visualising and mapping stakeholder influence. *Management Decision*, **43**(5):649–660.

Department of Sustainable Environment (2007) Stakeholder Analysis. From www.dse.vic.gov.au.

Fletcher, A., Guthrie, J., Steane, P., Roos, G. and Pike, S. (2003) Mapping stakeholder perceptions for a third sector organization. *Journal of Intellectual Capital*, **4**(4):505–527.

Glasser, W. (1998) *The Quality School: Managing Students Without Coercion*. New York, Harper Collins.

Imperial College London (September, 2007) Project Stakeholder Analysis. From www3.imperial.ac.uk.

Mitchell, R.K., Agle, B.R. and Wood, D.J.(1997) Toward a theory of stakeholder identification and salience: Defining the principle of who and what really counts. *Academy of Management Review*, **22**(4):853–888.

Moorhouse Consulting (2007) Beyond Conventional Stakeholder Management: Developing PRIME Intelligence on Complex Programmes. From www.moorhouseconsulting.com.

Murray-Webster, R. and Simon, P. (2007) Making sense of stakeholder mapping. *Project Management Practice*, **2007**(2):12–13.

Office of Government Commerce UK (2003) *Managing Successful Programmes*. London: The Stationary Office.

Rock, D. (2006) *Quiet Leadership: Six Steps to Transforming Performance at Work*. New York: HarperCollins.

Savage, G.T., Nix, T.W., Whitehead, C.J. and Blair, J.D. (1991) Strategies for assessing and managing organizational stakeholders. *Academy of Management Executive*, **5**(2):61–75.

Sbarcea, K. and Martins, R. (2003) The 'temporary knowledge organisation' as viewed from a complexity perspective. An enrichment of the traditional organisational project management paradigm. Retrieved 3 April 2004, from www.aipm.com.au.

Turner, J.R., Kristoffer, V. and Thurloway, L. (eds) (2002) *The Project Manager as Change Agent*. London: McGraw-Hill Publishing Co.

Walker, D.H.T., Bourne, L. and Rowlinson, S. (2008) Stakeholders and the supply chain. In: Walker, D.H.T. and Rowlinson, S. (eds) *Procurement Systems: A Cross-Industry Project Management Perspective*. London: Taylor & Francis.

Weaver, P. (2007) A simple view of complexity in project management. *Proceedings of the 4th World Project Management Week*, Singapore.

8 Strategies and Tactics for Managing Construction Stakeholders

Ektewan Manowong and Stephen Ogunlana

8.1 Introduction

The modern project manager critically needs to be attuned to the cultural, organizational and social environments surrounding the project. Understanding such project environment includes identification of project stakeholders and their ability to affect successful outcome of the project. The effective management of project stakeholders is considered as an important key to project success (Jergeas et al., 2000). If stakeholders are not managed effectively, the probability of successful project completion is reduced due to conflicts between stakeholders. Ineffective stakeholder management can also result in dissatisfaction with project outcomes and adverse disruption to budget and schedule. Besides, future works between project team and internal stakeholders may become more difficult and the community, as external stakeholders, may have negative reaction to the project.

In the stakeholder management process, the stakeholders' needs and expectations are evaluated with an important basis of stakeholder analysis (Olander, 2007). Stakeholder management strategies are applied aiming to increase the effectiveness of managing stakeholders' different interests and their disposition. However, finding the best resolution, which is acceptable for all stakeholders, to such differences remains a major challenge. Reaching the goals of effective management of construction stakeholders yet needs concerted efforts in attaining a formalized stakeholder analysis, improved stakeholders' relationship and effective communication, sustained stakeholders' commitment, and increased satisfaction of stakeholders.

As a compliment to the content described in previous chapters on stakeholders mapping and assessment, this chapter provides an insight on how to strategically manage and deal with project stakeholders. Besides, it seeks to support effective management of stakeholders in construction by providing a framework for analysis and understanding the construction stakeholders and their interrelations. Further, recommendations are made regarding useful strategies and tactics for engaging construction stakeholders more effectively so that the anticipated stakeholder related problems in construction can be managed effectively.

8.2 Project stakeholder analysis: identification and prioritization

According to PMI (2000), project stakeholders are defined as individuals and organizations actively involved in the project, or whose interests are positively or negatively affected by the project execution or the successful completion of the project. Project manager and team members have to deal with and manage other groups of project stakeholders. Theories of stakeholder management reveal ways to efficiently engage and manage stakeholders. The adopted management strategies may be formal, informal, detailed, or broad, depending on the project's needs, size, and complexity. Identifying stakeholders and assessing their interests and expectations throughout the construction project life cycle including the stages of inception, design, construction, commission, and operation of project are helpful for project managers to forecast project stakeholders' behavior and its effect on project outcome.

In the construction context, project stakeholders can be defined as individuals, groups, or organizations who are actively involved in the project including those having interests that are positively or negatively affected by the activities or results of the construction project (Olander, 2007). Attempts are then made to broadly differentiate the stakeholder groups such as, for example, internal and external stakeholders (Hill and Jones, 2001), or primary and secondary stakeholders. Internal stakeholders in construction project are comprised of project owner, clients, project leader, and core team members such as designers and contractors, suppliers and subcontractors. Meanwhile, external stakeholders include the regulators such as local and national authorities, the public and community groups, financier, the media, end users, and other independent concerned groups with special interests. Each group of stakeholders has different interests and objectives in the project. In addition, they also have different influences on the success of project. As such, it is important to know their different expectations and level of attention as well as to determine to what extent they could and would exert influences.

Since construction stakeholders can effectively contribute to success of project, it is also essential to prioritize the stakeholder to which project management team needs to pay more attention to. Stakeholder management is the process to identify stakeholders and win their support. For the purpose of planning management strategies, stakeholders can be defined as 'key' or 'nonkey' (Tasmanian Government, 2005). If the project is to be successful, the interests of key stakeholders, who will be positively or negatively affected during the project or on successful completion of the project, must be recognized. By contrast, the nonkey stakeholders will also be identified, but their needs do not have to be recognized for the project to be successful. Nevertheless, successful completion of project deliverables inevitably depends on the skills of managing relationship with stakeholders. As such, a critical skill of successful project managers is the ability to understand various stakeholders' power and influence which may be hidden or apparent (Bourne and Walker, 2005b). Understanding the stakeholders' nature and contributions on project is achievable via a useful technique of stakeholder analysis.

Stakeholder analysis is an approach used to identify and describe stakeholders based on their attributes, interrelationships, and their interests related to a given issue. It is used for understanding the system by identifying the key actors or stakeholders in the project and assessing their respective interest in that project (Grimble and Wellard, 1997). Stakeholder analysis is used for a variety of purposes, including management of change, risk, project's key issues, and promotion of project activities. Stakeholder analysis is therefore carried out to empirically discover existing patterns of interaction and analytically improve interventions. It is also used as a management tool in policy-making and as a tool for conflict prediction. As stakeholder analysis seeks to differentiate and study stakeholders on the basis of their attributes, the criteria appropriate for prioritizing stakeholders of specific situation may include relative power and interest of each stakeholder (Mendelow, 1991); their collaborating and threatening potentials (Savage et al., 1991); and importance and influence they have (Grimble and Wellard, 1997).

Stakeholders' power refers to their actual ability to influence the project, while the interest refers to their desire to influence. A stakeholder with both higher power and interest is considered to have more influence than one with lower power or interest. Stakeholders with low potential for threatening but high potential to collaborate means that they are supportive, while the nonsupportive stakeholders have low potential to collaborate but high potential to affect the project.

Stakeholder influence is defined as the extent to which a stakeholder is able to act on project operations and therefore affect project outcomes. Assessing the influence and importance of each individual or group of stakeholder is a significant step in stakeholder analysis. This assessment helps project managers to prioritize the project stakeholders in order to appropriately deal with them. As influence is a measure of stakeholders' power, the extent of control over the project resources and the extent to which stakeholder informs the decision-maker are considered as factors likely to lead to higher influence. Comparatively, stakeholders' importance is defined as the extent to which stakeholders' problems, needs, and interests are affected by project operations or desired outcomes. If important stakeholders are not assisted effectively, then the project cannot be deemed as successful. Based on these definitions, primary stakeholders are regarded as both important and influential. On the other hand, those who are prioritized as secondary stakeholders are either important or influential, but not both.

Stakeholder analysis can then help in analyzing and managing the project stakeholders effectively. That is, stakeholders' objectives and behavior are understood, their power is recognized, and strategies implications with stakeholders are anticipated. Furthermore, stakeholder analysis also helps in determining alternative strategies that are likely to contribute to project success. Besides attaining the established strategies' effect on the success of each group of stakeholders, the analysis of stakeholders also provides mapping of relationships between stakeholders. Stakeholder management is then associated with stakeholder analysis in the process of strategies formulation and control of the ensuing activities (Harrison and Caron, 1996).

In summary, strategies and tactics for managing stakeholders start with defining and identifying stakeholders. Generally, there can be various definitions of the term 'stakeholders' of the project. Such variations can lead to difficulties in ensuring involvement of all appropriate groups into project planning and operation. According to Cleland (1998), the internal or primary stakeholders are those directly involved and having contractual relationship to the project, while the external or secondary stakeholders are all interested groups but not directly involved to the project or contractually engaged into the project's transactions. However, regarding the project management purposes, the classes of stakeholders should make little practical difference as they have to be identified, their needs must be understood, and their potential effect on the project should be assessed.

8.3 Relating and communicating with stakeholders

Relationships in construction projects are those relationships between stakeholders and project managers as well as the relationships among the project stakeholders. Perceptions and expectations of stakeholders, the nature of the stakeholders' relationships with the project team, and the project manager's capability and willingness to effectively manage such relationships can strongly have influence on the project's success and failure (Bourne and Walker, 2005a). As such, the stakeholders' relative influences need to be assessed and their expectations need to be understood in order to define appropriate procedures of stakeholder engagement. In addition to the effective engagement of stakeholder which helps in anticipating and solving the problems, understanding stakeholders' relationship is very important in implementing the established stakeholder management strategies.

Stakeholder relationships are related to two common situations. Firstly, harmonious situations are those in which stakeholders share enough consensuses around an issue to collaborate. Secondly, stressful situations are those in which stakeholders may be in conflict. Since project stakeholders have different interests and different influences in the project, the project manager has to monitor and manage them effectively to ensure that their adverse actions do not affect project success. Management of stakeholder relationships induces a set of changes that require proactive change and development of relationships within the project (internal stakeholders' impact) and at the interface with other parties involved with the project (external stakeholders' impact).

Stakeholder relationships may be contractual or noncontractual. The corporate activity links some stakeholders through explicit contracts, while other interested parties are noncontractual or involuntary. Nevertheless, some parties involved may be yet unaware of their relationship until they experience some specific events or impacts such as economic benefits or environmental harms as a result of construction operations. The project manager cannot, therefore, ignore this aspect of corporate relationship with external stakeholders. To really understand the relationship with key stakeholders, a good project manager will attempt to build good relationships with the key stakeholders. Successful managers develop stakeholder relationship plans that include relationship matrices and communication charts. In fact, project managers must always ensure clear communication channels with stakeholders. As such, emphasis should be placed on communication channels being made available for all stakeholders.

The success of construction project is linked to the strength of the stakeholders' relationships that can be created by effective, regular, planned and *ad hoc* communication with all groups of stakeholders (Bourne and Walker, 2005b). Erroneous information conveyed to stakeholders can also result in serious problems to the project. Managing communication between construction stakeholders is therefore considered as an important process. Building and maintaining relationships with stakeholders can be successfully achieved through effective communication, which also creates higher possibility of maintaining stakeholders' support and commitment (Bourne, 2006). Besides, active communication channel will effectively maintain ongoing relationships with stakeholders and prevent risks of miscommunication, interference without adequate consultation, insufficient support, and poor delivery of project information. As such, it is essential that information generation, storage, dissemination, and disposition on construction projects are carried out timely and precisely in an appropriate manner.

The most effective method of ensuring adequate communication with stakeholders is to develop a communications management plan that outlines the project's periodic meetings or other methods of communicating project data to the stakeholder. The plan should facilitate conveying of stakeholder opinions to project team members. Besides, it should also be noted that face-to-face meetings are still the best way to communicate with construction project stakeholders (Tasmanian Government, 2005). The project manager will be more capable of reading the body language of stakeholders in order to better gauge their individual levels of satisfaction with project activities.

In summary, the key to monitoring stakeholder relationships within a project environment is to firstly identify the key stakeholders. Having identified stakeholders through the stakeholder matrix created during the project's initiation, the project manager should then utilize the attained stakeholder map to better understand the relationships among various stakeholders on specific problems. When the type of relationship is understood, the manager will know what strategies to be used in managing and dealing with stakeholders.

8.4 Strategies for managing construction stakeholders

Stakeholder strategy provides a means and direction to accomplish the goals, objectives, and mission of the stakeholders. The project management also needs strategies in dealing with and managing different types of stakeholders. Knowing the interest of the stakeholder, whether it is supportive or adverse to the project, is useful for establishing stakeholder management strategies. As aforementioned, prior to development of project strategies, identifying stakeholders is the primary task because all important decisions will be affected by the groups of internal and external stakeholders. Stakeholders' strengths and weaknesses should be assessed in order to determine the efficacy of their strategies. Then, managers can develop their own strategies to manage stakeholders accordingly. That is, potential stakeholders should be defined since the initial stage of construction project life cycle. The project management team then needs to discuss in order to find and modify strategic issues, which develop over time, and determine the changes of stakeholders' perception.

In the construction industry, different strategies should be used in order to manage different groups of stakeholders (Jawahar and McLaughlin, 2001). The project manager needs to analyze and focus on particular stakeholders at each stage of a project's life cycle. Stakeholders' influence strategies can be used to examine the relation among stakeholders. The type of identified relationship will then indicate which strategies should be used to deal with the project stakeholders. According to Weiss (2003), strategies and tactics developed to cope with stakeholders include:

- approach each stakeholder directly or indirectly;
- do nothing, monitor, take offensive or defensive with certain stakeholders;
- determine whether to accommodate, negotiate, manipulate, resist, avoid or wait, and see with specific stakeholders;
- combination of strategies.

Stakeholders with strong influence and interests need more delicate attention. To accommodate specific stakeholders, as an example, in an infrastructure development project, the owner/planner (the government agencies and/or the private organizations) may decide to initiate and employ the stakeholder participation plan to build good relationship with local communities and the general public. This plan addresses and involves stakeholders in both the project planning and the ongoing activities. The plan sufficiently describes how interested citizens can be involved in project proposal and approval or stay involved after the project execution is in place.

Negotiation with important stakeholders is needed when their interests are neglected or violated in many forms such as, for example, failing to conduct sufficient study on environmental impacts, providing incorrect financial information to the internal stakeholders, and offering unsatisfactory compensation to landowners. These practices of mismanagement inevitably lead to undesirable consequences in construction projects. To handle such situation, the management needs to prepare a structured procedure to negotiate with the affected important stakeholders. A good 'corporate governance' system can also be employed for defusing stakeholder conflicts and avoiding project protests.

Strategies to manipulate stakeholders are employed to deal with some groups of stakeholders who are analyzed as having 'less power' to the project. Manipulation encompasses tactics to influence or control stakeholders. In such situation, the project management has much power and can further increase such power. The stakeholders are not involved in active participation such as project decision-making. Oftentimes, the project management chooses to perform activities, such as public relations, to manipulate perceptions of the stakeholders. The management may also exert some degrees of resistance against stakeholder pressures. However, where stakeholders have more power, project management should concede to stakeholder pressure.

Where the power relationship between managers and stakeholders is more in balance, construction organizations generally avoid pressure generated by a certain group of stakeholder by reducing power of that group. This can be done by forming alliance with other groups of stakeholders or concealing facts about the project. Besides, stakeholder pressure can also be avoided by trying to escape from such pressure through stakeholder diversification (Oliver, 1991). For example, in a construction project, the project manager may consider supplier diversification in order to reduce the power of a highly influential supplier.

8.5   Developing and implementing stakeholder management strategies

8.5.1   Importance and influence of stakeholders

The information from stakeholder data analysis is collected to develop a strategy for handling stakeholders. In developing strategies, the managers have to make assumption regarding the stakeholders and then evaluate the assumption by rating its importance and certainty which has significant impact on the strategy. *Importance rating* is a guide to the amount of driving or resisting force on assumption exerts on the strategy. Meanwhile, *certainty rating* is a guide to the amount of knowledge the manager currently has about assumptions made on stakeholders. These ratings reveal stakeholders' level of significance and help in handling possible uncertainties such as inconsistent stakeholders' support resulting in risks that affect project decision-making. Therefore, the significance of each stakeholder should be analyzed based on their level of interest, influence, impact, and support. To effectively manage stakeholders, the key stakeholders should always be regarded as the driving factor in the adopted strategy. As such, the project manager needs to get them to share their views about the project, and communicate with them on a regular basis. Similarly, strategies can be implemented to defend against desirable actions or influences of key stakeholders.

To effectively employ strategies for managing stakeholders, primary stakeholders must be fully engaged in the governance and steering of project needing to succeed, while secondary stakeholders need to be actively managed during the project (Grimble and Wellard, 1997). As seen from Figure 8.1, stakeholders with high influence and importance are considered as the first priority for management focus. By contrast, stakeholders with low importance and influence require the least management effort.

If stakeholders are found to be supportive, having high potential for cooperation, the 'involve' strategy is used. Supportive stakeholders include the firm's employees, suppliers,

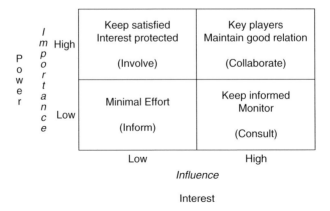

Figure 8.1   Stakeholder power/interest and importance/influence matrix (adapted from Grimble and Wellard, 1997; Newcombe, 2003).

board members, vendors, and the company. If the stakeholders are nonsupportive, the 'defend' strategy is used to retain interest and reduce dependence on that stakeholder. If the stakeholders are found to be mix-blessing, i.e., having high potential for threat and cooperation, the 'collaborative' strategy is used. If the stakeholders are marginal, i.e., having low potential for threat and cooperation, the recommended strategy is to monitor, wait, and see until they move to other categories.

8.5.2   Power and interest of stakeholders

Besides the technical issues, project managers also need to know how to use their power to strive for the project's success. It is also necessary to evaluate to which extent the stakeholders will exert their power on the project. Johnson and Scholes (1999) and Newcombe (2003) suggest that stakeholder mapping technique, using the power/interest matrix (as shown in Figure 8.1), is a useful tool for evaluation of the impact of particular stakeholders. It analyzes the stakeholders' level of interest in impressing their expectation on the project's decision as well as their willingness and power to follow their intention. Grouping stakeholders with the power/interest matrix therefore enables construction project managers to create clearer perspectives on the effects of communication and relationships between construction stakeholders upon the project implementation.

If stakeholders neither have a high own interest in project plans nor do they have power to exert much impact, the project managers should keep them informed to the necessary extent but should not invest too much effort in them. Meanwhile, if the stakeholders have high interest in the project and its actions but are having limited means to influence things, it is advisable to keep them informed about the issues they are interested in, as they could be valuable allies in important decisions. If it is found that the stakeholders have high power to make decisions or impact on the project but they behave passively and show low interest in corporate affairs, it is therefore necessary to analyze their potential intentions and reactions in all major activities and to involve them according to their interests. In this case, the relationship with stakeholders could be difficult. The final category is the stakeholders with high interests and power. This group has to be involved in all relevant activities. A government agency granting public–private partnership (PPP) concession is an example of a high power/high interest stakeholder.

8.5.3   Stakeholder coalition and participation

The stakeholder coalition is not static because time and event can change stakeholders. As such, it is also essential to monitor the shifting of coalitions. Today's highly involved stakeholders may be less involved tomorrow. Issues that are most salient at one time may be replaced by other issues at another time. The monitoring task requires periodically reviewing project stakeholders, drawing stakeholder maps, and revising the project's stakeholder–issue matrix. Managers need to be aware of and responsive to stakeholders' interests and influence shifts. The first sign of a rising public issue is conflict or confrontation with stakeholder groups. The project manager's political skills will be a useful attribute to assure maximum satisfaction among construction stakeholders (Wit, 1988).

8.5.4   Levels of stakeholder management

Although there are many different approaches to stakeholder management, they can be categorized into four levels to manage construction stakeholders.

1. *Inform.* This level expresses minimal effort of stakeholder involvement in the project. Secondary stakeholders with lower influence but higher importance need to be kept informed of decisions taken that may affect them directly. It is unlikely that they would play an active role in making those decisions. However, were they to highlight a particular issue with a decision, it is likely that serious consideration would be given to refining the decision made.
2. *Consult.* This is the way to keep stakeholders informed about the project. Since the secondary stakeholders with higher influence but lower importance need to be 'kept on board,' they should be consulted in order to seek their opinions and input for key decisions that directly or indirectly affect them. It is unlikely that the strategy will be altered as a result of such consultation, but tactics may be well adjusted to maintain higher levels of commitment.
3. *Involve.* Despite their low influence, stakeholders with high importance essentially need to be involved in all activities in the project according to their interest since they have power to make decisions that impact the project. The management should work directly with these stakeholders to ensure that their concerns are consistently understood, considered, and reflected in the developed alternatives. As long as their interest is achieved, they are kept satisfied and retain passive rather than active interest in the project.
4. *Partner/collaborate.* Since the primary stakeholders have high influence and importance to project success, they are likely to provide the project 'coalition of support' in planning and implementation. As such, they should be treated as partners to increase their engagement and commitment. This can be achieved by revising and tailoring project strategy, objectives, and outcomes if necessary to win their support.

These four levels of stakeholder management should be applied, according to the stakeholder management objectives, as a result of stakeholder analysis. When appropriately applied to stakeholders, the relationship is strengthened with more committed and satisfied stakeholders.

8.6   Useful tactics for construction stakeholder management

Essentially, project managers should exert their utmost leadership and managerial capability as well as tactical expertise to strive for coordination and commitment from stakeholders. There are tactics that are useful for deployment alongside various strategies of stakeholder management for mapping and analyzing stakeholders, managing stakeholder relationships and communication, achieving and sustaining stakeholder commitment, and satisfying stakeholders. The most important stakeholder must be identified and given the highest priority to pursue strategies that satisfy their needs through stakeholder analysis. Since stakeholder management activities can consume project resources, these activities should therefore concentrate on what will contribute to the project's success or where lack of communication will lead to failure.

8.6.1   Tactics for stakeholder analysis

The starting point is that those entities that have an interest in a project must be identified and the nature of their interests analyzed. Stakeholders can be identified, categorized, and prioritized by using several techniques including interviews with experts, brainstorming with project team members, and the use of checklists (Karlsen, 2002). A common mistake is committed by thinking that stakeholders are only those individuals who are decision-makers within the construction project. Nowadays, it is very risky and professionally unacceptable to undertake the project without a thorough understanding of every interest that is held in the project.

Stakeholder analysis is best carried out by the project team in consultation with potential stakeholders or representatives of potential stakeholder groups. The stakeholder analysis should be in the forms of foundation analysis (at project initiation), regular updates (at the end of every phase), and *ad hoc* updates (whenever there has been a change to stakeholder environment). In the stakeholder analysis, the created stakeholder matrix should contain information about each stakeholder such as role in project, role in organization, communication style or approach, support of project, relationship to other stakeholders, and any other pertinent information that will enable the project manager and project team, to effectively manage the stakeholders. Concerted effort is required to control and direct the management of stakeholders within and across the project management structure.

Application of the stakeholder mapping approach has valuable implications for project managers in assessing whether the political/cultural situation is likely to undermine the adoption of a particular project strategy, whether careful consideration is needed to pursue project strategies to reposition certain stakeholders, and the extent to which it is necessary, in order to ensure successful project implementation, to assist or encourage such stakeholders to maintain their level of interests and power. That is what is meant by keeping stakeholders informed and satisfied according to their identified power and interest in the power/interest matrix. The prioritization approach therefore enables the project team to ensure that the key stakeholders' expectations are understood, acknowledged, and managed.

Concisely, the process of stakeholder analysis comprises major steps including:

1. make a list of all stakeholders across the project establishment (i.e., identifying and prioritizing stakeholders as well as mapping their relationship and coalition);
2. determine stakeholder interests, importance and influence (i.e., assessing the nature of each stakeholders' attitude, interest, confidence, and power);
3. assess their ability to participate and find ways to involve them (i.e., constructing a matrix of stakeholder priorities, integrating stakeholders information obtained into the establishment of mission and goals, developing specific strategies and tactics, and monitoring shifting coalition).

In summary, tactical stakeholder analysis should include stakeholder identification, classification, and prioritization in order to review and analyze the nature of each group of stakeholders. These tactics help to confirm whether they are key or nonkey stakeholders. Besides, it reviews what is required to engage the key stakeholders in the project and gain their commitment, and how to communicate with them.

8.6.2 Tactics for management of stakeholder relationships: communication approach

Having identified the right stakeholders to engage, it is essential to understand their capability and willingness in order to manage relationships among stakeholders. Understanding that different stakeholders have different expectations of the project is the key to formation of successful project relationships. Bourne (2006) suggests that an important aspect of managing the project environment is to understand the direction of influence in which the project manager and management team must operate to realize the project successfully. The influential directions include managing forwards (anticipating and planning); backwards (developing and maintaining control systems, historical records, and the explicit and implicit knowledge of others); upwards (developing and maintaining relationships with senior managers whose support is vital and important to project success); downwards (managing the team); inwards (seeking feedback from stakeholders about the project, project management matters, and practitioner reflection and learning); sidewards (managing project manager's peers to ensure collaboration rather than competition); and managing outwards (addressing the needs and impact of a large group of the project's external stakeholders.

Thus, useful tactics for successfully managing stakeholder relationships are to develop plans that include internal relationship matrices and communication charts along with standard supporting communication plans. The communication strategy is therefore aimed at ensuring ongoing commitment and support by all key stakeholders for all aspects of the project. Communication, formal or informal, may be in the verbal, written, and electronic forms. Project meetings, project plans and reports, informal discussions, and formal presentations are considered appropriate ways of communication. With respect to the project objectives and expected outcomes, the purpose of understanding the current situation of each stakeholder can be achieved by undertaking a series of stakeholder interviews and opinion surveys. Particularly, it is essential that the relationships between construction project managers and stakeholders must be well understood in order to successfully engage stakeholders. A study on relationship management of construction project partnered by stakeholders from private and public sector also suggests that the proactive management of stakeholder relationships should be greater given strategic and tactical consideration, specifically focused at the client interface (Smyth and Edkins, 2007). Such action helps the project managers to manage poor relations as they emerge.

8.6.3 Tactics for achieving and sustaining stakeholders' engagement and commitment

A construction project should ideally have engaged and informed stakeholders who actively support the project's objectives and outcomes. Stakeholders' engagement can explain how well they understand the project's challenges and the strategies being used to overcome them while stakeholder commitment indicates level of the stakeholder support. As such, lack of understanding results in low engagement and the hostile stakeholder certainly provides weak support. To maintain stakeholder engagement, an active response to stakeholder's needs is essential. The project manager needs to involve the relevant stakeholders in the project planning processes. Strategies must be defined for each stakeholder and translated into action.

The Tasmanian Government (2005) recommends the following tactics for achieving and sustaining stakeholder commitment:

1. Provide active involvement of all stakeholders who can affect, and be affected, by the project early in the project definition and planning stages.
2. Legitimize the project manager's action in the realization of the project's benefits and outcomes. Creditability and trust should be engendered by establishing good personal relationships, illustrating that project actions are being seriously driven by the stakeholders' needs, using consultant's recommendations or the established formal methodologies to support the project, and involving senior executives as project champions in lending the project authority.
3. Implement early communication and persuasion. The communication strategy should appreciate stakeholders' differences and cater for their requirements.

It can be seen that project stakeholder management requires enthusiasm and commitment. Leung et al. (2004) investigated the impacts of commitment amongst major project stakeholders and reported that high affective commitment induces high performance and satisfaction while the continuous commitment provokes intention to resign from the firm. Thus, due to the dynamic nature of construction project and its stakeholders' relationships, the project management team should regularly update changes and assessment of the stakeholders' relationship and level of engagement and commitment throughout the construction project life cycle.

8.6.4 Tactics for increasing stakeholders' satisfaction

In reality, all stakeholders are interested in satisfying their expectation (Doyle, 1994). They feel dissatisfied when the organizational performance or resource transactions are insufficient to meet their minimum expectations. The external stakeholders may be satisfied if only they are adequately informed about the project (Manowong and Ogunlana, 2006). It is already recognized that project management process plays a critical role in ensuring successful delivery of construction project. Besides measuring project's success by comparing project's established goals with the project's final outcomes, evaluating the satisfaction of project stakeholders is another way to measure success of the project (Long et al., 2004). As such, it is essential to identify factors that critically affect the construction stakeholders' satisfaction with the project management process, because the management mechanism directly affects the satisfaction of construction project stakeholders (Leung et al., 2004).

Serving and satisfying stakeholders' expectation is not an easy task. It can be difficult to satisfy all groups of project stakeholders as the project supporters and opponents may have different levels of satisfaction with the project management depending on their process and outcome objectives during participation in the project activities (Manowong and Ogunlana, 2006). Managing to meet expectation of one group of stakeholders may bring dissatisfactions to other groups (Wheelen and Hunger, 2000). The project managers should therefore try to acknowledge the project's relevant concerns to all stakeholders as much as possible in order to satisfy every party or at least cater to their minimum requirements. Stakeholders' expectations can be fulfilled and satisfied if they are known early. In doing this, the opposing stakeholders may become more satisfied. As such, always keeping the key stakeholders informed of project information and decision-making is a useful tactic to satisfy construction project stakeholders, particularly in a construction project that has much impact on the public. Using open and trustworthy communication with the media and the affected stakeholders is also an essential tactic to make these groups satisfied with information given (Olander and Landin, 2005). Therefore, appropriate management mechanism is needed to avoid, manage, or alleviate conflicts on the project, to ensure stakeholder commitment and to improve stakeholder satisfaction.

8.6.5 Tactics for establishing stakeholders' common goals

It is also essential to investigate interrelations between stakeholders' commitment and satisfaction with the project goals and management mechanism, because the properly established project goals ensure that the stakeholders' requirement are well understood and satisfactorily fulfilled. It should be noted that, only when stakeholders are committed to the goals that are mutually accepted, will their high performance be achieved. Common goals refer to common interest of project stakeholders and they can be related to societal, economic, and environmental interests. Such common goals can be established through participation and dialogue. For example, in the development of rapid transit system in a city, stakeholders may have agreed on their common goals of economic development and environment protection.

The affected local citizen gives the developer their consent to construct infrastructures but the developer is obliged to comply with environmental protection requirements. In design and construction phases, project owners can establish a procedure for stakeholder participation at agreed stage to fulfill external stakeholders' need to have influence in expressing their views on the project. At the same time, with an established common goal of information exchange, it is an opportunity to obtain valuable inputs from the external stakeholders. It should also be noted that bad timing of stakeholder involvement may lead to stakeholders' frustration and less capability to deliver win-win solutions to stakeholders' interests.

Since contributions from project stakeholders are essential and needed to ensure success, the project manager needs to balance the interest of stakeholders and address them harmoniously. The stage before awarding the contract or the concession is the most appropriate to arrange intensive stakeholder involvement and balance stakeholders' interests. For example, in a PPP project, the government's interests are in terms of the amount of project subsidies, while the private sector's interests are in terms of potential benefits from the project. At the same time, the public interests are in terms of project's usage charges and benefits from using the project facilities while the local community's interests are in terms of preventive measures to alleviate project impacts on themselves and the surrounding environment. Balanced interests of stakeholders can be achieved by consultation and negotiation process. The interest groups attending the meeting should also be balanced according to their priority obtained during the stakeholder analysis. However, if the multiple parties cannot be balanced accordingly for stakeholder involvement in the cycle of project development, the manager needs to clarify the stakeholders' roles and responsibilities toward the project and try to enforce them through some form of contractual obligation. Again, to prevent conflicts, methods of negotiation, particularly with the group leader or representatives, can be employed to facilitate such action. It should be kept in mind that, when establishing the common goals, none of the stakeholders be excluded from the related project management activities.

Stakeholders' needs and opinions can be obtained through participation process, such as a public hearing, that is appropriately arranged during the project's life cycle. Public hearing is a participation mechanism that allows a large community of stakeholders, particularly the general public, to be involved in construction projects. It is regarded as one of the *Interaction* and/or *Information exchange* techniques of public participation. At public hearings, project managers have the opportunity to explain the purposes and constraints of projects and receive inputs from the public, especially the affected persons and potential users, before making major decisions regarding the project. Public hearings, therefore, provide opportunities to project stakeholders to present information, express concerns, present facts, and voice opinions, both in support of and in opposition to construction projects. As such, a public hearing is a useful venue for stakeholders to mutually communicate their desired goals. Through constructive discussions and mutual agreements, stakeholders' common goals can be established through public hearings.

However, public hearing can prove ineffective if it is not thoroughly planned and well conducted such that it could fail to constitute meaningful participation. To achieve effective public participation, project managers must devote time and resources to prepare for and execute public hearings. For example, the hearing must be conducted before major decisions on the project are made, and project information must be adequately provided to project stakeholders (Manowong and Ogunlana, 2006). If differences in opinion exist and the common goals are unlikely to be achieved at the hearing, negotiation with stakeholders must be carried out. To avoid damages to project's image and delays due to protests, the manager should make the effort to consult key project stakeholders in order to plan and conduct public hearing that is acceptable to all. Although conducting public hearing can be costly and time-consuming, public hearing is a good channel to transparently inform and adequately educate stakeholders to have broad understanding of the necessity for and the impacts of a project. The goal of public hearing is to encourage all the parties that are influential in and/or affected by a construction project, to contribute meaningfully into defining and realizing the project's objectives, ensuring sustainable benefits, and establishing common goals with mutual respect accorded to the interests of all stakeholders.

8.7   A compendium of tips

Construction project stakeholders express various needs and expectations about the project. These interests are often in conflict and it is highly unlikely that they will all be fulfilled. To effectively manage stakeholders in construction, stakeholders' need and expectations are evaluated based on the concept of stakeholder analysis that should include formalized identification and prioritization of project stakeholders. Identifying stakeholders helps the manager to know which groups of stakeholders are directly and indirectly involved in the project, to understand the interests and expectation of each group, and to forecast project stakeholders' behavior and its effect on project outcome. Prioritizing stakeholders enables the project managers to appropriately deal with stakeholders according to their power and influences to the project. Important and influential stakeholders are given priority such that their needs and expectation are adequately fulfilled. However, the rest of stakeholder groups must be handled appropriately according to their level of power and interests by applications of four major strategic management approaches, i.e., inform, consult, involve, and collaborate. Whatever approach is chosen, it must be conducted very early in the project life cycle.

Strong relationship with stakeholders is linked to project success. Effective communication helps the project managers to build and maintain good relationships with stakeholders. In order to define an appropriate procedure for stakeholder engagement, the stakeholders' relative influences need to be assessed and their expectations need to be understood. A good project manager will attempt to build good relationships with the key stakeholders in order to really understand the needs and expectations of the stakeholders. Stakeholder relationships plans should include relationship matrices and communication charts. Besides, the availability of communication channels should be emphasized. In particular, the generation, storage, dissemination, and disposition of information in construction project must be essentially carried out timely and precisely. Without effective communication, key stakeholders could miss out on vital information and may not understand why change is needed. Periodic meetings or other communication methods should be clearly planned and outlined in the project's communications management plan. Although the expectations have been clarified with project stakeholders, it is still necessary to carry out communication management in order to ensure that stakeholders are kept up-to-date on the project's status. Further, the issues arising during the project evolution need to be addressed and resolved with the respective stakeholders.

Stakeholders with good understanding on the project objectives are more engaged in the project. Similarly, stakeholders that are more committed to the project provide strong supports to the project. To improve stakeholders' engagement and commitment, the relationships with stakeholders should be regularly monitored and assessed. Involvement of appropriately prioritized stakeholders must be done early and proactively, with established trust and creditability as well as project authority lent by senior executives in the project. Stakeholders' requirements should be catered and differences among them should be appreciated by communication strategies. Stakeholders are satisfied when their minimum expectations are met. However, satisfying all groups simultaneously can be difficult. As the stakeholders may have different aspects of satisfaction, the managers should acknowledge concerns relevant to the project to all stakeholders as much as possible. If the stakeholders' requirements are known early and effectively fulfilled, the stakeholders then become more satisfied, particularly those stakeholders who oppose the project. Constructive use of the media is also a useful tactic to increase stakeholders' satisfaction. Figure 8.2 presents a chart showing the objectives and strategies for effectively managing construction stakeholders.

It is important to be aware that different stakeholder management approaches are appropriate for different types of stakeholders. Stakeholder management varies depending on project stages. During the planning and implementation phase, stakeholder management is mainly at 'Inform' and 'Consult' levels. During design and construction phase, the environmental impact study should also 'Involve' the external stakeholders. This stakeholder involvement can extend to the 'Collaborate' level. For example, the management levels at the operational phase are 'Inform' for general issues and 'Collaborate' for highly technical issues. Similarly, key stakeholders are required to 'Collaborate,' while other stakeholders with less significance need only to be informed.

There are particular and important concerns for construction stakeholder management strategy on international development projects in which stakeholders belong to different cultures, political risks, and demanding local constraints. For example, the foreign managers have to first build trust and create good relationship with the local employees. It is also essential to study the politics and bureaucracy of local government. It is possible that local government follows excessively bureaucratic procedure for government approvals. In addition, the managers should also consider the competency and commitment of local suppliers to deliver quality construction materials on time as required.

To foster collaborative working, it is primarily recommended that greater strategic and tactical consideration be given to the proactive management of relationships. Interpersonal relationships and communication with key stakeholders may be differently critical or significant, depending on the required specific knowledge, skills, and competencies involved. In addition, the project managers should also be constantly aware of the key factors to project success. Stakeholder management plan should be reviewed periodically. Any changes occurring should be updated on timely basis. If the strategy is not working, it should also be reviewed and updated. All development stages should be transparent and accountable especially where external stakeholders are excluded from the decision-making.

Formal forms of communication such as reports of progress, constant or frequent calls or meetings will make stakeholders feel that their interests are being met. However, allowing dissatisfied stakeholders to intervene can lead to disaster. That is, actions from low-interdependency stakeholders comprising local communities, councils, and other interest groups can be problematic when they express their dissatisfaction by engaging in harassing moves toward project. Such harassing moves include exposing conflict issues widely to the public through formal media such as newspapers, television, radio, and probably organizing protests and demonstrations. Subsequently, other groups of stakeholders are likely to be triggered to pay attention to those issues. As a result of such harassing actions, it takes longer time to secure wider public support in order to carry on the project. Project delays due to court cases are also undesirable consequences.

Regarding stakeholder communication, both good and bad news regarding the project to various members of the stakeholder groups should be communicated in meaningful ways. Informal communications such as conversations, discussions, and informal memos are generally not included in the project's communication plan. Nevertheless, the informal communication can be very effective in promoting unity of purpose and action within project teams and also improving understanding between project teams and other stakeholders. Informal communication channels such as interpersonal relationships are quite powerful such that it

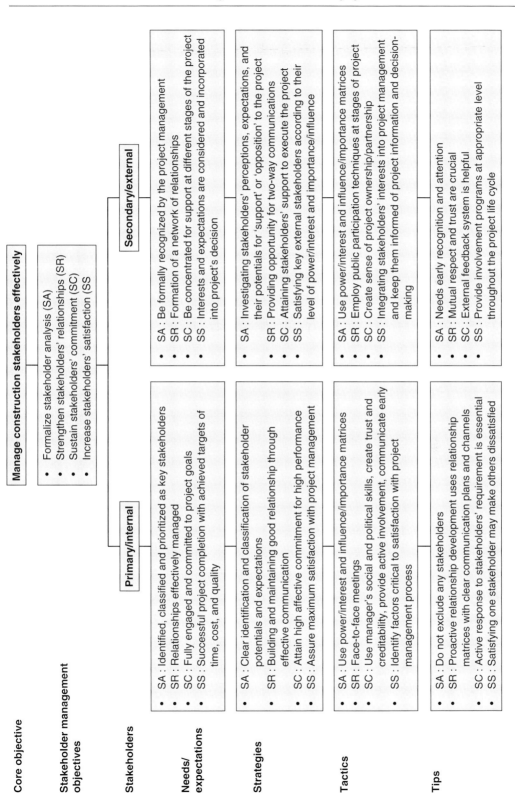

Figure 8.2   Chart of strategic stakeholder management.

can 'make or break' the success of the project. As such, planning and executing communication tactics is critical and, hence, informal stakeholder communication strategy should be active at all times.

8.8 Summary

Since the construction project managers have the responsibility to manage the construction sites and deal with many internal and external parties, they have to learn how to properly utilize stakeholders' power needs in order to avoid possible conflict. Effective and proactive stakeholder management strategies rely on the project manager's exerting power, decision-making power, referent power, and reasoning for selection, consultation, and friendliness. Balanced and well-managed relationships and power can then bring positive outcomes to the project. Through informal networking, the project manager can obtain invaluable insights through informal feedback that is more candid and emotive than written or meeting reports. Bridges across divergent groups of stakeholders can be built assuring that each group is gaining through their agreement to go forward. The most important things to do are to proactively reach out to stakeholders, actively create good relationships and mutual trust, and consistently keep stakeholders satisfactorily informed, involved, consulted, and collaborated with.

References

Bourne, L. (2006) Project relationships and the stakeholder circle. *Proceedings of PMI Research Conference*, July 16–19. Centre Mont-Royal, Montreal, Canada.

Bourne, L. and Walker, D. (2005a) Visualizing and mapping stakeholder influence. *Management Decision*, **43**(5):649–660.

Bourne, L. and Walker, D. (2005b) The paradox of project control. *Team Performance Management Journal*, **11**(5/6):157–178.

Cleland, D.I. (1998) *Stakeholder Management. Project Stakeholder Management Handbook.* San Francisco: Jossey-Bass.

Grimble, R. and Wellard, K. (1997) Stakeholder methodologies in natural resource management: A review of principles, contexts, experiences and opportunities. *Agricultural Systems*, **55**(2):173–193.

Harrison, J.S. and Caron, H. St John (1996) Managing and partnering with external stakeholders. *Academy of Management Executive*, **10**(2):46–60.

Hill, C.W.L. and Jones, G.R. (2001) *Strategic Management: An Integrated Approach.* New York: Houghton Mifflin.

Jawahar, I.M. and McLaughlin, G.L. (2001) Toward a descriptive stakeholder theory: An organizational life cycle approach. *Academy of Management Review*, **26**(3):397–414.

Jergeas, G.F., Williamson, E., Skulmoski, G.J. and Thomas, J.L. (2000) Stakeholder management on construction projects, *AACE International Transactions, PM.12*, AACE International, Morgantown, WV, 12.1–12.6.

Johnson, G. and Scholes, K. (1999) *Exploring Corporate Strategy.* London: Prentice Hall Europe.

Karlsen, J.T. (2002) Project stakeholder management. *Engineering Management Journal*, **14**(4):19–24.

Leung, Y.M., Chong, A., Thomas, S. Ng. and Cheung, C.K.M. (2004) Demystifying stakeholders' commitment and its impacts on construction projects. *Construction Management and Economics*, **22**(7):701–715.

Leung, Y.M., Thomas, S. Ng. and Cheung, S.O. (2004) Measuring construction project participant satisfaction. *Construction Management and Economics*, **22**:319–331.

Long, N.D., Ogunlana, S., Quang, T. and Lam, K.C. (2004) Large construction projects in developing countries: A case study from Vietnam. *International Journal of Project Management*, **22**(3):553–561.

Manowong, E. and Ogunlana, S.O. (2006) Public hearings in Thailand's infrastructure projects: Effective participations? *Engineering, Construction and Architectural Management*, **13**(4):343–363.

Mendelow, A. (1991) Stakeholder mapping. *Proceedings of the 2nd International Conference on Information Systems*, Cambridge, MA.

Newcombe, R. (2003) From client to project stakeholders: A stakeholder mapping approach. *Construction Management and Economics*, **21**(12):841–848.

Olander, S. (2007) Stakeholder impact analysis in construction project management. *Construction Management and Economics*, **25**(3):277–287.

Olander, S. and Landin, A. (2005) Evaluation of stakeholder influence in the implementation of construction projects. *International Journal of Project Management*, **23**:321–328.

Oliver, C. (1991) Strategic responses to institutional processes. *Academy of Management Review*, **16**(1):145–179.

Project Management Institute (PMI) (2000) *A Guide to the Project Management Body of Knowledge*. Newtown Square, PA: Project Management Institute Standards Committee.

Savage, G.T., Nix, T.W., Whitehead, C.W. and Blair, J.D. (1991) Strategies for assessing and managing organizational stakeholders. *Academy of Management Executive*, **5**(2):61–75.

Smyth, H. and Edkins, A. (2007) Relationship management in the management of PFI/PPP projects in the UK. *International Journal of Project Management* **25**:232–240.

Tasmanian Government (2005) *Stakeholder Management*. Tasmanian Government Project Management Guidelines Version 6.0.

Weiss, J.W. (2003) *Business Ethics, A Stakeholder and Issues Management Approach*. OH: Thomson Learning, South-Western.

Wheelan, T.L. and Hunger, J.D. (2000) *Strategic Management Business Policy*. Englewood Cliffs, NJ: Prentice-Hall.

Wit, A.D. (1988) Measurement of project success. *Project Management Journal*, **6**(3):164–170.

9 Constructing Negotiations: Bargaining, Learning and Fighting

Jeroen Warner

9.1 Introduction

Like politics, negotiation doesn't have a great reputation. It smacks of wheeling and dealing, of unscrupulous confrontation and unethical concessions, carried out by cynics who would sell their mothers to get a result. Negotiation is also scary: if others are more skilful than you, you may look back and have given away the family jewels.

On closer scrutiny, however, many everyday social situations turn out to be negotiation by another name. 'Joint planning', 'consensus building', 'consultation' can prove euphemisms for a great deal of bargaining. At some point or other, there will be a moment in which you will need to strike a deal on the terms of a project definition, the rules of engagement or the project's implementation. These deals may be implicit and be understood without being explicated and formalised.

Negotiation doesn't come easily to everyone. Managers used to top-down management may not be used to negotiation, accustomed as they may be to a patrimonial style, where the patron sets the terms and hands out favours. Poor people may reinforce this image, as they feel intimidated by the presence of someone in authority.

Because planners tend to assign rationality and expertise to themselves, they may be inclined to see stakeholders as irrational nitwits, even as obstacles. Planners often seem to believe that the plan and its planning is so self-evident that only the uneducated will not see its logic. Public participation, if included at all, then becomes an exercise of selling the project to the people.

But 'the people' are not so rational. The seeming irrationality of stakeholders can make the planner averse to negotiating. While negotiation *isn't* about point scoring, finding out who is right or pressing one's point forever, this may be the reality in a meeting.

Activists on the basis of their moral stance may be used to polarisation, to expose and denounce the wrongdoing of a company or public authority. Politicians may grandstand without saying anything definite. But if each of these wants to achieve or change anything, at some point, they too will have to negotiate. Therefore, it is important to know some basics of bargaining.

The first part of this chapter will address negotiation, irrespective of whether this takes place within or outside a formal or informal setting. After explaining types and strategies of negotiation, Section 9.2 zooms in on dealing with power relations. Section 9.3 discusses what a successful negotiation might involve and whether that includes a consensus outcome. Section 9.4 will explore what can be influence in a negotiation process.

With kind inputs and feedback from Dr Sandra Inês B. Granja, University of Sao Paulo/FUNDAP

9.2 Why negotiate?

Negotiation is a way of trying to get what you want through dialogue with others. You enter into negotiations because:

- someone has something you want;
- its value is not fixed (there's not a single set price for it);
- you can't force the other to give you exactly what you want.

In everyday life, skilful negotiation is often needed to get what we want: to get our kids to bed without too much fuss, to trade shifts with our colleagues because we want to see the game, to get that fetching tie and cuffs thrown in with an expensive suit. In a construction project, you will need the cooperation from many stakeholders without twisting their arms too much, so it makes a lot of sense to develop techniques and strategies that help lead to successful negotiation.

Let's first establish when *not* to negotiate, though (Box 9.1). Mack and Snyder (1957) are representatives of a traditional approach that defines negotiation as a social interaction process between parties with mutually exclusive or incompatible values. This definition focuses on differences, zooming on something that is hard to change – people's values. Therefore this approach is most likely to lead to 'zero-sum' outcomes; one gains, the other loses. If Mack and Snyder were entirely right, very few disputes would be satisfactorily settled and deals brokered. But if two (or more) parties have a dispute, problem or want to do something new, they need to seize an opportunity they cannot seize without the other. We agree therefore with the Consensus Building Institute (CBI) when they say that *'In simplest terms, negotiation is a discussion between two or more disputants who are trying to work out a solution to their problem'*.

If parties are willing to settle for what the other will give, there is little need to negotiate. If parties are not willing to take the other party's interest into account and can get away without loss, there is little need to cooperate. The missing link here is that negotiation starts from actor *interdependence*: if the sense of interdependence is weak, there is little need to negotiate. If you can choose between hundreds of equally good suppliers, you make a take-it-or-leave-it offer and that's it. Likewise, you don't bargain in the supermarket over the price of butter – they will sell it to someone else. But if you need a particular favour, material, price, quantity, location, and they need you to clinch a deal, you are in business

> Negotiation is based on interdependence. Negotiation defines or redefines the terms of interdependence (Walton and McKersie, 1965).

Negotiation can work because actor interdependence is usually not symmetrical: one party doesn't want exactly the same thing as the other at exactly the same time. This gives opportunity for exchange, conceding on one aspect while gaining on another (CBI, 2000). But the interdependence is rarely as strongly developed both ways – dependencies tend to be asymmetric, and bring different interest positions. The present contribution will specifically discuss the issue of power differences.

9.2.1 Collaborative negotiation

Another conventional pitfall is that negotiation is often assumed to be 'one-shot' (one round of negotiations) and without communication. Negotiation theory has benefited a lot from 'game theory' which models negotiators in an ideal-typical setting and predicting their strategic behaviour.

BOX 9.1   When not to negotiate

In addition to negotiation, learning and fighting, a fourth response is: non-engagement. It can be 'non-entry' into negotiations, or 'exit' during negotiations, which does not preclude informal communication and later (re)-entry.
   In which cases is this an option? It is:

   – When you don't care
   – When you have everything to lose
   – When you have nothing to gain (e.g, You are already running at full capacity)
   – When it compromises your character or reputation
   – When they act in bad faith
   – When you don't have enough time
   – When you have too much time, e.g. waiting will improve your position
   – When you are not prepared (Lewicki et al 2007)
   – When you prefer your BATNA. A key aspect of negotiation (80% of negotiation is preparation) is building the Matrix below. This Matrix show the points of contact between 2 different actors, what is possible to negotiate and what is not. The matrix should be built with the BATNA of each actor. You can do a role playing game to simulate different negotiation scenarios to identify BATNAs in different situations.

The 'prisoner's dilemma' and similar games like 'stag hunt' and 'chicken' are famous examples of negotiation games. In 'prisoner's dilemma', each of two suspects of a major crime are given the option to go free if they squeal and denounce the other, who gets 10 years in prison, while both suspects will serve 1 year each if neither squeals. The outcome of such simulations is usually non-cooperative.

The assumption is that the two cannot talk with each other, and will not want to work together once they are released from prison. Indeed, this looks a lot like the first phase of engagement in a negotiation, where people do not know each other well. The first stage is mostly confrontational. People do not feel comfortable together, and want to secure maximum clarity: 'these are my interests, these are my non-negotiables'.

However, analysis of *iterated* (repeated) game shows that repeatedly playing the same game brings significantly more cooperative behaviour between the players (Cardenas and Ostrom, 2004). Practice makes perfect, but this perfection appears to include benefit for the other as well as yourself. In practice, we can note that as people meet more frequently, they often cannot help but bond over common interests and reduce mutual confrontation (Poncelet, 1999). This has to do with empathy and appreciation, but also with the very real joint gains from cooperation, as there inevitably are some common interests. Conflict escalates with non-communication and stereotyping – it becomes harder to pursue when the opponents has a face and personality. In the course of negotiation, parties tend to adjust to each other, seek to influence and pre-empt the other's bargaining position. After a breakthrough, things can become surprisingly cooperative and constructive, as the bargaining sides look for common interest and similarities.

Psychologists and others have noted that psychological factors influence the players. Negotiators tend to be lazy: they 'leave value on the table' rather than going for the optimum outcome for themselves or for all parties involved. One reason is that people will not push things to breaking point when they are tired. The influence of 'satisficing' (Simon, 1947) behaviour should not be underestimated.

Much recent literature on planning and governance extols the virtue of collaboration and Alternative Dispute Resolution. It promotes integrative negotiation (baking cakes together) over distributive bargaining) arguing over 'how the cake should be cut'. Integrative negotiation is 'the process by which the parties attempt to increase the size of the joint gain without respect to the division of the payoffs' (Walton and McKersie, 1965:13). Table 9.1 lays out the steps towards integrative negotiation.

A collaborative strategy is especially expected when you value the longer-term relations with the other party. A more 'sustainable' outcome is one in which stakeholders share and add to information, learning about each other's interests, explore options jointly and ultimately find an appropriate balance among goals. This makes it possible to

Table 9.1   A methodological framework for facilitating integrative negotiations

Task 1: Preparation

- exploratory analysis of conflicts, problems, relations, practices, etc. in:
  - historical perspective;
  - selecting participants;
  - securing participation by stakeholders;
  - establishing relations with the wider policy environment.

Task 2: Agreeing upon a process design and process protocol

- creating an agreed-upon code of conduct and provisional agenda;
- reaching agreement about procedures, methodologies, etc.;
- process management and maintenance of process agreements;
- securing new process agreements as the process unfolds.

Task 3: Joint exploration and situation analysis

- group formation;
- exchanging perspectives, interests, goals;
- analysing problems and interrelations;
- integration of visions into new problem definitions;
- preliminary identification of alternative solutions and 'win–win' strategies;
- identification of gaps in knowledge and insight.

Task 4: Joint fact-finding

- developing and implementing action plans to fill knowledge gaps.

Task 5: Forging agreement

- manoeuvre: clarifying positions, making claims, use of pressure to secure concessions, create and resolve impasses;
- securing agreement on a coherent package of measures and action plans.

Task 6: Communication of representatives with constituencies

- transferring the learning process;
- 'ratification' of agreement by constituencies.

Task 7: Monitoring implementation

- implementing the agreements made;
- monitoring progress;
- creating contexts of.

Source: Leeuwis (2000).

Table 9.2   Learning vs. fighting in negotiation

Fight	Debate/learn
Tit for tatw	What is vs. what should be
Driven by fear	Driven by values
Emotional	Rational
Tends to escalate	Tends to compromise/consensus

Source: Rapoport (1960).

transcend the limits of distributive bargaining, 'politicking'. *If the outcome accommodates both interests, but nothing more, there is question of 'distributive' deal making. But if the deal brings a surplus by optimising the outcomes for both it is 'integrative'.* It is about maintaining good, mutually advantageous relations without yielding everything to the other. A focus on integration is the basis for what Fisher and Ury (1981) call 'principled negotiation' and Lax and Sebenius (1986) 'contingency bargaining'.

While a necessary, civilised correction to the earlier confrontational perspective, I would like to nuance the view that the learning approach is also more effective. Table 9.2 illustrates some tactics we can adopt while learning. Even knowing that we stand a better chance of constructive negotiation over time, most of us will have experienced in educational or work situations that a measured use of confrontation and muscle-flexing may be extremely effective. Negotiation is supposed to be rational, as opposed to conflict, which involves emotional language, shouting, threats, avoidance. Sometimes, being difficult works.

In highly adversarial situations, the stakeholders generally use 'hard bargaining' (setting out extreme positions, withholding information and making concessions grudgingly) in order to achieve a settlement, distributive 'hard' bargaining approaches may be efficient if there is little time to prepare and deal. We cannot, and should not *expect* constructive negotiation.

There may be good tactical reasons not to give in any more and drag one's feet. Both parties always have the alternative *not* to negotiate any further if they see no advantage in it. In the back of their minds, each negotiator should have a more or less clearly defined '*BATNA*' (*Best Alternative to a Negotiated Agreement*), that is, the point at which no deal is better than a bad deal?

> **BATNA** (Best alternative to a negotiated agreement) is the course of action that will be taken by a party if the current negotiations fail and an agreement cannot be reached. A party should generally never accept a worse resolution than its BATNA.

9.2.2   So, learn or fight?

If you stick with a confrontational, unyielding position for too long, there is a danger of stagnation. Axelrod (1981) has called this 'one tit for two tats': when one negotiator makes a serious offer and gets nothing, (s)he will make only one more concession. If the other party fails to play the 'give and take' game, there is no incentive for the first party to compromise.

In the long run, conflictive stances are almost never economically, environmentally or socially sustainable. But alternatively, when things are too cooperative from start to finish, there is a danger of losing acuity and clarity, inviting complacency and coming out with something that is neither fish nor flesh. A generous proposal at the start can almost force the other party to 'cash in' and bargain hard to see if there is any more 'value on the table' for him/her. A nuanced position at the start can create confusion and makes it hard to compromise in clear terms.

Which of the tactics then should one opt for, or prepare for?

There is no *a priori* 'better' choice between distributive and integrative bargaining, between competitive and cooperative negotiation. Each may be useful at any stage of the process, and often we find both of them, even combinations played out simultaneously as complementary strategies, in a balancing act. This is not so easy to get one's head around. It is attractive to plump for one of the polarities, rather than accept the tension and ambiguity of a balancing act, but maintaining the balance is more likely to be effective over time.

We shall follow the insight of Mastenbroek (1992) who argues that the most effective negotiation is situated *in between* the two polarities of conflict and cooperation, drawing from both, depending on the circumstances. Negotiation is the middle ground between learning and fighting, between bargaining on how the cake is cut and baking the cake together. Negotiation relies on interdependence as well as difference of interests. If the sense of interdependence clearly prevails over difference of interest, cooperation will be the most likely form. Fighting is the most logical when difference in interest prevails over interdependence between the parties involved.

A level-headed assessment of these two variables therefore can help you assess at which point you, or the other party, might shore up negotiation, and at what point you look for an aggressive, dominating style, driving for a distributive deal and when you might look for an integrated, collaborative result.

When there are two negotiators per side, one can play 'good cop, bad cop': one negotiator is hardliner, the other more flexible. For each of these tactics, dosage is important to prevent an impression of irresolvable conflict.

Determine problem type

To know what can be gained from different strategies, it is useful to determine the type of problem you want to tackle: is it mainly about information and uncertainties, is it about values and interests, or both? (see Table 9.3). When it is not clear who the actors are (Table 9.4) and what the definition is, the 'learning' view of negotiation can prevail. Value differences and uncertainties are only to be expected, the issue here is whether they are divisive. This needs to be analysed and verified early on – you may be sure there is agreement on basic issues or disagreement but others may not.

Strategy: avoid/explore?

Apart from conflict or cooperation, Mastenbroek (1992) identifies another polarity: evasive (or avoiding) versus explorative negotiation. The former is more oriented to seeking security and protection (cards close to chest), the latter courts risk and adventure (cards on the table).

Table 9.3   Problem type and strategy

Values ↓ Facts →	Agreement on values	Disagreement on values
Low uncertainty	Technical problem: solved by coordinated knowledge	Political problem Dominated by 'fighting'
High uncertainty	Untamed technical problem	Wicked problem
	Learning and pacification prevails	Dominated by 'learning' but also a degree of 'fighting'

Source: After Hisschemoller and Hoppe (1996) and Kok (1998)

Table 9.4   Example of multi-actor analysis of bargaining positions

Actors	A	B	C	D	....Z
A		B and A	etc.	etc.	
B	What are points of contact/negotiation Between actor A and B	–			
C	A and C	B and C	–		
D	A and D	B and D		–	
Z	...	...			–

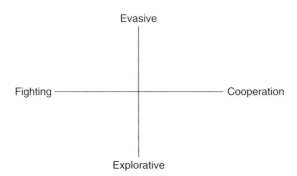

Figure 9.1   Mastenbroek's two strategic dimensions (see also Fisher and Ury, 1981).

An *evasive* (avoiding) strategy is non-confrontational, but also non-accommodating, for example by repeating the same point, without bringing any new insights or benefiting from new information. This passive style may be expressed in an active, noisy way or in a calm voice and an inscrutable smile, but the position is the same. It may be a very effective tactic, but self-defeating in the long run as people lose patience and break off the talks.

On the other hand, an *explorative* strategy is flexible and accommodating, constantly on the lookout for new avenues. Too much flexibility makes the negotiator seem untrustworthy: any position may be changed the next day. Again, Mastenbroek strongly prefers a balance between the two: place some cards on the table, not all – show some flexibility, but avoid a flip-flop pattern, which only conveys an impression of fickleness. The foregoing can be visually represented as two dimensions (see Figure 9.1).

9.3   Recognising and influencing the power element

It is tempting to overlook or deny the power element in planning. Like money, those who have it do not like to talk about power (as it exposes them to accountability), while those who are not, may feel that 'the powers that be' will always win. Collaborative planning approaches have been faulted for negating the crucial operation of power asymmetry. *Bargaining power* is the capacity to produce an agreement on one's own terms, and those who lack this power may not even be fully aware that the facts are stacked in a certain way (Lukes, 2005). In any planning and implementation process (where 'idea meets reality'), a struggle for power is always present. 'Fighting' is only the more overt expression of this struggle. However, the centrality of the power element in negotiation does not by definition have to prevent reaching satisfactory outcomes for each side involved.

First, power differences may be perfectly legitimate: we don't all want to make life and death, war and peace decisions every day; that is why we delegate this power to presidents, army generals and hospital surgeons. Normally, technical knowledge and resources are not equally distributed among stakeholders who do not have the time to read up on everything – that is why we have experts. If legitimacy (the justification of power differences) is undisputed, there is little problem with power.

Just like interests are normally dissimilar between the parties, power is always asymmetrical: it is impossible to find two actors who have the same power, or the same resources (cognitive, political, etc.). Project initiators face stakeholders who are much stronger than them (e.g. banks, governments) but also those who are much weaker.

A national government has crucial authority relations with local government, executive directors over middle managers but also dependency relations with banks and electorates.

Great power disparity may seem comfortable for the stronger party, but can be counterproductive. If power asymmetry and distrust between disputants is great, there is a tendency for either party to resort to physical violence and intimidation, civil disobedience, litigation, etc. rather than productive negotiation.

Coercion and authoritarian decision then may seem the logical option. In practice, these are rarely decisive with respect to the overall outcome, though. This has a lot to do with the limits to coercion. It proves hard to make a success out of a purely coercive strategy because coercion must frighten and reassure the opponent at the same, using threats without motivating them to resist. Coercion only works if the other party cooperates! (Jakobsen, 2007). In top-down situation, there would appear to be little flexibility, no discretion for lower-level negotiators to make their own deals, unless some 'residual space' is left. Yet if lower-level participants are seriously unhappy with this situation, they will 'erode' the project. Scott (1985) has called attention to the 'weapons of the weak' – they include discrediting the project, putting obstacles on the flow of the paperwork, paralysing or obstructing the decisions, talking about the coordinators, consulting, etc. There are many informal ways to sink a project.

In multi-party negotiation, both the very powerful and very powerless tend not to negotiate (Warner, 2007). Non-participation by the strong comes from a feeling they only stand to lose from bargaining. Non-participation by the weak comes from an impression they have nothing to gain.

'Fighting' is the tendency and expectation from the constituency of the less powerful party in negotiation, but it tends to worsen an already weak negotiation position. From a weaker position, it is tempting to claim a 'bottom line' to satisfy one's constituency (Mastenbroek, 1992). From a stronger position, there is a tendency to try and control the process and outcome. From the perspective of the more powerful, keeping control is the central challenge, for the less powerful, to prevent being encircled.

Depending on the context, negotiators or facilitators may find themselves in horizontal (equality, 'deliberative') level or vertical (inequality, 'tough') modes. Projects often have a decision in horizontal mode, but the execution is on vertical, because the institutions which have money more often work on 'vertical' relations. In practice, you may work on horizontal and vertical levels at the same time. One option to crosscut levels in a 'vertical situation' is to have an expert platform working in parallel with a decision-making platform (see below), with overlapping memberships, or a 'mirror' group of social actors who follow the process and interact with some decision-makers (but not all).

Thus in negotiation, say over building in the flood plain, experts might bring their wisdom to bear, citizens organising their cunning, NGOs their moral authority and a sponsor or investor their financial strength.

This points at the parallel existence of different styles of power being brought to bear in negotiation. One amusing yet serious power categorisation is the following metaphorical zoo (Scott, 2001):

Box 9.2   Three forms of power

'Power over' (dominance) can be exercised in a benign as well in an exploitative way: A's power *over* B may be used to the benefit of A, but also of B, or of A and B. If the stronger party exercises power through force and coercion, it is unlikely to last for long; if A however uses his/her power over B such that both benefit (if differentially) B will be more likely to accept A's power, because A's power also increases B's 'power too'. (Haugaard and Lentner, 2006).

'Power with' on the other hand, refers to a certain form of getting things done **together**, that is, collaborative endeavours (Woehrle, 1992).

Both 'power over' and 'power with' can enhance the total 'power to'; a reference to the capacity to get things done. This is true at the level of states and citizens, but also of experts and layman, donors and recipients.

    LION strength
) hard power   inducements
    FOX cunning
    BEAR authority
) soft power   persuasion
    OWL wisdom

This categorisation shows that there are multliple sources of power, which can be countervailing with respect to the other.

In another categorisation, from Conflict Resolution Info, a negotiating party has three basic ways to induce adversaries to move towards the position it desires: It may try to coerce reward or persuade the opponents, that is, use sticks, carrots or hugs (http://crinfo.beyondintractability.org/essay/Power/   CR Info The Conflict resolution Informatkon Source). These reflect the difference between the exercise of 'power over', 'power to' and 'power with' (see Box 9.2).

9.4   What is successful negotiation?

Having talked a lot about process, how about outcomes? Where do you want to end up? Should there be a consensus outcome or is residual conflict acceptable? Again, there is more than one way to evaluate a successful negotiation. Apart from the direct benefits, other gains are important as well. A special think tank on negotiation at Harvard Law School called the Program on Negotiation (PON) has carried out a lot of research over the past 15 years or so to look at the questions: 'What is a successful negotiation?' and 'What kinds of strategies and behaviors lead to success?' This research has involved watching people role-play different kinds of negotiations encouraging them to adopt first one approach and then the other and watching the results:

– Satisfying outcome: 'Even if the deal is not terrific for you, if it is better than anything else you might have received elsewhere, then the outcome can and probably should be considered successful'.
– Sustainability: The agreement is only likely to last if negotiating partners can live with it.
– Acceptable cost: If you spent much longer negotiating than you had time for, or expended more resources than your organisation could afford, the settlement reached is not efficient.

- (No) 'value' on the table: 'If there was more that could have been squeezed out of the agreement, the negotiation was not as successful as it could have been'.
- (No) damage to the relationship: 'If your negotiating partners never want to see you again because they feel that their needs were not met by the agreement or because they feel taken advantage of, then you haven't left yourself in a very good position to negotiate with those people again. The question to ask is: Did this negotiation put us in a better or worse position to deal with each other the next time?'

A mutual gains approach to negotiation assumes that 'the best way for me to ensure that I get what I want is to listen carefully to what you need'. People may be unsure about what they want, and what they have to offer. The approach can improve all stakeholders' capacity to meet their goals by *training* them, both about the subject matter and about negotiation. It is quite a hard skill to learn to focus on needs and interests, not positions, let alone on persons (play the ball, not the (wo)man).

An example of this is the NEGOWAT project in peri-urban areas in Sao Paulo, Brazil. New arrivals often are illegal or semi-legal settlers without basic services such as potable water and sanitation. When a group of such settlers was invited to talk with local authorities, they were divided and could not make their wishes and needs clear to their counterpart. Training helped them enormously, while games helped everyone visualise the physical and social interdependences and consequences of different policy scenarios (Ducrot, 2007).

9.4.1 Should negotiation lead to consensus?

The above has not insisted on consensus as an outcome of successful negotiation. In many cultures, consensus building and mutual gains are social desirables. Securing a majority vote still leaves a dissatisfied minority, while a consensus decision[1] is one that all unanimously support (the word 'consensus' confusingly has multiple interpretations). The assumption is that if people know each other well, they will overcome political, organisational constraints and reason beyond their self-interest, even if the consensus is suboptimal for any one party involved.

Reasons for not pursuing consensus building

Coglianese (s.d) sees at least reasons for not pursuing consensus building:

- Consensus building is time consuming.
- Consensus does not lead to better decisions.
- Consensus takes time, it does not save time.
- Consensus does not reduce conflict on other issues.

Organise your own resistance?

It takes a lot of dedication and energy to get a project going. To maintain constructive energy, a 'coalition' championing the project gets going, and when it feels it has a solid case, it confronts the outside world.

After all, it is an unattractive prospect to meet actual or potential opponents right from the start and negotiate with them even about the project definition. It can dissipate the energy and morale. But this reasoning only obtains if the opposition is united, determined and unchanging. In practice, stakeholders are rarely united and may well be open to a change of perspective. Waiting too late before opening the floor to stakeholders invites antagonism.

In a one-shot negotiation, or the early phase of any first-time negotiation round, the natural inclination of most stakeholders will be to present one's interests as non-negotiable, to 'draw the line' especially if there is a general feeling that the project is already decided, even when it's not.

If you start painting the situation in black and white, you leave space for shades of grey later.

But if negotiators meet early, most often, trust, confidence and mutual understanding can grow as the details and parameters of the project become clearer.

9.5   What can you influence?

As you negotiate, your mind will be focused most on a favourable *outcome*. But to stack the cards in your favour, there are more elements of negotiation you can influence.

What can you influence during negotiation? Mastenbroek (1992) identifies five aspects negotiation can try and influence:

- substance
- atmosphere
- constituency
- procedures
- power balance

The subsections below will go into each of these in some depth.

9.5.1   Framing and reframing substantive issues

If the challenge is how to distribute a pie, the options for fair division are limited. When buying a rug at the Persian market, you can haggle over the price, the seller can throw in an extra rug, but that's about all the options you have. But most problems are more complex than pies and rugs, and can be seen in a variety of ways.

Let's look at two concrete examples of dilemmas in civil engineering and planning.

Reframing

In the early years of extension studies, communication used to be seen as a message sent by a sender, received by an audience. If the audience did not get the message, there was question of 'distortion'. We now have come to understand that people filter information

---

Example 1

Geldof (1994), a trained civil engineer himself, recounts a session in a newly built estate where hurried construction led to wholly predictable groundwater problems. The water engineer called in models the problem in his computer and suggests building drains. But this is not how the neighbourhood perceives the problem. To them, the problem is one of survival as a group, to prevent their estate being demolished. This would necessitate them to move to higher-rent houses, maybe causing their closely knit community to fall to pieces. This makes damp cellars only a minor element in a tangle of poverty, rents, etc. Their voice is 'noise' to the engineer, but to stakeholders this complexity constitutes their life-world.

Example 2

Take local residents faced with a construction project of new housing and recreation facilities by the river. The residents may not be against housing itself, but the houses take away their view of the river, produce noise from construction works, unsightly stacks of building materials, bulldozers and trucks tolling on and off which may be dangerous to small children. Meanwhile, the tourists attracted by recreation facilities bring more noise and pollution, and may disregard local customs. From the perspective of these stakeholders, there is almost no gain in the project. A straight buyout or compensation strategy can go some way in appeasing the residents. Some families wanted to move elsewhere anyway, others need the cash. But you can also seek to reduce the concerns of the residents, bringing in noise-reducing nuisance from building, or bringing in new collective facilities for the area (a nature education centre, a playground, a cycle path) that everyone will benefit from.

Do the different sides to the dispute live on different planets? Of course not, but they process their information through different filters. Both rationalities lead to 'correct' interpretations of the situation, but not necessarily compatible to begin with. The two sides may ignore each other, shout at each other.

A different representation of the problem, the state of the situation and the actors, can shed a completely new light on a problem, which may not even look like a problem anymore. In the course of the bargaining process, negotiators constantly seek to influence the 'frame' of the issue, and expect the other to do the same. Whether aware of it or not, framing is a kind of tactical manipulation of information and images, but it can be manipulative in a positive (constructive) and negative (deceptive) sense, such that others cannot easily see beneath it.

To understand and to some degree predict the frames the other brings to the table, a proces of learning about the other's frames is important at the start of the negotiation. If it is really not possible to meet face-to-face (mutual learning) and tease out and compare 'frames', the project initiator can at least research the background of the other stakeholders, to understand their concerns, and assess what they can realistically offer. Conversely, a stakeholder can learn about the constraints and degrees of freedom the project manager has to work with, so as not to make unreasonable, impossible demand, but also to discover new possible linkages. As all negotiation gurus maintain, *Preparation*, finding out what you can bargain for, the resources and goals the other party has, really is everything. Learn about your counterparts to find out what is possible and what is impossible for them, what they might accept and what they will reject. This pre-assessment is not set in stone, but it gives an orientation.

A frame that is seemingly about a distribution issue, is often related with identity and belonging. A farmer, a taxi (or rickshaw) driver may hang on to their self-image even if they are only part-time farmers or taxi drivers, because it gives them clear place, role and status in their community, establish them as competent breadwinners. A project that forces them to relocate brings *uncertainty* of place, role and status, that cannot simply be bought off with a handsome handout.

Some frames may seem undebatable: for example when the issue is about 'x' or 'non-x', and one of the stakeholders proposes 'y'. The speakers of accepted, 'hegemonic' 'x-talk' may find it impossible to see the point of discussing 'y' (Haugaard, 1997). This is a frequent source of frustration between professional disciplines or (sub)cultures.

'Fans of straight batting may feel this approach to be a manipulative and dishonest'. But we are not necessarily talking about covering up injustice, but, as in everyday life, issues of cognition and tact.

Imagine a discussion between hard-core civil engineers who see the problem of linking two river banks in clear-cut terms; for them the issue is whether to build a bridge or a viaduct. Their discussion can then touch on construction designs, materials, budgets, but they may not be very open to a person who fears nuisance from building works and asks 'y' questions such as: 'How about a ferry? Can't people use their own rubber boats or swim? Why cross a river anyway? Can't we just relocate the village to the other bank?'

It may be beyond the cognitive frame of the others, or simply be inopportune and time-wasting for them, to accept the y-frame. The y-framer therefore stands to gain from framing his/her interests and proposals at least in terms of the vocabulary of the x-framers, co-opting dominant frames without subscribing to all the ideas that go with that frame. The word 'uncertainties' itself was a taboo for many Dutch engineers and politicians in the 1990s, as it seemed to make the engineers look, or feel, incompetent. But calling them 'failure factors' or 'challenges' did the trick, such that, according to one engineer, a report on flood management has to mention them to be taken seriously. A final example of this is the currently 'hot' discourse of governance. Suppose you badly need to address issues of corruption in the project consortium, but calling it thus would sound accusatory and imperil good relations between partners. A call for 'transparency' in the name of 'good governance' has proved a convenient strategy to put corruption on the agenda.

Careful reframing of issues and offers is therefore a good strategy to make progress on touchy or impasse-prone issues. Drake and Donohue (1996) argue that each move in a negotiation frames the issue in a specific way; thus, a frame is an interaction mode, which can be accepted or rejected by other participants through maintaining or altering the frame in their response respectively.

according to interpretative frames, from which people construct meanings. To a child a dolphin is a funny, cuddly animal that can perform amazing tricks, to a tourist entrepreneur an asset to attract customers, to a biologist a highly intelligent mammal, to an Inuit hunter a healthy meal and to a fisherman a competitor for fish catch. These meanings assign very different values to the dolphin. The good news is that these frames are not immutable. Through interaction, people can learn to see an issue, and their interests with respect to that issue, quite differently. This does not mean that frames are 'engineerable'. People may agree to represent an issue differently (*consent*), without agreeing that this is the whole story now (*consensus*). The pragmatic concept of 'equifinality' means that actors accept the same goals, but for very different reasons. This leads to vague, multi-interpretable, but often quite workable agreements.

People's interpretative frames are more flexible than it appeared in the first round of negotiation. Re-framing is the 'transformational process where one or more parties in a dispute actually change their frames – they develop new ways of understanding and interpreting either their own perception of the "issue", another party's view or the issue at stake' (Leach and Wallwork, 2003).

Gray and Donnellon (1990) distinguish between top-down framing, that is disputants interpreting the negotiation according to their original frames, and bottom-up framing, that is observing one another's (linguistic) behaviour and making inferences about what is the most appropriate frame for the situation. In this approach, issue framing is a dynamic process, responsive to preceding moves and anticipatory of possible reactions. This suggests strategic frames, which can be changeable. (An excellent overview of frames and types can be found online in Dewulf *et al.*, 2005).

Frames often include 'applause-generating' concepts that others can hardly be against. Who could be against participation, peace, animals, employment opportunities, the environment? In such case, opponents may feel with their backs against the wall. But opponents can analyse carefully whose interest the 'applause-generating' frame benefits, and if the frame can be reframed such that the same issue is more 'democratic', that is, opens alternative voices and avenues.

There is always a selectivity at work, that includes some aspects and excludes others. A more inclusive, 'umbrella' frame can seem to marry different frames, but runs the risk of being so general that it becomes meaningless (*depoliticised*).

By extension this pertains to the presentation of information. Parties will seek to gain and give information in such a way that it influences expected costs and benefits. Lying and deceiving can often be exposed, but selecting and silencing relevant information is less evident.

When you start negotiating, you may not at all be sure what kind of result is feasible and desirable. Therefore, an exchange of information (mutual learning) is required. But people do not generally lay all the information they have on the table, they pick and choose what best suits them. Both parties know that putting cards too close to one's chest makes negotiation impossible, but too much openness exposes one to advantage-taking. It can be expected that the other would do the same.

Tactical information serves to reduce the minimum demands on the part of the other party and to present one's own demands as realistic and inevitable. If 'softer' techniques don't elicit relevant information from the other party, 'harder' tactics like temporary breaking off negotiation or posing a deadline, bluffing are possible. This may be justified by the need to get vital information on the table. But as this is risky, one should be aware not to supplement a tough stance for a lack of preparation – always consider if you might not obtain the information in another way.

9.5.2   *Influencing atmosphere: avoiding unnecessary conflict*

Negotiations are often so much about substance that it is easily forgotten how much the process and environment can influence the success or failure of a negotiation process. You do not need to perfume the room, but stand a good chance of producing easier results by avoiding tactics that spoil the show even before it started.

A key pointer is to talk about situations and solutions rather than problems. It may sound counterintuitive, but identifying a problem at the start of the process can be counterproductive. It may be clear to you 'what the problem is' but others may take a different perspective. A problem mostly has an implicit or explicitly attribution of a cause, a finger pointed at what or who 'did it'. This makes the indicated party feel extremely uncomfortable and defensive. In complex situations, a simplifying reduction of cause and effect chains may not do justice to the situation at hand either.

The meeting's chairman, mediator or facilitator if involved has an important role to play in avoiding unnecessary antagonism. Making suggestions, asking questions and using phrases like 'What if' and 'Why don't we…' can ease the pressure, taking tension off relations as it makes it possible to explore avenues without committing to them upfront. You can also reduce tensions by making little jokes, paying compliments when the other has a good point or nice turn of phrase, respecting their line of argument even if you feel it is all wrong and stressing interdependence. In practice, this also involves – avoiding characterising one's own views as 'honest, fair and generous' which imply the opponent's views are unreasonable and dishonest. Finally, take regular breaks. Recess is not a sign of weakness! Staying up all night to broker a deal is a war of attrition that makes most people irritable.

> Say you can't accept the offer rather than say you won't accept the offer.

Clarity and openness?

As integrative bargaining brings uncertainties, people will only agree to do this if there is a clear and visible problem, interdependency and a foreseeable result that the environment will value (Leeuwis, 2000). Lack of clarity, or lack of clear boundaries in the policy network brings uncertainty and conflict.

In a new project, people come to the arena who may never have worked with each other. The 'project team' may be composed of people from different organisations, professional backgrounds and world view. It takes some time before they can work together well, before they have found common values and goals.

Therefore, it is not so surprising when they are not immediately eager to open up the planning process to other stakeholders. Prins *et al.* (2005) found that a planning group working on a river valley rehabilitation in Belgium sought to reduce rather than enhance interdependence with other stakeholders. While not desirable, this is quite understandable, as a project task group is often fragmented, seeking collective mission and identity. Therefore initiators minimise interdependence and openness with other participants. A task group tends to seek clarity and irreversible decisions. It helps them build a culture of trust and cooperation. Interactive decision-making requires a lot of flexibility to accept ideas and opposition from stakeholders. It is therefore not useful to strive too quickly for consensus (premature closure), as it can silence or kill important objectives and alternatives. Techniques like joint problem finding and iteration improve democracy but slow down effective progress form the perspective of the task group.

Planning engineers and local (private and civil society) stakeholders very much appreciate clarity. What is the budget, how many hectares will the project affect my back garden? Political decision-makers however tend to prefer – keeping options open until the last moment. This difference of focus can lead planners and stakeholders to focus on details too early (which makes it hard to deviate from them later) and politicians on details too late (so that stakeholders feel they have been 'played with').

Clarity between participants however should not be confused with transparency vis-á-vis the wider project environment. As Leeuwis (2000: 951) notes, the actors involved have to be able to communicate *freely* for a successful negotiation. If there are too many binds in saying what you want to say and what you can talk about, promising options will remain unexplored. This can mean an avoidance of too much limelight. A creative bargaining process requires a level of confidentiality that may clash with maxims of transparency and accountability. Options may be tested and rejected without a constant need to account for each avenue explored. As secrecy may raise suspicion with one's constituency and the general public, one should not forget to report whatever it is safe to report and to have good and thorough *process documentation* so that more insight can be shed after a deal has been struck. This is all the more important where public or private money is involved. It may also prevent legal action for non-transparency later on.

Some authors recommend maximum formalisation once agreement has been reached. It is true that bargaining on the last iota can slow down progress and erode momentum. Yet one reason to be careful with secret deals is that a code of confidentiality however cannot be taken for granted. To a negotiating party, openness may be their best alternative to a negotiated agreement ('BATNA'). One should always be on the lookout for opportunism: 'Forum shopping' is the strategy of a social actor trying all available avenues for negotiation with different agencies and partners to see which yields the most for them. If the agreed negotiation setting is not working out for them, they may go to the press with their story, or lobby a politician or public official. They may break the confidentiality by leaking to the press or opponents when the negotiation doesn't bring a good enough outcome for their interest.

It is preferable to keep informal communication channels open even when formal communication breaks down. This kind of informality may hardly be palatable to one's constituency but has proved very practical, most famously the 'hotline' between American and Soviet leaders in the Cold War, for example in its clash with Royal Dutch Shell over the decommissioning of the Brent Spar platform in 1995. Even action groups like Greenpeace, who present themselves as fighting alone against the powers that be and generally refuse to negotiate, have informal contacts with their adversaries, outside the limelight.

This brings us to the next area of influence: relations with one's constituency.

9.5.3   *Negotiating with your constituency*

Multiple stakes in negotiation

Seemingly incompatible goals may not only be pursued between bargaining parties, but even within the same person! Do negotiators suffer from Multi-Personality Syndrome? Rarely, but it is useful to remind oneself that the negotiation table has a mix of three agendas:

- *Personal interest*: For example fame, the gratification of 'winning', personal principles, office seeking (being the chairman or secretary), developing a professional network, etc.
- Group interest: For example a positive, constructive outcome, a nice working atmosphere, a good reputation, a good fight.
- *Constituency interest*: For example being well represented, not losing acquired rights, …

Each of these three agendas needs to be satisfied to some degree for the negotiator to be able to face the world after negotiation.

This may cause a negotiator to send contradictory signals: both confrontational and conciliatory. We have already seen that reframing can take the sting out of conflict situations and be a help in combining seemingly contradictory goals within the bargaining setting. The same goes for contradictions between the bargaining parties and their environment.

Representing a constituency does not mean agreeing with the constituency. It is attractive to try and please constituency members all the time because they are powerful, have high expectations and can withdraw their mandate. Too much independence from constituency can create distrust.

Yet it is important to secure flexibility in the mandate, otherwise it will be impossible to make deals with the partners in negotiation. You can modify your constituency's expectations by steadily informing the constituency about what is feasible. Make much of the concessions made by the other party. Mastenbroek notices that negotiators on both sides of the table often have an (im- or explicit) Gentleman's agreement:

- Don't make each other look bad in public.
- Allow the other some public 'theatre' now and then.
- Don't concede too quickly so as not to create non-realistic expectations.

Construction Stakeholder Management

9.5.4 Influencing procedures

An explorative style can promote a conducive environment for negotiation. Some elements you can consider are:

- Informal pre-meetings.
- Explorative ('open') phase first before argumentation and bargaining: wide range of alternatives with a range of actors.
- Ask lots of clarifying questions throughout.
- Agree on criteria (starting points).
- If this is impossible, start with a global 'platform proposal' on which others can build.
- If you can't reach a definitive bargain, make a provisional one.

9.5.5 Influencing power relations

Those who are obviously in the weaker position, or feel they only stand to lose, can frequently still work out a deal that will benefit them as well as the other, those in the stronger position. You can think of deals that create value (benefit) to all involved (CBI, 2000). An actor is unlikely to be weaker on all aspects: however, you can be weaker on one issue and stronger on the other. Through smart linkage and bargaining, power disparities in one issue-area can be compensated by another. In the United Nations, thus, moral leadership helped the small island state of Malta realise a convention on the seas.

Likewise a weak citizen's organisation can mobilise the press or sympathetic experts and funders.

In that sense, the power position for each party is not absolute, it depends crucially on their 'best available alternative to a negotiated agreement' (BATNA).

Whichever strategy you choose, a confrontational or collaborative use of power, Pruitt (1981) notices a certain 'path dependency' in negotiation. At several points in negotiation, a bargainer must choose among three basic strategic options:

- conceding unilaterally;
- standing firm and employing pressure;
- collaborating with the other party in search of mutually acceptable solution (Pruitt, 1981).

There will remain multiple crossroads in the course of the bargaining in which you can change strategy, but a choice at one point diminishes the likelihood of other strategies in subsequent phases.

Linkage

The above ideas on linkage between issues or actors leads us to the issue of multi-stakeholder negotiations. So-called Multi-Stakeholder Processes (MSPs) are currently popular modes of accommodating multiple interests, often at difficult scales, to focus on a specific project or policy. Take for example, a housing development project in a river flood plain. When building in the flood plain, some actors may worry about increased flood risk, encroachment on an area of scenic beauty or 'building for the rich'. Others will see opportunity for social cohesion and control in a neglected area, economic growth, environmental enhancement or architectural opportunities. Balancing environmental, planning, economic and construction concerns ('floating houses') turns out to enable previously inconceivable multi-functional land use.

While one-on-one negotiation about a single topic appears the most 'pure' and open form of negotiation, it is also the toughest kind of negotiation. It very easily has a win–lose (zero sum) outcome, plus the additional 'loss of face' for the 'loser'. It has a large conflict potential, with topics only painted in black and white. Another characteristic of 'conflict-type' negotiation is a minimum of communication and cooperation and a maximum of emotive language and grandstanding. People often resort to this mode if they are anxious – a black and white world reduces ambiguity, uncertainty and obscurity, you know what you are getting, who wins and who loses. The arena for such 'negotiation' is the court, a traditional arbitration authority or even the press, as is the Greenpeace strategy mentioned earlier. Such a process is highly moral and value-laden, quite independent of 'the facts'.

Negotiations often consider one issue at a time. The negotiation process can break down when the bandwidth is too narrow, that is: when there is too little to 'trade' (e.g. in 1977, the Dutch negotiation to form a multi-party administration broke down when only names of Ministers were at stake, and nothing else, so no package deal was possible).

Broadening the range of topics that are open for bargaining enough keeps the bargaining space open. When issues are not so black and white ('digital': one or zero) – you have more options for linkage and package deals, so you never only lose or only win. Unfortunately it is often not easily determined what the bandwidth is. Negotiations are almost always about more than one issue, thus achieving mutual gain. This allows you to give in one aspect and win on another. In the early stages, sometimes nothing seems possible, as positions ('frames') are fixed. To resolve this, one has to meet often and communicate a lot to build trust and overcome the conflict mode, so that 'reframing' the problem and range of options becomes possible.

Apart from broadening the range of topics for negotiation and widening the frame, it is also possible and often preferable to broaden the *range of actors*. One reason for this is tactical. Zartman (1993; in Faure and Rubin, 1994) relates the story of failure in negotiations between Israelis and Arabs. The Israelis insisted on bargaining face–to–face, while the Arabs wanted a third party. The face-to-face (one on one) option means it is very clear who are winners and who are losers. The 'third party' option is preferable as nobody directly looses face (you can always blame the third party).

If you increase the number of actors and broaden the range of topics, more subtle ways of engagement become possible, with more 'integrated' outcomes. So-called multi-stakeholder processes and multi-stakeholder platforms (Warner, 2007) facilitate more nuanced bargaining. This complexity increases the number of possible (combinations of) alternatives and hence, 'shades of grey'. A three-or n-way negotiation on a broader topic, gives more opportunities to make deals and to 'link'. A linkage strategy connects two seemingly unrelated topics (issue-areas). Because they are not immediately connected, chances are high that the preference structure differs among the two, you take a big interest in Issue Area A but not so much in Issue Area B. That makes it easier for you to compromise on B, so as to gain more on A.

9.6 Conclusions

This chapter has sketched two extremes of relations with stakeholders; a conflictive and cooperative style. In practice, however, fighting and learning with stakeholders are two sides of the same coin, alternating within the same project, sometimes even on the same chessboard. They constitute two forms of negotiation, *distributive* (hard) and *integrative* (soft) modes, and their combinations; as there is no fighting without learning, and no learning without a degree of fighting (struggle).

As negotiation in building projects and large-scale spatial interventions can be expected to involve multiple stages, we should look at them as 'iterated games'. The chapter has stressed the importance of taking power differentials into account. Even the powerful are ultimately interested in some levelling of the playing field to arrive at a constructive, durable result. If weaker actors themselves are not well positioned to improve their own bargaining position, *facilitation* can help. The Negowat project (Ducrot, 2007) supported peri-urban actors with little technical and political background to identify and formulate their position and help voice their concerns in a language that their technical counterparts could relate to. Edmunds and Wollenberg (2001) list opportunities to 'level the playing field', including the type of facilitation, stakeholder training in technical aspects of the project or debating techniques, choice of non-threatening, non-technical environment location and availability of interpretation service.

Of course the stronger actor can be inclined to develop and impose a 'hegemonic concept of control', a frame or agenda that presents its own interest as the common good, while co-opting some of the agenda items of his opponents. The others can (possibly with some training) however analyse who benefits from this frame and seek to change the odds.

Endnote

1. The word 'consensus' confusingly has multiple interpretations.

References

Axelrod, R. (1981) *The Evolution of Cooperation*. New York: Basic Books.
Cardenas, J. and Ostrom, E. (2004) What do people bring into the game? Experiments in the field about cooperation in the commons. *Agricultural Systems*, **82**(3):307–326.
CBI (Consensus Building Institute) and Land Use Law Center. (2000) *Conducting Conflict Assessments in the Land Use context: A Manual*. Pace University School of Law; Cambridge, MA.
Coglianese, C. (s.d.) The limits of consensus http://www.ksg.harvard.edu/prg/cary/limits.pdf.
Dewulf, A., Gray, B., Putnam, L., Aarts, N., Lewicki, R.J., Bouwen, R.and van Woerkum, C. (2005) *Disentangling approaches to framing: mapping the terrain*. Paper given at the 18th IACM Conference. http://papers.ssrn.com/sol3/papers.cfm?abstract_id=728203.
Drake, L.D. and Donohue, W.A. (1996) Communicative framing theory in conflict resolution. *Communication Research*, **23**(3):297–322.
Ducrot, R. (2007) Building capacities to tackle the infrastructural and environmental crisis in São Paolo: Role-playing games for participatory modelling. In: Butterworth, J., Ducrot, R., Faysse, N. and Janakarajan, S. (eds) *Peri-Urban Water Conflicts. Supporting Dialogue and Negotiation*, Technical paper Series No. 50, Delft: International Water and Sanitation Centre (IRC), pp. 54–84.
Edmunds, D. and Wollenberg, E. (2001) *Disadvantaged groups in multi stakeholder negotiation*. CIFOR report, www.cifor.cgiar.org-publications-pdf?files-Strategic?Negotiation?report.pdf.
Fisher, R. and Ury, W. with Houghton M. (eds) (1981) *Getting to YES: Negotiating Agreement Without Giving in*. Boston, MA: Penguin Group.
Geldof, G.D. (1994) *Adaptief waterbeheer* (Adaptive Water Management), Deventer: Tauw.
Gray, B. and Donnellon, A. (1990) *An Interactive Theory of Reframing in Negotiation* (unpublished manuscript). Pennsylvania State University, College of Business Administration.
Haugaard, M. (1997) *The Constitution of Power: A Theoretical Analysis of Power, Knowledge and Structure*. Manchester: Manchester University Press.

Haugaard, M. and Lentner, H. (2006) *Power and hegemony*. New York: Rowman and Littlefield.

Hisschemoller, M. and Hoppe, R. (1996) Coping with intractable controversies: The case for problem structuring in policy design and analysis. *The International Journal of Knowledge Transfer and Utilization*, **8**(4):40–60.

Jakobsen, P.V. (2007) Coercive diplomacy, frequently used, seldom successful. *Küngl. Krigsvetenskapsakademiens Handlingar Och Tidskrift*, **4**:29–40, available online at http://www.kkrva.se/Artiklar/074/kkrvaht_4_2007_3.pdf (Last accessed on 20 December 2008).

Kok, A. (1998) *Internationaal Onderhandelen. Problemen bij international zakendoen* (International Negotiation: Problems when conducting international business). Deventer: Kluwer.

Lax, D.A. and Sebenius, J.K. (1986) *The Manager as Negotiator*. New York: Free Press.

Leach, G. and Wallwork, J. (2003) *Enabling Effective Participation, Negotiation, Conflict Resolution and Advocacy in Participatory Research: Tools and approaches for Extension Professionals – APEN 2003 National Forum*

http://www.regional.org.au/au/apen/2003/refereed/083leachgwallworkj.htm.

Leeuwis, C. (2000) Reconceptualizing participation for sustainable rural development: Towards a negotiation approach. *Development and Change*, **31**:931–959.

Lewicki, R.J., Barry, B. and Saunders, D.M. (2007) *Essentials of Negotiation* (4th edn). Boston, Mass.: McGraw Hill.

Lukes, S. (2005) *Power: A Radical View* (2nd edn). Basingstoke: Palgrave Macmillan.

Mack, R.W. and Snyder, R.C. (1957) The analysis of social conflict: toward an overview and synthesis. *Journal of Conflict Resolution*, **1**(2):212–248.

Mastenbroek, W.F.G. (1992) *Onderhandelen* (3rd edn). Utrecht: Het Spectrum/Marka.

Poncelet, E. (1999) A Kiss Here and a Kiss There. Environmental Partnerships in Belgium. http://dlc.dlib.indiana.edu/archive/00000132/00/poncelet.pdf.

Prins, S., Craps, M. and Van Rossen, E. (2005) Managing 'psychological' boundaries in a collaborative effort: The dynamics of inclusion and exclusion in an integrated study of a river valley. In Gössling, T., Jansen, R.J.G. and Oerlemans, L.A.G. (eds) *Coalitions and Collisions*. Nijmegen: Wolf Publishers, pp. 255–264.

Pruitt, D.G. (1981) *Negotiation Behavior*. New York: Academic Press.

Rapoport, A. (1960) *Fights, Games and Debates*, Michigan: University of Michigan.

Scott, J. (2001) *Power*. Cambridge: Polity Press.

Scott, J.C. (1985) *Weapons of the Weak: Everyday Forms of Peasant Resistance*. New Haven, CT: Yale University Press.

Simon, H.A. (1947) *Administrative Behavior*. New York: Macmillan.

Walton, R.E. and McKersie, R.B. (1965) *A Behavioral Theory of Labor Negotiations*. New York: McGraw-Hill.

Warner, J.F. (ed.) (2007) *Multi-Stakeholder Platforms for Integrated Water Management*. Aldershot: Ashgate.

Woehrle, L.M. (1992) *Social Constructions of Power and Empowerment: Thoughts from Feminist Approaches to Peace Research and Peace-making*. Syracuse: Syracuse University Press.

Zartman, I.W. (1993) A skeptic's view. In Faure, G.O. and Rubin, J.Z. (eds) *Culture and Negotiation*. Newbury Park: Sage, pp. 17–21.

Webliography

http://www.cbuilding.org/: Consensus Building Institute: In partnership with Corporate Mediators to promote the use of alternative dispute resolution.

MIT: http://web.mit.edu/publicdisputes/ : Since 1983 PDP has been part of the interuniversity Program on Negotiation at Harvard Law School. PDP is also affiliated with the MIT Department of Urban Studies and Planning, the Tufts University Fletcher School of Law and Diplomacy and the not-for-profit Consensus Building Institute.

http://www.democraticdialoguenetwork.org/index.pl
http://www.negotiations.com/definition/agenda/
http://www.kellogg.northwestern.edu/drrc/index.htm: The Dispute Resolution Research Center
   (DRRC) The DRRC office is located within the Kellogg School of Management, Northwestern
   University.
http://www.negotiations.com/definition/

10 Communication in Stakeholder Management

Amir W. Al-Khafaji, Douglas R. Oberhelman,
Wayne Baum and Bernie Koch

10.1 Introduction

The survival of any organization depends on its ability to develop and maintain effective and continuing relationships with its stakeholders. In business, communication in its various forms is the primary tool used for building long-lasting relationships and partnerships. Communication is the art and science to structure and transmit information in a manner that can be easily understood.

The content of this chapter is focused on how stakeholders within an organization can communicate with each other. The most frequent or major approaches used by the bigger companies will vary from those used by smaller firms. Also, the way one would communicate with different constituents might vary. These intricacies are captured in the chapter. In this regard, case study examples are drawn from a big multi-national organization (Caterpillar Inc.), a top 100 companies (CORE Construction), and a medium firm (River City Construction). These case studies illustrate the array of communication techniques that are useful for stakeholder interaction. How different departments and personnel communicate with each other is presented through the case studies. Top-down, bottom-up, and lateral approaches to communication are covered. The case studies provide unique perspectives on how different communication patterns are employed by real companies. Their insights provide a clear set of guidelines that transcends the academic realm of communication.

10.2 The stakeholders in context

Stakeholder communication is critical to the success of every project in every organization. A well-planned communication process helps maintain good relationships between the organization and all its stakeholders. An organization has its internal and external stakeholders. In recent years, the term 'partnering' has entered the vocabulary of construction management. Partnering is simply getting along with the people you work with and getting the project done with quality, efficiency, and profitability for all parties involved. It creates a truly win/win situation for the owner, and all other stakeholders. A stakeholder is someone who is potentially impacted by the project and/or who has an impact on the project as shown below.

Stakeholders

1. Owners and sponsors
2. Customers and local community
3. Subcontractors
4. Project managers and superintendents

5. Project team members
6. Project end users
7. Architects and consultants providing services to project
8. Material and product suppliers
9. Government regulators and public agencies involved in issuing permits
10. Insurance and bonding companies
11. News media

Although stakeholders may be both organizations and people, ultimately communication is between people. Irrespective of the size of the project and the company involved, stakeholders can be grouped into the following categories:

1. internal to the project
2. internal to the company
3. external to the company

Once the stakeholders are identified and a project team is assembled, it is essential to identify the possible resistance level to the project of each stakeholder. Attention should be focused to better leverage stakeholder power and influence and to identify possible risks.

10.3   Communication networks

Communication can be classified as:

1. verbal (projected as oral, written, and/or electronic);
2. nonverbal (expression, expressive behaviors, and/or body language).

Communication flow is classified as formal where the message moves along regulated pathways otherwise it is informal. The pathways and flow patterns of messages and information can be:

- downward
- upward
- horizontal

Downward communication is where the top management of an organization gives orders to a middle manager who conveys this order to the first-level manager. *Upward communication* begins at a lower level of the organization and moves to a top level. In this case, the management in an organization receives feedback from the employee. *Horizontal communication* is between employees, between owners, managers, subcontractors, etc. In all cases, effective communication requires appropriate timing, simplicity, clarity, relevance, credibility, and style.

Communication networks are another aspect of direction and flow of communication. Such networks may affect the group's completion of the assigned task on time, the position of the leader within the group, or may affect the stakeholders level of satisfaction in occupying certain positions in the network.

The process that a given organization utilizes to communicate and interact reveals much about its group dynamics, quality of leadership and stakeholders' satisfaction. Bavelas (1950) suggested five basic communication patterns employed by organizations. These include (a) the wheel, (b) the chain–Y, (c) chain, (d) circle, and (e) all-channel, as shown in Figure 10.1.

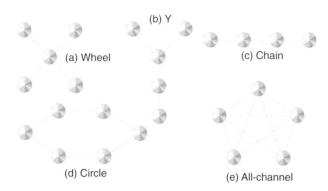

Figure 10.1   Patterns of communication.

The five communication patterns have distinctive characteristics. The *wheel* pattern is centralized around an experienced leader. This pattern is typically found in an autocratic organization and limits stakeholders' participation. Organizations with a wheel model are usually most efficient for simple tasks where problems are solved quickly. Such a model is not suitable for large organizations with complex projects.

The *Y and chain patterns* are centralized with information flowing along a predetermined channel with little interaction among members of the organization. This pattern requires a strong leader and produces a low level of satisfaction within members of the organization. The *Y and chain* models represent the hierarchical pattern of information flow as is the case in military and some business organizations.

The *circle* pattern is the most decentralized form of communication. It requires low level of leadership skills and originality. The group team is usually disorganized and performance tends to be erratic.

The *all-channel* network is analogous to the free-flow of communication in a group that encourages all of its members to become involved in making a decision. The *all-channel* pattern may be considered as an informal communication pattern and produces a high level of satisfaction amongst members.

10.4   The stakeholders map

Effective project management requires that the stakeholders involved in a project are identified and their relative power and levels of interest clearly identified. As part of this process, consider all the people who are affected by the project and have an interest in its successful completion.

The first step required in the development of an effective communication program is to produce a list of people and organizations that are affected by a project. Subsequently, one needs to identify those with power to block or advance the project. It should be clear that some may be interested in the project while others may not. The second step is to map out the stakeholders using the power/interest grid as shown in Figure 10.2.

In the power–interest map (Gardner, et al., 1986), the owner will have high power and influence over the projects and high interest in its successful completion, while the general public may have high interest, but is unlikely to have high power over the project. Once a stakeholder map is developed, one should identify stakeholders who are blockers or critics, and which stakeholders are advocates and supporters of the project.

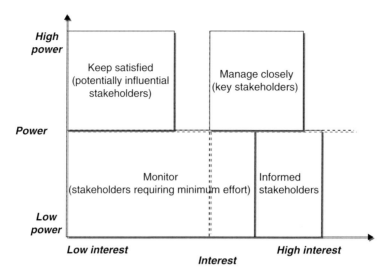

Figure 10.2   The power and interest map.

10.5   Case studies

This section provides three case studies that illustrate good practice in terms of communication between stakeholders. The application of the concepts discussed in the foregoing sections is also demonstrated in the case studies.

10.5.1   Case study 1: Caterpillar Inc., USA; annual revenue - $51 billion

Corporate structure

For more than 80 years, Caterpillar Inc. has been making progress possible and driving positive and sustainable change on every continent. With more than 100 000 employees in 50 countries, Caterpillar relies on a robust communication process to ensure employees are informed of key company news, events, and information.

  Caterpillar's Chairman and Chief Executive Officer (CEO) leads a team of group presidents who have responsibility for the company's 31 business units. Each business unit is structured as an independent organization, with a vice president who acts as the 'CEO' of his or her business. Employment in a business unit can range from several hundred to several thousand and may include both office and production employees. Often, a single business unit's employees are spread across multiple facilities and multiple countries.

Internal communications

Just as each business unit produces specific products and services, so too are its communications needs and challenges. Therefore, every business unit has a dedicated communications staff, ranging from one individual to a team of professionals. Business unit communicators are responsible for knowing their organization's unique audience, culture, job functions, facility services, and offerings. They create and publish value-added communications tied to the enterprise strategy, ensuring relevant, customized information reaches the right individuals in a timely manner.

Communications flow

Many of Caterpillar's communications to employees follow a 'top-down' approach. They are created within the company's Corporate Public Affairs department with the guidance of Caterpillar's executive office or a global process owner. Public Affairs communicators are the 'Associated Press' of the corporation, drafting and distributing corporate news to business unit communicators – the 'local reporters,' who in turn customize and distribute information to their employees. In some instances, business unit news deemed strategic and relevant to a global audience follows a 'bottom-up' approach and is distributed via corporate communications vehicles.

Communications channels

Caterpillar's many communications vehicles include traditional outlets such as newsletters, home mailings, posters and facility signage, videos and more. The company's corporate internet portal, Cat @work, is a primary source of information for employees worldwide, with sections devoted to corporate news, values and strategy updates, messages from the chairman, compensation and benefits information, and more. Each business unit has its own news 'portlet' within Cat @work to distribute organization-specific information to employees. Caterpillar also considers its annual and sustainability reports to be key employee communication tools.

The number one channel for employee communications at Caterpillar, however, is direct communications from leaders to employees. Surveys indicate this is the preferred and most trusted method for employees to receive information. For that reason, great emphasis is put on regular all-employee meetings and staff meetings. Talking points documents are a key resource for leaders, with succinct messages that can be customized for use with a supervisor's immediate team during staff meetings or between shift changes.

As the company grows, so does its communication challenges. An increasingly global workforce – more than half of Caterpillar's employees are now located outside the United States – requires communicators to be sensitive to cultural differences. Determining what materials to translate is also a challenge.

Enterprise values and strategy launch

Caterpillar's internal communications processes were put to the ultimate test in 2005 when the company launched a new enterprise strategy and updated Worldwide Code of Conduct, *Our Values in Action*. Developed by the company's Strategic Planning Committee, a small group of high-level leaders, the new strategy, and values were disseminated first to the company's top 100 leaders during the annual Strategic Review Conference in August, then to the top 600 leaders during global off-site meetings in early October, and finally to all Caterpillar employees during a 2-day global launch in late October.

Working closely with the Strategic Planning Committee and business unit communicators, a team in Corporate Public Affairs developed a comprehensive communications plan to ensure the right media and messages were in place to reach the global leader and employee audience. Public Affairs also worked with Caterpillar University to create a 'leaders as teachers' approach to communications. This method was used throughout the strategy launch as leaders at various levels developed 'teachable points' to articulate key messages as they cascaded information to their direct reports – an approach that also helped build the communications capability of leaders across the enterprise.

One critical requirement was ensuring every employee could understand the strategy and connect to the values, so launch materials were translated into the 14 most common languages spoken by employees. On launch day, every Caterpillar employee received an individual strategy manual and a Worldwide Code of Conduct book in his or her spoken language. Public Affairs also distributed a range of supporting information to business unit communicators – including facility signage, leader talking points, leader videos, learning journals for meeting participants, and more. Regular articles were also distributed in all of Caterpillar's corporate news vehicles. An emphasis was placed on consistent messaging and graphics to help ensure employees worldwide developed a common language and understanding of the strategy and values.

Accountability was key to the success of the launch. A detailed plan and timeline were distributed to leaders and communicators to ensure message and materials were launched on schedule. In addition, questions about the strategy and values were added to Caterpillar's annual employee opinion survey to track employee understanding and the correlation with employee engagement.

To keep strategy and values messages top of mind following the launch, Caterpillar restructured its strategic communications team within Corporate Public Affairs and added a team responsible solely for strategy and values communications. This team works with other corporate and business unit communicators to ensure strategy and values messages are integrated into all Caterpillar communications.

Customer communications

In 1925, Caterpillar's founders established a first-class engineering and manufacturing operation, committed to quality, innovation, and customer value. To distribute their products, they enlisted the support of independent dealers and aggressive entrepreneurs who were responsible for managing customer relationships at the local level and adding value to Cat products through service excellence.

Throughout its 80-plus-year history, Caterpillar has chosen to interact and communicate with its customers primarily through this dealer network, which today stretches across six continents, with 180 dealers employing more than 130 000 people. Each dealer is the 'face' of Caterpillar in a particular geographic region, and dealers serve as Caterpillar's 'eyes and ears' with customers. The enduring partnership between Caterpillar and its dealers is often seen as the company's number one differentiator in the marketplace.

The Caterpillar–dealer relationship
Caterpillar's global marketing organization is structured to support its worldwide dealer network. Employees in four marketing organizations – representing the Americas, Asia-Pacific and Europe, Africa, and the Middle East – are responsible for building and maintaining dealer relationships and supporting dealer sales efforts. These marketing organizations serve as the liaison between Caterpillar product managers and Cat dealers.

Each marketing organization is structured into regional and district offices that provide on-the-ground support for dealers. Most communications flow between Caterpillar and dealers through these district offices, with Caterpillar keeping the district offices aware of all marketing strategies and growth initiatives, and the district offices in turn updating dealers.

Caterpillar also provides services to help dealers support customers, maintaining a ready supply of high-value, competitively priced parts and providing service tools, training, repair processes, and technical information. Caterpillar also works closely with dealers to serve global and regional customers who demand the corporation's direct presence in the relationship.

In return, dealers keep Caterpillar informed about what is happening in the marketplace. As local experts and trusted advisors to Cat customers, dealers represent an ideal source of business intelligence.

The dealer–customer relationship

As the critical link between the company and its customers worldwide, Cat dealers understand local markets, provide custom support, and develop long-term relationships with customers. They work to understand customers' businesses and needs, when and where it is needed, to keep productivity up and costs down. A dealer's chief responsibility is to take care of Cat customers – adding value to Cat products by bundling the company's collective resources and capabilities into integrated solutions that help customers succeed.

Caterpillar's global dealer network has long been recognized as the best in the world at distributing parts and equipment. But Cat dealers do much more than sell products. They also offer rental and used equipment, information products, business planning services, financing, insurance, and more through dealer main stores, branch stores, Cat Rental Stores, and the web-based Dealer StoreFront and PartStore.

While dealers rely on traditional and new-media advertising techniques to reach customers, the business is primarily relationship-driven. Face-to-face meetings, customer events, and site visits are the key methods through which dealers communicate with customers. The use of the Internet to purchase Caterpillar equipment, particularly for smaller customers, is growing – providing new challenges and opportunities for Caterpillar and dealers.

Customer events and visits

Caterpillar and its dealers also work in partnership to host customers at a variety of locations and events, showcasing the value of Caterpillar and dealer solutions.

- *Factory tours*
  Caterpillar's Corporate Customer Services group provides customers worldwide with in-depth tours of various Cat facilities, reaching approximately 12 000 dealers and customers each year. The goal is to show first-hand why Caterpillar is a world leader and allow customers to interact with the employees who are producing their products.
- *Demonstration centers*
  Several Caterpillar Demonstration and Learning Centers worldwide are designed to demonstrate the value of Caterpillar products and services through the operation and application of our equipment. Dealers bring customer groups to these facilities to operate equipment in a real field setting, then step into a high-tech classroom or meeting room for additional demonstrations or discussions about the company's people and products.
- *Trade shows*
  Trade shows provide an opportunity for Caterpillar to showcase its products, services, and people directly to customers in various industries. Through the use of interactive displays, audio-visual media, and personal contact with industry experts, Caterpillar develops an experience for visitors to reinforce its industry leadership position, solidify the Cat brand, and showcase the strength of its dealer network.
  Caterpillar participates in a variety of trade shows each year. One of the largest, which occurs every 3 years, is CONEXPO-CON/AGG. This show is a major event for Caterpillar with more than 125 000 customers and potential customers attending. Caterpillar's exhibit in 2008 featured four displays, more than 60 000 square feet of exhibit space and over 400 employee show workers. The displays and messaging at this

event delivered a fully integrated marketing appeal to customers regarding the products, services, solutions, technologies, and value offered by Caterpillar and its dealers.

External communications

The reality of business is that every company must contend with a number of external factors, and successful organizations generally make the choice to embrace and participate in select issues that have the biggest impact on their business. To do so, Caterpillar and its leaders stay engaged with key industry, government, non-governmental, and community organizations.

Industry leadership

As a leader in a variety of industries, Caterpillar actively seeks opportunities to develop a network of business executives, policymakers, and scholars to develop innovative solutions to key business issues. In 2004 and 2006, for example, Caterpillar leaders participated in the International Construction Innovations Conference (ICIC) sponsored by Bradley University. This regular conference brings together global leaders in education, industry, and government to discuss issues facing the construction industry. A small group of Caterpillar industry experts and customers also participates in a full-day, roundtable ideation session – a creative brainstorming process in which industry stakeholders, policymakers, and scholars discuss future trends and current challenges in the industry. The following represent just a few additional examples of Caterpillar's industry involvement:

- Caterpillar is actively involved in developing International Standards Organization (ISO) criteria and chairs the committee that develops global standards for earthmoving equipment, including regulations for visibility, rollover protection structures, and braking.
- Since 1993, Caterpillar has worked closely with the Tropical Forest Foundation to promote sustainable forest management, demonstrating how reduced-impact logging techniques and technology can increase harvesting efficiency and provide environmental benefits.
- In 2002, Caterpillar was a major sponsor of the Global Mining Initiative's 'Resourcing the Future' conference, where the industry discussed key issues raised by the Mining, Minerals and Sustainable Development project. At MINExpo 2004, Caterpillar hosted 'Today's Partnerships, Tomorrow's Practices,' drawing hundreds of industry representatives to hear from experts and discuss best practices.
- In 2006, Caterpillar signed a letter of intent with China's National Development and Reform Commission (NDRC) to help China as it establishes policies related to growing its remanufacturing industry. Caterpillar is providing expertise to help the NDRC and Chinese research institutions pursue China's 4R initiative: reduce, reuse, recycle, and remanufacture.
- Caterpillar is a founding member of the United States Climate Action Partnership (USCAP), which is pressing for the establishment of a 'cap and trade' system that will put a price on carbon emissions and drive significant reductions in an economically efficient manner. The company also holds a leadership role in the Council on Competitiveness's Energy Security, Innovation and Sustainability Initiative and is an active participant in the UK-based Energy Technologies Institute, the Asia-Pacific Partnership on Clean Development, and Climate and other regional, national, and international organizations.

Policy advocacy

There is no product or service Caterpillar provides that is not impacted by federal and state legislation. Policy affects every part of the Caterpillar organization – facilities, products, and even employee benefits. Caterpillar's governmental affairs team works with public officials to help ensure legislation does not negatively impact the company's operations.

- Issue advocacy
  Issue advocacy is another name for lobbying – meeting with elected officials and their staffs to share Caterpillar's concerns on an issue. Cat lobbyists try to persuade contacts that the company's position on an issue is the correct one and urge them to support that point of view.
- Association memberships
  An association is a permanent, formalized group that represents a broad interest. An example is the National Association of Manufacturers (NAM), which represents the interests of US manufacturers. Caterpillar has representatives on several association boards to ensure the company's voice is heard on all fronts. Caterpillar executives are active in the US Chamber of Commerce, the Manufacturers Alliance, the US–China Business Council, the Business Roundtable, the Business Council, and the Council on Competitiveness, among others.
- Coalitions
  Unlike an association, a coalition is not permanent. It is a group organized to address a specific topic, such as the National Alliance Against Blacklisting. The coalition lasts for the duration of the issue, and when it is over, the coalition dissolves. When appropriate, Caterpillar participates in or leads coalitions related to topics important to the business.
- Grassroots coordination
  Grassroots activities range from members of Caterpillar's political action committee writing letters to legislators to the company's CEO writing a letter to the President. In 2007, Caterpillar used grassroots tactics to promote support for House passage of the US–Peru Free Trade Agreement.

Corporate Speakers Bureau

To encourage leaders across the company to speak out on policy issues critical to its business, Caterpillar has established a Corporate Speakers Bureau. Leaders can access speeches, presentations, and talking points on globalization and trade, sustainable development, climate change, and global workforce issues, via the web. Leaders are encouraged to customize these materials and seek out internal and external opportunities to educate stakeholders on issues critical to Caterpillar's success as a worldwide leader.

10.5.2 *Case study 2: Core Construction, USA; annual revenue > $600 million*

Company overview

The CORE Construction Group (CCG) companies have a history dating back 71 years when Otto Baum, a 26-year-old German immigrant bricklayer, began building masonry barn foundations in central Illinois. His tools, a mortar box, mortar hoe, shovel, and a two wheel trailer towed by a very old Chevrolet, along with his strong work ethic, core values, and a strong Christian faith, provided a solid foundation for what is now the CCG. The CCG maintains successful operations in five states with an annual volume of over $600 000 000. The separate divisions include CORE/IL, CORE/AZ, CORE/FL, CORE/NV, and CORE/TX. The founder's vision is still evident in all of the CORE divisions. Communicating the founder's principles to those within the organization, and demonstrating their value to their clients, subcontractors, suppliers, and to external stakeholders, has allowed the CCG to grow into national prominence.

The CCG is a multifaceted company providing general contracting, design-build, and construction management services to the construction industry ranked number 94 in the Engineering News Record's top 100 contractors in the United States. The meaning behind the name CORE is very important to all of their employees as well as the clients they serve. CORE represents the core values of its founder, Otto Baum, and expresses the culture and foundation of the entire organization. These core values are honesty, integrity, fairness, and nurturing personal growth.

Management style and information flow

At CORE Construction, it is essential to the management of the organization that its, mission, vision, culture, and goals are enumerated, understood and accepted, and communicated to everyone involved in the company. This requires an orientation and explanation of how the company operates when hiring new employees. This orientation needs to describe all facets of the operation, including the communication among those who are stakeholders in the organization. This orientation is both job specific and general, so members of the organization understand not only what is their job expectation, but how they interact within the company. They understand that they are representing the company to clients, subcontractors, suppliers, and the outside community. CORE Construction has an open 'top-down and bottom-up' communication program. Ideas and concerns can be shared with top management without going through intermediary individuals.

Specific methods used by CCG including, but not limited to, estimating, contract writing, purchasing, project management, scheduling, field operations, and safety are utilized as standard practices. CCG provides extensive training in these specific areas, both internally, and through the use of outside consultants. General operational policies, compensation, benefits, incentives, and strategic planning goals are also communicated throughout the organization.

The CCG established a constitution under the federalist system somewhat mirroring the US's Federal and State governmental system. The CCG operates as the 'federal' government providing certain services to all companies, such as bonding, bank credit lines, all insurance including business and employees' medical, fleet, and equipment purchases, data processing, and internal accounting oversight. Policies and services are provided to the individual companies by the CCG. This is determined by a Board of Directors which consists of the President of each division and a Chairman of the Board. An assessment (similar to a tax) is charged by the CCG to each of the divisions to pay for the cost of these provided services which is budgeted at the beginning of each fiscal year and approved by the CCG board. The CCG operates with a fulltime staff consisting of a president, secretary, Certified Public Accountant (CPA), and risk manager. The individual divisions operate as the 'states,' providing their own individual management of all the primary construction service operations required by their local markets and geographical locations. Local area business development, estimating, project procurement and purchasing, project management, scheduling, field operations, job cost control, and invoicing are all responsibilities of the individual divisions. The individual divisions are monitored quarterly by a Decision Making Board (DMB) consisting of the President and Director of Operations of the specific division and three individual shareholders employed by one of the other divisions. There is a standard reporting format of all phases of the individual divisions which is the agenda for the quarterly DMB meetings. Each of the individual divisions has a separate DMB.

Delivery methods

CORE uses three primary construction delivery methods to obtain the largest percentage of their work. These methods are competitive lump sum bid, Construction Management (at Risk), and Design Build. Business development is both provided by the individual divisions and through the use of national advertising provided by CCG. The CCG companies place a major emphasis on client relationships, and, therefore, negotiate approximately 35–40% of their work. For example, CORE/IL has, over the past 20 years, constructed seven major buildings for a single college campus. Repeat clients are a testimony to the core value culture of the CORE companies of fairness, honesty, and integrity when dealing with their clients. Preconstruction services, preliminary budget estimates, and value engineering are provided on many CCG projects to assist the owner and Architect/Engineer (AE) in using efficient construction methods to reduce the project costs and control the budget. Maintaining good subcontractor relationships is essential to the CCG companies. In order for the CCG companies to obtain the lowest cost for the client, they must receive all of the lowest qualified subcontractor bids on a project. It is CORE's policy and responsibility to qualify subcontractors for their ability to adequately maintain the workforce and have the financial viability required to maintain a project's schedule.

CCG encourages the individual CORE divisions to be involved in their local communities through their involvement in local governments, charitable organizations, and to lead by example, that everyone has either time, talent, or financial resources to provide support to these organizations. CORE/IL has established a foundation for their corporate giving which is administered by their internal foundation board. Other CORE divisions have established 'CORE CARES' to assist community organizations.

10.5.3   Case study 3: River City Construction, USA; annual revenue > $178 million

Introduction

Over its first 23 years, River City Construction has grown continuously, radiating outward as it gains relationships with new customers, subcontractors, and suppliers. Their management and administrative group now numbers over one hundred with an additional 400 field personnel. Over the past 5 years, River City Construction has averaged $178 million in annual revenues and plans to grow to an average of $225 million over the next 5 years. The vision of leadership has grown beyond Central Illinois and Illinois into the surrounding states. Services have grown to include design-build and turnkey projects. River City Construction has also expanded by opening three branch offices in Benton, IL; Huntley, IL; and Ashland, MO.

Communication tools

The ability to communicate as an integrated project team is becoming increasingly critical on construction projects. As the construction industry demands that projects be completed in abbreviated periods of time, project teams face the complex task of disseminating information to all project stakeholders in the most efficient and effective (timely) manner possible.

The basic and most commonly used communication vehicles run the gamut from the telephone and fax to email and instant messaging to the use of portal sites. These modes of communication can make most project information readily available to both internal and external project stakeholders. Yet, even with these communication tools, the most often overlooked impediment to effective communication is not deciding the mechanism to share information among project participants but more importantly, communicating the information in a manner that tells people what they need to know, how it will impact them and allows for appropriate feedback. Still, communication is often one-sided and feedback is typically limited or nonexistent.

Partnering workshop

It is for that very reason that River City Construction begins all major projects with a Partnering Workshop and holds quarterly follow-up partnering sessions to ensure that the lines and means for appropriate communication that were established at the original workshop are working effectively. Partnering, in its simplest form, is a semi-formal structure that establishes working relationships, on an equal basis, through a mutually developed strategy of communication, coordination, and cooperation.

Partnering is about a common sense approach to building a project. It is getting along with the people you work with and getting the project done with quality, efficiency, and profitability for all parties involved. It creates a truly win-win situation for the owner, architect, engineer, contractors, and all participating entities. The key elements in partnering include:

1. efficient and effective communication
2. commitment
3. equity
4. trust
5. development of mutual goals
6. continuous evaluation

The River City Construction philosophy of partnering creates ownership for everyone involved in each aspect of the project. It is a simple idea, but an incredibly powerful tool. It allows every individual a chance to be heard. River City Construction believes that the partnering process creates opportunities for mutual trust, coordination of effort, and expedited problem solving through open communication. Working as a focused unit, each team and every team member has an incentive to deal fairly and resolve problems quickly and equitably, as well as develop a deeper respect for one another.

River City recognizes that inadequate communication on projects, particularly as it relates to the expectations of the project team, contributes to frustration and a resistance by some team members to participate openly and honestly in discussing issues critical to the success of the project. Some team members may believe that it is beneficial to keep difficult issues from the rest of the team for fear of the consequences. This, however, is absolutely the wrong approach and invariably delays the ability of the team to work toward a consensus solution and address those issues in a timely manner. Information that is critical to the work of others and to the overall success of the project must be discussed in an open and honest forum. Thus, it is imperative that each project team member feels comfortable acknowledging and discussing any issues that would have an impact on the project.

Through the partnering process, River City is able to address those impediments to appropriate communication. At River City, the common expectations of the project team are first identified, then the concerns of the team are examined and lastly impediments to the communication process are explored. Below are examples of this process used in a recently partnered project.

Expectations

The regular expectations of the stakeholders are highlighted below.

Owner/user group
Meet expectations

- Deliver project on time and budget
- Conflict resolution up front with quick resolution
- Minimal change orders

Exceed expectations

- Deliver project ahead of projected date with quality construction
- Communicate and communication
- Minimal change orders

River City Construction
Meet expectations

- Complete on schedule
- Under budget
- Timely response from all parties
- Safety – zero lost time
- Open/honest communication
- Participation in weekly meetings
- Profitable project for all

Exceed expectations

- Ahead of schedule
- Final completion – 30 days from substantial
- Zero recordable incidents
- Same team used for future projects
- Quick execution of project paperwork – Request For Information (RFIs), submittals
- Fun – enjoyable project for all

Subcontractors/suppliers
Meet expectations

- Finish on time
- Work as a team (respect)
- Stay within budget
- Build relationships – long term
- Safety participation
- Quality end product
- Provide good leadership – set example
- Communicate
- Coordination
- Sufficient manpower
- No surprises

Exceed expectations

- Finish on or ahead of time
- Stay under budget

Table 10.1 Decision making, communication and issue resolution (POC, point of contact for communication.)

Organization	Level 1 Day-to-day	Level 2 Time or money	Level 3 Buck stops here
Owner/user group	Rick (POC) to owner and user	Rick	Board of managers
Architect	John & Jerry to consultants, John (POC)	John & Jerry	Willie S
River City Construction	Kent to subcontractor/supplier	Cody (POC)	Bernie

- Owners' expectations exceeded
- Zero lost time incidents or accidents
- No punchlist
- Satisfied customer
- All are profitable
- Paid on time

Greatest concerns of the project team

- Schedule-liquidated damages – 3
- Poor communication – 3
- Weather – 3
- Gaps in coordination – 3
- Site congestion – 3
- Permit and regulating – 3
- Decision-making – 3
- Cost – 2
- Submittals – 2
- Cooperation – 2
- Safety – 1
- Underground conditions? – 1
- Closeout – 1

Given this array of stakeholders, it is sometimes worthwhile to adopt a structured approach to communication where a group is involved. For instance, the group could be reached via a named individual who will serve as the channel of communication. Table 10.1 illustrates how named individuals have represented groups within the context of this case study. Using these representatives has been quite effective and has minimized the aspect of noise in communication.

The communication processes utilized by the project team during the course of a project can dramatically influence, either positively or negatively, the degree of success achieved in that project. The challenge is to communicate in a manner which respects the views of others, while at the same time, provides an opportunity for constructive discussion and feedback.

10.6   Conclusion

This chapter has discussed the concept and relevance of communication. Whether an organization is big, medium, or small in size, communication is useful to it. Without communication, many things will not be done. Particularly, an organization cannot engage its stakeholders effectively without communication. The case studies in this chapter illustrate the diversity of stakeholders and their different expectations. It is via dialogue with stakeholders that their respective needs can be met. It is also via good communication that differing stakeholder requirements can be addressed. Communication is certainly not the only solution to meeting stakeholders' requirements but it certainly plays a major role. The significance of communication in stakeholder management should thus not be undermined.

As there are different stakeholders, the means to communicate with them will vary. Even with a particular stakeholder, the circumstances on hand should dictate the means of communication and engagement. It is thus worthwhile to be conversant with how to communicate in diverse ways and to use this circumspectly.

References

Bavelas, A. (1950) Communication patterns in task-oriented groups. *The Journal of the Acoustical Society of America*, **22**(August): 725–730.

Leavitt, H. (1958) *Managerial Psychology*. Chicago: University of Chicago Press.

Gardner, J.R., Rachlin, R. and Sweeny, H.W.A. (1986) *Handbook of Strategic Planning*. New York: John Wiley & Sons Inc.

11 Culture and Leadership in Stakeholder Management

Abbas Ali Elmualim

11.1 Introduction

The UK construction industry is widely viewed as fragmented and diverse due to its project-based nature. According to Wild (2002), 80% of construction projects involve one-off clients and are non-recurrent. There is mounting pressure for construction to collaborate. The discourse of change towards collaboration espouse improved contracts, communications and management (Egan, 1998). The recommended change, however, remains aspirations of the policy makers (Wild, 2002). The inception, design, construction and operation of any construction project require the participation of various agents according to their professional knowledge, experience and their required input into the specific project. By nature of the diverse number of participants in construction, the delivery of projects requires collaboration by these participants. However, each constituency of participant or participants has its own agenda and interests. Each constituency of participants will mobilise its resources, knowledge and practices, as part of the project, to meet its interests. These interests are mainly financial to achieve a competitive edge over their competitors (Elmualim *et al.*, 2006). Hence, there is a stringent need for collaboration between various organisations from architectural practices, consulting engineers, general contactors, specialised sub-contractors, to manufacturers and material suppliers as well as management firms in the delivery of construction projects. This intra-organisation collaboration is mirrored by an inter-organisational practice. It is argued that individual and organisational behaviour is highly influenced by national culture (Hofstede, 2001, 2003). The conflict arises when an individual, an organisation or a group of individuals or organisations meet their own interest and ignore the interests of others. The dichotomy is that many firms and individuals while engaging in collaborative practices are seeking competition to advance their interests through competitive practices (Elmualim *et al.*, 2006). Such practices of collaboration and competition at both levels of intra- or inter-organisations have caused the emergence of a culture of adversarial attitudes (Egan, 1998) that are pertinent within construction and decried by all (Green and May, 2003). The most compelling and pressing question is why organisations and individuals develop such paradoxical practices of competition and collaboration? And what are the implications of such behavioural traits on stakeholder management in construction? No doubt the role of culture and leadership to further advance mutually collaborative projects through stakeholder management needs no elaboration. To attempt to address these paramount questions requires adopting a multidisciplinary approach, particularly a psycho-socio structural approach that seek to understand culture and leadership and advance mutual relationships in construction through stakeholder management.

11.2   The role of culture and leadership in stakeholder management

The introduction of stakeholder management as a concept in construction attempts to redefine construction organisations by understanding how individuals or groups can affect or be affected by the realisations of construction firms' objectives. According to Friedman and Miles (2006), an organisation should be viewed as a grouping of stakeholders. The role of an organisation therefore is managing the interest, needs and viewpoints of all stakeholders. The conventional strategic view of stakeholders focuses solely on their management as part of the process of pursuing profits for the shareholders. Recently synthesised and integrated views for stakeholder management are emerging such as the multifiduciary approach where stakeholders are viewed as more than mere parties to wield economic or legal power (Carroll and Buchholtz, 2006). The multifiduciary approach states that an organisation has fiduciary (trust and duty) to stakeholders as well as shareholders (Carroll and Buchholtz, 2006).

Friedman and Miles (2006) investigated and synthesised many definitions of the concept in strategic management particularly building on the work by Freeman (1984). The concept was further elaborated on by Starik (1994) by adding that stakeholders are individuals or groups with which organisations interact and who have a stake or a vested interest in the organisation (Starik, 1994). Furthermore, the concept included the principles of corporate legitimacy and the stakeholder fiduciary principle (Evan and Freeman, 1993, Friedman and Miles, 2006). The principle of corporate legitimacy aims to safeguard the rights of groups and those groups have to be participating in decision-making processes that substantially affect their vested interests. The corporation should be managed in a way that benefits its stakeholders. The fiduciary principle views the organisation as an abstract entity and its managers/leaders being entrusted with acting in the best interest of stakeholders and the organisation. The fiduciary principles require honesty, adequate care, transparency and trust in avoiding personal gain or harming the interests of the stakeholders (Friedman and Miles, 2006). According to Friedman and Miles (2006), stakeholders may include all or some the following: shareholders, customers, suppliers and distributors, employers, local communities, stakeholders representatives, NGOs, competitors, Governments/policy makers, financier, media, public in general, non-human aspects such as the earth and natural environment, business partners, academics, future generations, past generations, archetype or memes.

Much of the literature on stakeholder management lies within the strategic management arena. In strategic management, the stakeholder theory and practice is about segmentations of stakeholders to enable organisation to investigate various ways of dealing with each stakeholder segment or group (Friedman and Miles, 2006). Friedman and Miles (2006) articulated that the literature treated managers (or leaders as the focus of this chapter on managers as leaders) in many ways. Managers/leaders are regarded as part of the stakeholders with access to organisation resources. Managers/leaders are viewed in the literature as referees or the go between investors and employees.

It is argued that there will be different approaches to stakeholder management depending on the cultures of organisations, industry or society as large. Most of the literature is solely focused on the Anglo-American profit-seeking organisations. Other human cultures might have less emphasis on profit returns. Indeed organisations, industry and human cultures vary enormously (Elmualim, 2007a). It is evident that the style of managers/leaders and aspect of cultures whether at organisation, industry or society unit of analysis will have a profound implication on the practice of stakeholder management. The current terrain of construction – in the western world with the adversarial and conflictual attitudes as the norm – will hinder the application of such a concept.

There are various norms at organisational or societal level. The argument is that in the circumstance where there are conflicting norms, the hyper-norm should be adhered to (Friedman and Miles, 2006). Managers and leaders of organisations are considered to be the central group steering the role of stakeholder management (Friedman and Miles, 2006).

It is vital to readdress and redefine not only the role of organisations in the construction industry but also the roles of culture and manager/leaders that best harness and exploit the practice of stakeholder management.

11.3   The reality of construction terrain

11.3.1   Fragmentation of the UK construction industry

According to the figures from the Department of Trade and Industry (DTI), the UK construction sector employs more than 1.6 million people with a third registered as self-employed (DTI, 2003). The UK construction is dominated by small and medium enterprises and with output of more than £83.5 billion. The sector is highly fragmented with low levels of workload continuity, little interdependence and communication and lack of trust. According to Egan (1998), this sector's fragmentation led to the extensive use of subcontracting and prevented the continuity of efficient and effective team work. However, this fragmentation assisted organisations in having greater flexibility to deal with highly varied workloads. Eventually, the sector's overriding practice is characterised by adversarial relationship, low cost, short-term profits and opportunistic behaviour (Green *et al.*, 2004). Best practice initiatives such as teamwork, collaborative work and the call for higher level of trust are failing to bring about dramatic changes that are sought (Egan, 1998).

Large construction firms in the UK are typically seen as hollowed-out organisations in that very few of them actually carry out the work themselves. Practically all their work is carried out by subcontractors or by subcontract labour. The large firms are thus more involved in managing the processes on projects rather than doing the work themselves. This has implications in that if collaborative models are to be formed to promote knowledge transfer at all levels closer to the workface and users, then this will probably have to take place with people from different organisations. Secondly, it means that large contracting firms have limited influence on promoting learning at the levels that work actually gets done. However, they have a great deal of influence in promoting learning among 'white collar' workers. A significant factor explaining the negative examples of inter-organisation communication is the relatively low level of trust that exists between these organisations and individuals representing them. The importance of trust in intra- or inter-organisations is increasingly being recognised, with a lack of trust likely to inhibit the extent to which people are willing to share knowledge and experiences with each other (Hislop, 2005). Trusting relations are based on, and develop from, an expectation of reciprocity. However, the nature of intra- and inter-organisation social relations, where the consensual knowledge base is limited, or where people have a limited sense of shared identity, makes the existence of trust less likely and the development of trust more difficult. Coupled with this, construction projects run to very tight timescales and thus coming together to meet, for the purpose of sharing knowledge and developing mutual interests, is not seen as a priority to a number of project staff. Getting staff to collaborate online also poses its own difficulties in that, on many projects, the IT infrastructure which houses collaboration software is not always available (Egbu, 2004).

The industrial recipe of construction has been defined as the overriding culture that typifies a particular industry. Culture has been defined as the deeper level of basic assumptions and beliefs that are shared by members of an organisation, that operate unconsciously and defined in a basic 'taken for granted' fashion, as an 'organisation's view of itself and its environment. The industry recipe is thought to be particularly strong in the construction industry, the flavour of which is captured in the perceived view of the traditional design–tender–build method of procurement, in which design is separated from construction and the system characterised by fragmentation, friction and mistrust. Newer forms of procurement, such as construction management and partnering, are being used although slowly as vehicles to change the industry recipe (Fernie *et al.*, 2004; Green *et al.*, 2004).

11.3.2 *Dichotomy of competition and collaboration in construction*

The need for a greater degree of competition was undoubtedly one of the guiding threads of public policy during the 1980s and 1990s. During this era, governments repeatedly looked at competition as the means of achieving efficiency.

Much of the current discourse of competitiveness is derived from Michael Porter's typology of competitive advantage (Porter, 1980, 1985, 1990a, b, 1998). While Porter's work initially focused on guiding firms whether or not to enter a particular industry, it has since become more commonly used to analyse relative market power as a means of informing strategic action. The analysis is often known as 'Five Forces', referring to the factors used. Porter proposes three generic strategies:

1. cost leadership
2. differentiation
3. focus.

Cost leadership strategies are deeply embedded within construction, largely because firms frequently struggle to differentiate their products. More recently, there is evidence of some companies adopting a 'focus' strategy by concentrating on particular market niches with a view to providing excellent services to a limited clientele. However, the assumption that such strategies are mutually exclusive has been repeatedly challenged in the literature. Furthermore, Porter notably offers little advice on *how* organisational resources are to be aligned in the effective implementation of the adopted strategy (O'Shaughnessy, 1996; Flint, 2000; Klein, 2002; Connor, 2003).

Within the strategic management literature, the term competitive advantage remains ambiguous and is frequently criticised for being divorced from context. Competitive advantage is seemingly a relative term that defies universal definition. Whilst the discourse of competitive advantage remains attractive, there is little agreement on what it means in operational terms (Klein, 2002; Connor, 2003).

The broader community of strategic management researchers are highly critical of Porter's perceived lack of precision (O'Shaughnessy, 1996; Flint, 2000; Klein, 2002; Connor, 2003). Flint (2000) is especially dismissive: 'The terminology used in the field of strategic management that might possibly garner the prize for the most overworked and least understood catchphrase is "competitive advantage". The extension of that phrase into "sustainable competitive advantage" is currently an elaboration of ambiguity'.

The continued ambiguity of the discourse of competitive advantage has profound implications on the adopted managerial practices (Thomas, 2003), particularly, in the construction industry. The issue of which narrative of competitiveness is mobilised depends in part on the vested interests of different institutionalised groups. Managerial practice in construction can be categorised into identifiable constituencies (assembly of people, materials and practices that are governed by power, social relations/interactions, material practices, institutions and rituals, beliefs and values and discourse) (Thomas, 2003) and sub-constituencies of interests as ascribed by industry participants. The identification of different constituencies together with the need to contrast the 'competitiveness discourse' as mobilised by: (i) construction academics, (ii) construction consultants, (iii) construction practicing managers, (iv) the policy makers and (v) clients.

The terms competitiveness and competitive advantage are widely used within the strategic management field and also in the UK construction. In its broadest sense, the term competitive advantage means nothing more than success or superior financial performance (Klein, 2002). Within the strategic management literature, generally or in the context of construction, it is evident that the term competitive advantage is very ambiguous and becoming divorced from the context within which one operates (O'Shaughnessy, 1996; Flint, 2000; Klein, 2002; Connor, 2003; Thomas, 2003).

However, more recently sub-discourses have been mobilised to direct attention towards the damaging side-effects of destructive competition and the inefficiencies of adversarial relationships (Egan, 1998; Blockley and Godfrey, 2000). Perhaps the so-called adversarial attitudes of the UK construction sector are a direct product of the mantra of competitiveness and the broader ideology of the 'enterprise culture'. Certainly the holy grail of labour market flexibility and the casualisation of the workforce has exacerbated the UK construction problems (Green and May, 2003). Within the domain of public procurement, emphasis has switched away from competition towards partnership. Current notions of 'partnering' and 'collaborative working' are indicative of this same shift in discourse (Latham, 1994; Egan, 1998).

The supposed dichotomy between competition and collaboration/partnership however continues to attract much debate within both the public and private sectors (DTI, 1998; Egan, 1998; Hartfield, 1999; Fernie *et al.*, 2004; Green *et al.*, 2004; Karlberg, 2004). The distinction between competition and collaboration is something of a red herring. Firms seek to engage in collaboration as a means of securing *competitive advantage*; that is, they seek collaboration as a means of securing long-term returns. Clients are likewise attracted to collaborative relationships on the basis of the alleged benefits of innovation and continuous improvement. Paradoxically, effective collaboration can therefore make construction firms and clients more competitive. Similar arguments apply throughout the supply chain, although there is frequently a significant inertia to be overcome before firms can switch to relational contracting. It must also be recognised that firms frequently have to operate collaboratively and competitively at the same time, thereby eroding any prospects of an associated 'culture change'. A further complication is introduced by the fact that some firms espouse the rhetoric of collaborative working whilst continuing to behave opportunistically.

11.4 Mutual and adversarial relationships

In today's western-liberal culture, competitiveness and conflictual relationships are ubiquitous cultural practices that are hegemonic in nature serving the interests of a few. Such competitive and conflictual relationships are becoming the norms that are engraved in a conventional wisdom such that conflictual relationships are inevitable expression of human nature (Karlberg, 2004). According to Karlberg (2004), 'throughout the contemporary public sphere, competitive and conflictual practices have become institutionalised norms. Indeed, contest models of social organisations and protest models of social change have become so ubiquitous that they tend to appear normal, natural and inevitable to those raised in western-liberal cultures'. It is further argued that the norm makes it difficult to envisage alternatives (Fellman, 1998).

Advancing collaboration through mutualistic relationships requires understanding of power sharing and distribution amongst individuals, across organisations and society at large. Indeed 'in the long history of mankind those who learned to collaborate and improvise most effectively have prevailed' (Darwin, 1859). Our existence as human beings will depend greatly on mankind to act together, though we might be thinking differently (Hofstede, 2001).

Archaeology and anthropology suggest that evidences of both modes of behaviour are as two strands woven in the human culture. The paramount questions are: is the culture of contest and protest a reflection of human nature? Are humans having the potential for either adversarial or mutualistic behaviours? What are the implications? Which culture promotes social justice and ecological sustainability?

11.4.1   Adversarialism

Adversarialism is the pursuit of mutually exclusive interests by individuals or groups working against one another. Adversarialism appears in the form of contest, competition and confrontational relationships. According to Karlberg (2004), 'though not all of adversary practices and forms of expressions may not be problematic, these forms of cultural codes are maladaptive, maladjusted and maladroit'. Adversarialism creates a division between winners and losers.

11.4.2   Mutualism

On the other hand, mutualism is the pursuit of mutually inclusive gains by individuals or groups working with one another. Mutualism enhances the chances of all members of the group to have the benefit of being winners, though the sense of winning might not be of a tangible form (Karlberg, 2004). Many schools of thought such as feminism, systems theory, ecology and environmentalism, communication theory and alternative dispute resolutions advocate the vitality of mutualistic relationships to advance human progress. However, mutualism is difficult to evaluate and to a greater extent is being marginalised because of the norms and therefore appear naive and idealistic (Karlberg, 2004). To fully appreciate the benefit of the application of mutualistic relationship in real cultural communities, there will be a stringent need to have all embracing vision established through the understanding of structures, practices, institutional structures and strategies and collective decision-making processes.

11.5   A Tripartite axiological system

Construction organisations are shaped by divergent objectives and the specific interests of different individuals, groups or departments; each one with their own power base and discourse to legitimise their interest to the wider organisation (Hardy *et al.*, 2000). This specifically identifies the fragmentation of the construction industry as a whole and organisations operating in the construction market. Hence, the construction organisations can be described as pluralistic organisations that are characterised by multiple objectives, diffused power and knowledge-based work processes (Denis *et al.*, 2007). According to Denis *et al.* (2007), pluralistic organisations throw three types of problems for those who wanted to promote concerted organisational actions:

- individual autonomy;
- inflationary consensus;
- layering of structures.

Individual autonomy is associated with diffuse power and emphasis on knowledge-based work. Pluralistic organisations endeavour to provide a broad scope for individual action, encouraging local development and flexibility. This can become a barrier to the integration and coordination of the organisation as individuals have the freedom to dissociate from the common objectives of the organisations. The second problem is the 'inflationary consensus'. Under diffuse power, in practice, building a consensual decision may be achieved at the expense of realism. The third is that decisions and strategies are produced by the same people and processes which can result in producing a various layering of structures (Denis *et al.*, 2007).

In any construction project, individuals are required to take decision in relation to other parties in the project. These decisions might be enacted in a mutualistic or adversarial manner to promote the interest of the whole entity or oneself or organisation limited interest. Those individuals are part of the organisation and part of the society as a whole. There is a tripartite axiological system involving the interaction between society, firms and individuals. This system is inherently 'irreducibly complex system' (Karlberg, 2004). It is evident that there is interplay between psychological and sociological behavioural dynamics (Lips-Wiersma, 2002). 'Psychological theory is fundamentally concerned with constellations of personal beliefs, attitudes and attributes and based on the assumption that individuals potentially have a moderate degree of destiny or control in their career' (Hotchkiss and Borrow, 1990). While sociological theories emphasise the role of the environment based on the assumption that reality is socially constructed and institutionally influenced, culture and structure play an important role in shaping personal behaviour, experiences and opportunities. Individual shared stories show the duality of psychological and sociological influences and highlight a wide range of relationships and interactions. There is an assumption that in changing oneself, one can exert influence on his/her organisation and hence the proposition of one can change the world. Likewise individual choices, decisions and behaviours are greatly influenced or limited by organisational structure and culture (Lips-Wiersma, 2002).

In attempting to fully understand the drivers for mutualistic or adversarial behaviours, a different but complementary perspective is needed. This perspective is the notion of this chapter. It is argued that to understand the model of adversarial behaviours in construction and advance mutualistic relationship, a perspective of psycho-structural and socio-structural is needed. This perspective can be enlightened further by understanding power and its distribution. Such a joint perspective helps sheds lights on the reasoning underpinning adversarial or mutualistic behaviours.

11.6   Psycho and socio-structural perspective

Culture has a complex and different shades of meaning (Hofstede, 2001, 2003). It is a collection of systems of representations, meanings, beliefs and other ideological variations among particular social groups (Hofstede, 2001). It encompasses political, economic, legal and other structural variations among social groups (Karlberg, 2004). Culture can also be seen as social heritage. These phenomena are socially learned or constructed, relatively fluid and variable between populations and across generations, and hence there are infinite cultural representations (Hofstede, 2001). At organisational level, culture is the collection of relatively uniform and enduring values, beliefs and customs, traditions and practices that are shared by an organisation's members, learned by new recruits and transmitted from one generation of employees to the next; it is the way we do things around here; the way we interact and the way we cooperate or compete. Hofstede (2001) refers to the mental programming of humans that is physically determined by the brain cells. However, it is difficult to observe these mental programmes physically but are possible to be observed through human behaviour (Hofstede, 2001). According to Hofstede (2001), every person's mental programming is partly unique, partly shared with others at the three levels of individual, collective and universal.

In the broadest sense, culture is the entire social heritage of a community including material, ideological and structural expressions. These contingent expressions can be socially learned or constructed. For example our need to eat or drink is an innate characteristic of our human nature; however, all forms of our tastes in food including the diverse ways of producing and distributing food are expressions of our human culture. Indeed, and according to the anthropologist Clifford Geertz 'our capacity to speak is surely innate; our capacity to speak English is surely cultural' (Geertz, 1973). The interplay between human nature and culture is well established as our need for culture itself is an innate characteristic of human nature. Hence the distinction between nature and culture is very subtle and very complex (Hofstede, 2001; Karlberg, 2004).

Culture has a great impact on the managers' and leaders' thinking and decision making. However, with the potpourri of culture it is hard to summarise messages our culture is giving to leaders in terms of ethics and morality and community management (Carroll and Buchholtz, 2006). Conflicts are there in organisations working in foreign countries. A process and type of reconciliation and balance to honour both cultural and moral standards is required but not easily achievable.

Having said this, the distinction between natural and cultural practices is of paramount importance to the study of managerial practices in construction. As adversarial practices in construction have been internalised, they are then represented as natural and inevitable to those who have internalised them. The prevailing assumptions are that adversarial practices are, as commonly argued, part of the fundamental make up of human nature as we are intrinsically selfish and aggressive (Lewontin, 1991). However, Karlberg (2004) suggested that such assumptions are not only misinformed but socially oppressive and ecologically unsustainable and that there is a need to establish norms to regulate human nature (culture).

Cultures are not monolithic entities and are processes of physical and cultural evolution that are not distinct nor sequential. The fundamental issue is that cultural practices can become internalised and therefore appear natural, inevitable and impossible to change. In his study, Hofstede (2001) found variations of individual behaviour across the globe. These behavioural variations where individual can be cooperative or competitive depend on power distance (relating to human equality), uncertainty avoidance (related to stress level in unknown circumstances), individualism versus collectivism (related to the individual integration into the primary group), masculinity versus femininity (related to the division of the emotional roles of men and women in society) and long term versus short term that relate to the focus of people's efforts on the present or the future (Hofstede, 2001).

It is evident that to understand human behaviour, one needs to appreciate the two levels of culture: the socio-structural and the psycho-structural elements (Hofstede, 2001, 2003). The socio-structural element of culture (the culturally contingent models of social organisations) deals with groups and organisations of society and the relationship between labour and capital in capitalist economy, between first and third world nations in a post-colonial world order. Changes here are done through reforming underlying structures such as making of laws and regulations (Karlberg, 2004).

The psycho-structural dimensions of culture (understanding the contingent structure of the human mind) highlight subjective understandings of individuals. Changes here are by raising awareness, rethinking values or identities or reforming attitudes and beliefs (the subjective structure of the mind rather than the structure of social organisations). This can be appreciated at the early years of development of individuals, mostly in childhood and adolescents, although media campaigns for different age groups or specific groups can be used these days to achieve the desired changes (Karlberg, 2004).

It appears that there is a need for a dual strategy. Individual humans have the potential for adversarial as well as mutualistic behaviour. The degree to which we realise either of these potentials is a function of the cultures we are raised within (Hofstede, 2001; Karlberg, 2004).

11.7   Power to facilitate stakeholder management

Hislop (2005) argues that one of the major criticisms of the majority of the mainstream management literature is the neglect of issues of power and conflict. Greater account of 'unequal relations of power' within organisations has been neglected by subsequent writers, with the exception of the few (Lave and Wenger, 1991; Fox, 2000). Lave and Wenger (1991) discuss these issues of power and conflict and their appeal for future analyses. Fundamentally, individuals and organisations in construction have inherent tensions built into them which unavoidably results in them possessing an, 'unequal distribution of power'. The uneven distribution of power results from the reality of the diversity of construction and hence participants in the processes are not necessarily equal. This uneven distribution of power creates potential conflicts in the processes of construction projects (Wild, 2002). There are many spheres where power can be manifested and utilised. Economic and political power are often used but there are other subtle sphere of power in business such as cultural, technological and environmental power (Carroll and Buchholtz, 2006).

According to Karlberg (2004), there are four distinct power distribution models whether related to individual or organisations. The first model of power distribution is power over, to have power to and to have power over, that is derived mainly by culture of contest. This model of power distribution results in a state of inequality between parties coupled with adversarial relationships. The second model of power distribution is assisted power which results again in a state of inequality but might help to develop mutual relationships. The third model is the balance of power that results in equality between parties with probability of the parties resorting to adversarial attitudes. The fourth model is mutual empowerment or power with (mutualism). This model clearly advances mutualistic relationships on an equal basis.

Indeed, the model of contest can be seen in our economic, legal and political systems. In academia, the Greek-based competition – epistemological model of contest of ideas (intellectual contests) – is viewed as the best way to generate knowledge by structuring its production as a contest between competing ideas. This assumption views humans as aggressive and greedy by nature 'Natural selection' (Karlberg, 2004). However, cultural models of cooperation are seen as enablers for humans to survive rather than aggressive impulses. 'Rejecting the fundamentally self-interested and competitive understanding "homo economus", economist expand the understanding of human nature as cooperative, altruistic and even self-sacrificing' (Geertz, 1973). Such assumptions can also be seen in game theories and among economists. Furthermore, other schools of thought such as Feminism, System theory, Ecology/Environmentalist and Communication theories advocate the later models of culture, a culture of mutuality.

Feminism argues that aggressive and competitive structures serve as structures for male domination. In Feminism theory, empowering others is seen as rewarding and gratifying activities. Women stress power as a capacity rather than domination (Boulding, 1977). This paradigm of thinking is complemented by System theory. System theory is a relational paradigm that characterises the study of complex systems adapting a holistic or systemic focus that reveals the connections or interrelations within the system. It aims to understand the wholeness or interdependence of a system sharing an adaptive organic view of the world/system.

It further emphasises the reality of increasing interdependence and the need for a movement towards post-hegemonic cultures (Karlberg, 2004). System thinking requires that individual members of a system should be attached to the vital interest of the whole and not to hesitate to subordinate every particular interest for the sake of overriding interests of the whole (Sherwood, 2002). In a world of increasing interdependency, the interests of the parts are best reached by the advantage of the whole; as no lasting results can be achieved by any of the component parts if the general interests of the whole entity itself are neglected or marginalised by any of the parts. System thinking advocates an integrative power model. This integrative power model is seen as 'the capacity to build organisations, to create families and groups, to inspire loyalty, to bind people together, to develop legitimacy' (Boulding, 1990). Such is the underpinnings of the Ecology/Environmentalism model that provide an epistemological model for reconciling 'oppositional' viewpoints where the differing views are seen as complementary and necessary for the dynamics of the group (Karlberg, 2004, Sherwood, 2002).

Another school of thought that promotes mutualistic relationships is Communication theories. Theories of interpersonal communication suggest that adversarial modes are counterproductive within families, among friends and co-workers. Individuals can't afford to offend each other especially in long-term relationships. Communication theory advances the interest of the group by developing a collective discourse through dialogue. A dialogue opens up the channels that 'contribute to the thinking about an issue so that everyone involved gains a greater understanding of the issue in its subtlety, richness and complexity. An understanding that engenders appreciation, value and sense of equality'. Absent are efforts to dominate others because the goal is the understanding and appreciation of one another's perspective rather than the denigration of some simply because they are different (Foss and Griffin, 1995).

11.8   Leadership for stakeholder management

11.8.1   Leadership in construction

Leadership and practice in construction is continually developing with ongoing research mainly in other fields such as strategic management and organisational theories. Leadership models, capabilities and theories have developed over the past century to understand what makes a leader or a person with authority (Chan and Cooper, 2007). Within the realm of construction, leaders are criticised for being a technocratic totalitarian type with a 'lip service' given to contested themes of empowerment and teamwork (Green, 1998). Green (1998) further commented that all calls for change within construction are imposed on organisations to ensure control of the processes by the dominant power groups. He further added that 'whilst industry leaders are fond of calling for attitudinal and cultural improvement, the language of the accepted research agenda continually reinforces the industry's dominant culture of control and command'. In a study with in-depth interviews with leaders in the UK construction industry, Chan and Cooper (2007) found that leaders are highly motivated by pursuing personal agendas.

It is apparent that there are various business and professional moral and ethical challenges facing construction today. Many of these moral and ethical challenges are associated with the contemporary culture of construction such as conflict of interest, customer needs and service, fair competition and responsibilities. The fierce competition within construction is seen to engender the unethical behaviour which is prevalent in construction (Elmualim, 2007b). Such unethical attitudes will no doubt hamper the practice of stakeholder management. Although there are various codes of ethics governing the practice within construction, breaches in professional responsibilities are common (Bowen et al., 2007).

Recently, the role of leadership and entrepreneurship in construction is widely being examined due to the dynamic and changing nature of the industry. Although there is a wealth of management knowledge evolving around leadership in other disciplines, it is contested that leadership as a concept is not fully understood nor developed in construction. It is argued that leadership studies in construction management remains inadequate (Chan and Cooper, 2007). Toor and Ofori (2007) in their survey of authentic leadership in construction found that individuals are not satisfied with authenticity of their managers/leaders as well as their organisations. They further concluded that 'the overall perception of the ethical and professional environment of overall construction industry is not healthy'. This profoundly will have a great implication for stakeholder management. Having said that, leadership studies in construction focus solely on managerial functionalism and to a great extent ignoring the broader sociological, psychological, historical and political perspectives of leadership (Chan and Cooper, 2007), hence the difference between a manager and a leader. Current educational and management practices within construction focus on managerial aspects of leadership, hence, the industry has managers and being criticised of lacking aspiring moral and authentic leaders.

Having said that, in construction as in other disciplines leadership practice and study will ever remain a fascinating subject. According to Munshi et al. (2005), leaders are vital for the progress of organisations at all levels. Particularly important are the two distinct roles performed by leaders; firstly motivation of others particularly into new arenas and secondly instigating organisations, structure and environment that foster innovation and continual improvement (Munshi et al., 2005). Leaders are seen as individuals with vision; with the ability to take groups or organisations into new territories where there is lack of knowledge (Chan and Cooper, 2007). During the past half-century, it is estimated that over a thousand studies were conducted to establish the styles, characteristics or personality traits of great leaders. However, none of these studies have produced one style that individuals can imitate (George et al., 2007). 'No one can be authentic by trying to imitate someone else. You can learn from other experiences, but there is no way you can be successful when you are trying to be like them. People trust you when you are genuine and authentic, not a replica of someone else' (George et al., 2007).

Without a doubt the study of leadership can't be separated from the study of culture and power. However, there are five main clusters of theories describing leadership:

- contingency
- transformational
- transactional
- distributed
- structuralist theories.

Contingency theories assert leaders as taking action in difficult circumstances. The difficulty here is that any response or decisions taken by the leader are dependant on the context or the situation in which the response or the decision was taken.

The transformation and transactional theories examine the differences that exist between leaders and followers. Transformation theories place a significant emphasis on vision and effective communications through personal qualities such as charisma (Strange and Mumford, 2002). On the other hand, the transactional theories point to the reward and punishment system (carrot and stick) that will influence the attitude and behaviour of followers.

The rationale underpinning distributed leadership theories is that new forms of interdependence and coordination (a characteristic of the construction industry) give rise to distributed managerial and leadership practices. The emphasis of this cluster of theories is on self-management (Munshi *et al.*, 2005).

The structuralist leadership focuses on organisation systems and the inclusion of system thinking whereby leaders are facilitators and responsible for the creation of organisational structures that allow efficient and effective distribution of resources (Chan and Cooper, 2007).

Most of these early theories were mainly concerned with finding out styles and characteristics of leadership based on the assumption that these are what differentiate leaders from followers. Sceptics of these theories argue that having an individual with such characteristics will not necessarily mean that this individual can become a leader (Chan and Cooper, 2007). Indeed some theories according to specific characteristics describe the prevailing models of leadership as authoritarian, paternalistic, egoistical, know-it-all, manipulative and democratic leadership.

11.8.2 Aspects of moral leadership

Construction is being widely criticised for being slow in adopting leadership theories with more emphasis on organisational performance. Some research in leadership in construction identify four leadership styles: shareholder, autocrat, consensus and consultative. Other studies point out that leaders are technically competent entrepreneurs focusing on driving innovation (Chan and Cooper, 2007). However, the focus within construction remains on performance agenda and mainly achieving financial performance or competitive advantage over other competitors.

With the turn of this century, there is a growing interest in positive psychology that is contributing to the rising research in moral leadership and authenticity in general and in construction in particular. While most research work deals with the scientific aspect or reality (what), morality and authenticities focus on what ought to be (Walker, 2007). Indeed the concept of morality is one of the most complex and widely discussed concepts over the centuries. Research is widely being conducted on authenticity in various disciplines such as psychology, sociology, philosophy and managerial science (Toor and Ofori, 2007). However, according to Toor and Ofori (2007) in-depth understanding, measurement, development and sustainability of authenticity are some of the challenges facing researchers today. Of course 'the different perceptions of authenticity in different cultural contexts have resulted in a plethora of measurement challenges and research questions'.

For construction the need for moral leadership and authenticity is well established due to the poor social image, ethical and moral difficulties and fragile professional practices (Toor and Ofori, 2007). Toor and Ofori (2007) further emphasised the need for moral leadership in construction as corruption remains at a high level and the situation is worsening worldwide. They further articulated that there is a stringent need for raising the awareness of authenticity, positive organisational behaviour, positive organisational scholarship and authentic and moral leadership in construction.

Moral leadership is seen as the behaviour of leaders in addition to the moral character, values and programs (Toor and Ofori, 2007). According to Luthans and Avolio (2003), moral leadership is a process that integrates positive psychological capacities and highly developed organisational understanding resulting in greater self-awareness, self-regulating positive behaviours and fostering positive self-development. Hence moral leaders are genuine, reliable, trustworthy, real and veritable (Luthans and Avolio, 2003). Other authors described moral leaders as honest, open, transparent, compassionate 'and with a heart'. Further definitions for moral leadership is the possession of vision, the practice of a solid set of values, leading from the heart, establishing enduring relationships and demonstrating self-discipline in their lives (Toor and Ofori, 2007). Another complementary articulation of moral leadership is that the individual is fully aware of who they are, how they think and behave, confident, hopeful, optimistic, resilient and of high moral character (Luthans and Avolio, 2003).

According to George *et al.* (2007) authentic and moral leaders work hard at understanding and developing themselves and drive long-term results. Through their integrity, moral leaders sustain the organisation results through good times and bad times. They further introduced four main points in order for an individual to be a moral leader; firstly practicing values and principles. The values and ethics that motivate moral leaders are derived from their convictions and beliefs. It is suggested that one will not know what their true values are until tested under real life pressures. Secondly balancing extrinsic and intrinsic motivations. Extrinsic motivations are those gained when comparing with peers and outside world parameters such as promotion, recognition and status and financial rewards. Intrinsic motivations are mainly derived from the sense of meaning of life such as personal growth, helping other people develop, taking on social causes and ultimately making a difference in the world (George *et al.*, 2007). Such issues are vitally important in order to advance moral, ethic and authentic leadership in construction and further stakeholder management.

In his highly acclaimed book, Covey (1992) argued that growth is natural and comparison of oneself with others is dangerous. He further warned against knowledge without character, pleasure without conscience, science without humanity and finally against commerce (business) without morality and ethics. Having said that moral leaders throughout history, with all their varieties, helped bring us up all morally and can become part of a nation's moral fibre – a child, an adult, a person in politics or one quietly trying through a seemingly ordinary life (Coles, 2000).

It is apparent that moral leadership is a very complex and multidimensional concept containing elements from various domains. In his holistic framework for moral education, Clarken (2007) suggested a tripartite theory of the human mind. These three elements of the human mind depend and are conditional on cognition (related to knowing, intellect, epistemology, thinking), affection (which is related to valuing emotion, aesthetics and feelings) and finally conation (which is related to striving, volition, ethics and willingness) (Clarken, 2007). These three elements of a moral education framework are part of the self-development and self- awareness process of an individual in a variety of contexts such as individual, family, group/team, organisation level and society at large. This understanding has a great implication in construction. According to Toor and Ofori (2007) if a construction leader is morally oriented and authentic 'it is likely that the overall performance of the project will be superior'. Since authenticity and morality take into account the component of self-awareness and self-regulation, . . . It is likely that an authentic project leader would not suffer from personal bias and self-interest. Moreover, an authentic project leader will develop better interpersonal relationships which will result in reduced disputes and conflicts on the projects'. Such moral attitudes will benefit individuals and organisations. It is anticipated that employees will enjoy better interpersonal relationships, satisfaction and well-being and good working environments. Authenticity and moral leadership will benefit organisations with better chances to grow in challenging times which will result in higher standards of ethics, morality and professionalism of the wider construction industry. Toor and Ofori (2007) called for investing in academic curricula as well as professional development among other human resource management activities to advance the cause of moral leadership in construction in relation to stakeholder management as a concept, theory and practice.

11.9   Cultural renewal and leaders as facilitators

11.9.1   Altruism, philanthropy and strong reciprocity

More than two centuries prior to Porter's work on competitive advantage (Porter, 1980, 1985, 1990a), Adam Smith advocated market competition as the main driver for economic prosperity (Smith 1982 (1759); 1991 (1778)). Smith is famously quoted: 'it is not from the benevolence of the butcher, the brewer, or the baker that we expect our dinner, but from their regards to their own interest'. This has led to the ubiquitous belief that Smith was the champion of self-regarding or the homo economicus nature of human hypothesis, exclusively. However, Smith articulated the altruistic and philanthropic nature of man when he wrote 'how selfish so ever man may be supposed, there is evidently some principles in his nature, which interest him in the fortunes of others, and render their happiness necessary to him, though he derives nothing from it, except the pleasure of seeing it'. Such principles were advocated by Darwin (Darwin, 1859) contrary to the contemporary neoclassical economics which assumes solely the self-interested and self-regarding nature of individual (Gintis *et al.*, 2005). According to Darwinism 'in the long history of mankind (animal kind too) those who learnt to collaborate and improvise most effectively have prevailed'. Advocates of social justice such as Karl Marx called for the state ownership and control of the main elements of the economy to kerb the self-interested behaviour of individuals (Das capitalism, 1867). However, both liberal and social economic theory have their flaws (Huddleston, 1999) which contribute greatly to the adversarial cultural norms in western countries, economic stifle and social inequality. Such cultural norms, economic problems and inequality are relevant in the construction industry (Elmualim, 2007b).

It is difficult to understand the effects competition has on individual, groups or culture if concerns for fairness and strong reciprocity were ignored. Competition sometimes will have less impact than the prediction by model based solely on the self-interest assumption (Fehr and Fischbacher, 2005). Fehr and Fischbacher (2005) articulated that lack of understanding of cooperation is due to the interaction between self-interested strong reciprocal individuals. Such interaction is generally shaped by institutional environments (Fehr and Fischbacher, 2005).

In the evolution of cooperation in primate groups, it was found that competition generates powerful incentives for cooperation (Silk, 2005). Similar findings were established in other disciplines. For example engaging in group production with sharing provided the individuals with higher returns (Kaplan and Gurven, 2005). According to Kaplan and Gurven (2005) 'reciprocal altruism will only emerge among self-interested actors if there is repeated interaction that rewards cooperation and punishes defection'. They further proposed that 'multi-individual negotiations result in the emergence of social norms that are collectively enforced'. In our course of human evolution, natural selection shaped our psyche with certain characteristics such as sensitivity and taking advantage of potential gains from cooperation, motivation to avoid being victim of free-riding, take advantage of free-rides, motivation to negotiate social norms and motivation to obey and enforce social norms. Systems of cooperation and sharing need control.

In Kantian moral theory actions emanate from the sense of duty of the individual or obligations (good will). According to Friedman and Miles (2006), Kantian moral theory stresses the moral imagination whereby individuals imagine they are at the receiving end of the decisions taken by others. 'Governing rules should be formulated so that they are applicable to everyone within the organisation being governed'. The survival and existence of organisations specially in construction is largely dependant on their good and amicable relationships with their stakeholders (other organisations in the construction project supply chain and actors) (Pajunen, 2006).

The self-interest model is based on the assumption that self-interest is the ultimate goal of human actions. The self-interest is the main motive to take actions to satisfy individual desires and needs indifferent to the effects on others. Literature also indicates that the self-interest model includes altruistic behaviours as individual can compromise part of their interests for the common good of others (Rocha and Ghoshal, 2006). The critique of self-interest and the call for alternatives is based on the increasing number of people and their tendencies for counter examples. Particularly in cases where people walk away from highly profitable transactions as they believe the terms of such transactions are unfair or people help others without expectations of reciprocity (Rocha and Ghoshal, 2006). Rocha and Ghoshal (2006) argued for an integrative approach for self-interest and its critique and hence the self-love model. They articulated that the 'integrative approach contributes to a better understanding of the human potential to develop motives other than self-interest, which are neither anomalies nor intrinsically conflictive, but part of an integrated motivational system that guided by self-scrutiny and freedom drives different behaviours and outcomes'.

While there is a continuing development in stakeholder management literature, there are few generic models for stakeholder identifications (Pajunen, 2006). The self-love model could be traced back to the ancient Greeks – Aristotle and Socrates. The model is Aristotelian in nature where the assumption is that humans strive for their own good and perfection and a great tendency towards good. Good can include wealth, honour and pleasure as well as other attributes such as sentimental love, duty and excellence (Aristotle, 1984). To Socrates people find their real self and real happiness in accomplishments of perfection.

Rocha and Ghoshal (2006) criticised the self-interest view on human behaviour and proposed a framework of 'self-love view, that integrates self-interest and unselfishness and provides different explanations of the relationship between preferences, behaviour and outcomes. According to Rocha and Ghoshal (2006), the self-love model provides an understanding of motivational assumptions and their influence on individual behaviour. It is argued that such understanding and various designs of motivational contexts will help policy makers and managers/leaders rather than the assumption of the conflict between self-interest and unselfishness (Rocha and Ghoshal, 2006).

Society is the macro environment for any business. The macro environment provides the all-encompassing context for businesses including construction (Carroll and Buchholtz, 2006). Our modern society is considered to be pluralistic in nature where there is a diffusion of power among individuals, groups and organisations. Carroll and Buchholtz (2006) argued that we are taking pluralistic to the extreme which results in a special interest society each pursuing its own limited agenda. Such societies will prevent against the danger of a leader of any group or firm to be left uncontrolled. Indeed the rise of the discourse in corporate citizenship and social responsibility is the reply by businesses to the harsh criticism against them. Such responsibility includes the pyramid of responsibility: with economic responsibility as the base, legal and then ethical and philanthropic responsibilities on top of the pyramid (Carroll and Buchholtz, 2006).

Gintis et al. (2005) argue that 'strong reciprocity is a behavioural schema that is compatible with a wide variety of cultural norms'. And they extend this to say that strong reciprocators punish those who behave selfishly 'but the fairness and the nature of punishment are cultural variables'. There is a need for policies that limit social and economic inequality. Such policies are not only important from an ethical point of view 'but necessary to harness moral sentiments to future economic and social development around the world' (Gintis et al., 2005). Such policies should be steered towards influencing peoples' values and codes of behaviour by evoking, empowering and cultivating public spirited motives (Gintis et al., 2005).

11.9.2   Research and educational implications

Stakeholder management as a concept for construction change and renewal requires the reordering of industry norms through the transformation of individual leaders and organisations. Establishing the industry aspired milieu of trust, cooperation, mutualism and prosperity ultimately involves change in basic attitudes and values. Consultative mechanisms should be in place for individuals, groups and communities in order for them to become the principal actors in stakeholder management. Inclusiveness and cooperation in decision making are vital indicators for social and economical progress. A process of collective decision making and action that are devoid of adversarial posturing should be a coherent feature of stakeholder management. The ability and opportunity for individuals, groups or firms to consult within the framework of stakeholders is vital for the success of effective, just and efficient stakeholder management. This requires a new breed of leaders and managers to create and realise a new social reality for the construction industry. These leaders and managers will be acting as facilitators for the processes of decision making for stakeholder management. The role of leaders as facilitators is to foster unity among stakeholders' participants through equality and justice where the needs and interests of all are met and maintained. This will only be possible if leaders, within their fiduciary duties, understand the oneness and wholeness of their stakeholders. Such a call for cultural and leadership renewal based on philanthropic, altruistic and strong reciprocity will have great implications for research and education in construction. The renewal and cultural change that embraces a stakeholder approach will need to be embedded in all forms of construction educations and curricula. Research agenda also needs to be re-oriented towards a multidisciplinary sphere to further emphasize the benefits of the concepts and approaches such as stakeholders, strong reciprocity and philanthropy to be ripe for policy makers, management practitioners, the industry and society at large.

11.10   Conclusions

The construction industry is widely viewed as a fragmented and diverse industry. As it is a project-based industry, the delivery of construction projects requires the participation of various agents. Each agent comes with their own cultural values, modes of operation and interests. Such cultural values and interests produce a power distribution system amongst participants that leads to conflict and adversarial relationships that are strongly disapproved by all participants, particularly policy makers and academicians. However, organisations within construction while espoused to the collaboration rhetoric of the policy makers, exploit every opportunity to gain a competitive advantage over their rivals and other organisations in the supply chain. No wonder concepts such as stakeholder management resonate within construction to bring change in construction.

Current cultural, management and leadership practices are a hindrance to the actualisation of the benefits offered by the stakeholder management approach. For the realisation of the concept of stakeholder management, there is a stringent need to readdress the issues of collaboration and competitiveness in construction via a different yet complementary approach of psycho and socio structural study. A complementary approach seeks to discern the dichotomy of collaboration and competitiveness by understanding power distribution among social groups. Such an approach sheds light on the nature of power, its distribution and its influences on driving competitive or genuine collaborative behaviours as foreseen by individuals, organisations and society as a whole. It is argued that our culture of competition has become the norm, and norms make it difficult for individuals and organisations to see other alternatives. Contests are seen as normal and necessary models of social interactions and, hence, 'normative adversarialism' has become the predominant strand in contemporary western-liberal societies.

The psycho and socio structural approach shows that individuals are greatly influenced by their organisations. Both individuals and organisations are influenced by the wider society in a very complex and intertwined tripartite axiological system. Individual humans are intrinsically and culturally programmed towards competitive and collaborative behaviours. It is argued that in the post-hegemonic society, cooperation and mutualistic relationships will be best for serving the interests of the participants of construction projects delivery. Such a view is supported by various schools of thought such as communication theory, system thinking, feminism and environmentalists groups. Competition discourse is better seen as a means for training or acquiring cooperative skills in order to advance collaborative and mutualistic relationships in construction.

The application of stakeholder management will be accelerated via a cultural renewal that seeks to integrate aspects of our gene-cultural assumptions away from the exclusivity of the self-regarding hypothesis. Leaders with altruistic, philanthropic and strong reciprocity traits will drive the desired need through unity in diversity serving the interest of all across the construction supply chain and society as a whole. In addition to cultural change and need for moral leaders, there is a need for sound policies and regulation to be in place to advance cooperation and eliminate inequality in order for construction to reap the benefits of stakeholder management as a concept and further develop it as common practice.

References

Aristotle (1984) Nicomachean ethics. In: Barnes, J. (ed) *The Complete Works of Aristotle. The Revised Oxford Translation.* NJ: Princeton University Press.

Blockley, D. and Godfrey, P. (2000) *Doing It Differently – Systems for Rethinking Construction.* London: Thomas Telford.

Boulding, E. (1977) *Women in the 20th Century World.* New York: Sage Publications Ltd.

Boulding, K. E. (1990) *Three Facets of Power.* Newbury Park: Sage Publications Ltd.

Bowen, P., Pearl, R. and Akintoye, A. (2007) Professional ethics in the South African construction industry. *Building Research and Information,* **35**(2):189–205.

Carroll, A.B. and Buchholtz, A.K. (2006) *Business and Society: Ethics and Stakeholder Management.* Mason, Ohaio: Thomson South-Western.

Chan, P. and Cooper, R. (2007) What makes a leader in construction? An analysis of leaders in the UK construction Industry. In: *Proceedings of CIB World Building Congress.* 13–17 May 2007, Cape Town, South Africa, pp. 498–510. CIB (International Council for Research and Innovation in Building and Construction).

Clarken, R.H. (2007) Toward a holistic framework for moral education Annual Meeting. In: *The American Educational Researcher Association.* Chicago, USA: The American Educational Researcher Association.

Coles, R. (2000) *Lives of Moral Leadership.* New York: Random House.

Connor, T. (2003) Managing for competitiveness: A proposed model for managerial focus. *Strategic Change,* **12**(4):195–207.

Covey, S.R. (1992) *Principle-Centered Leadership.* New York: Fireside.

Darwin, C. (1859) The Origin of the Species and the Voyage of the Beagle. Vintage Classics, Vintage and the Oxford University Press, Oxford.

Denis, J-L., Langley, A. and Rouleau, L. (2007) Strategizing in pluralistic contexts: Rethinking theoretical frames. *Human Relations,* **60**(1):179–215.

Department of Trade and Industry (DTI) (1998) *Competitive Futures: Building the Knowledge Driven Economy.* London: Department of Trade and Industry (DTI).

DTI (2003) *Construction Annual Statistics.* London: Department of Trade and Industry.

Egan, J. (1998) *Rethinking Construction.* London: DETR/Stationery Office.

Egbu, C.O. (2004) Managing knowledge and intellectual capital for improved organizational innovations in the construction industry: An examination of critical success factors. *Engineering, Construction and Architectural Management*, **11**(5):301–315.

Elmualim, A.A. (2007a) Psycho-social approach for discerning the dichotomy of competition and collaboration to advance mutualistic relationships in construction. In: *CME25: Construction Management and Economics: Past, Present and the Future*. 15–18 July 2007, Reading, UK: University of Reading.

Elmualim, A.A. (2007b) A locus for competitiveness in the UK construction. In: *CIB World Building Congress: Construction for Development*. 14–18 May 2007, Cape Town, South Africa. CIB (International Council for Research and Innovation in Building and Construction).

Elmualim, A.A., Green, S.D., Larsen, G. and Kao, C.C. (2006) The discourse of construction competitiveness: Material consequences and localised resistance. In: *Joint International Conference on Construction Culture, Innovation and Management (CCIM)*. 26–29 November 2006, Dubai: The British University in Dubai, UAE and CIB.

Evan, W.M. and Freeman, R.E. (1993) A stakeholder theory of the modern corporation: Kantian capitalism. In: Beauchamp, T.L. and Bowie, N.E. (eds) *Ethicak Theory and Business*. NJ: Prentice-Hall, pp. 75–84.

Fehr, E. and Fischbacher, U. (2005) The economics of strong reciprocity. In: Gintis, H., Bowels, S., Boyd, R. and Fehr, E. (eds) *Moral Sentiments and Material Interests: The Foundation of Cooperation in Economic Life*. Massachusetts: Massachusetts Institute of Technology, pp. 151–191.

Fellman, G. (1998) Rambo and the Dalai Lama: The Compulsion to Wind and its Threat to Human Survival. Albany: State University of New York Press.

Fernie, S., Weller, S. and Green, S.D. (2004) Aspiration of collaboration: Integrated procurement and the mediating effects of context. In: *CIB World Building Congress 2004: Building for the Future*. Toronto, Canada: National Research Council Canada, Vol. CIB T6S6.

Flint, G.D. (2000) What is the meaning of competitive advantage? *Advances in competitiveness research*, **8**(1):121–129.

Foss, S. and Griffin, C. (1995) Beyond persuasion: A proposal for an invitational rhetoric. *Communication Monographs*, **62**(March):2–18.

Fox, S. (2000) Communities of practice foucault and actor network theory. *Journal of Management Studies*, **37**(6):853–867.

Freeman, R.E. (1984) *Strategic Management: A Stakeholder Approach*. Boston, MA: Pitman.

Friedman, A.L. and Miles, S. (2006) *Stakeholders Theory and Practice*. Oxford, UK: Oxford University Press.

Geertz, C. (1973) The Interpretation of Cultures: Selected Essays. New York: Fontana.

George, B., Sims, P., Mclean, A.N. and Mayer, D. (2007) Discovering your authentic leadership. *Harvard Business Review*, **85**(2):129–138.

Gintis, H., Bowels, S., Boyd, R. and Fehr, E. (2005) *Moral Sentiments and Material Interests: The Foundation of Cooperation in Economic Life*. Massachusetts: Massachusetts Institute of Technology.

Green, S.D. (1998) The technocratic totalitarianism of construction process improvement: A critical perspective. *Engineering, Construction and Architectural Management*, **5**(4):376–386.

Green, S.D. and May, S.C. (2003) Re-engineering construction: Going against the grain. *Building Research & Information*, **31**(2):97–106.

Green, S.D., Newcombe, R., Fernie, S. and Weller, S. (2004) *Learning Across Business Sectors: Knowledge Sharing Between Aerospace and Construction*. Reading, UK: The University of Reading.

Hardy, C., Palmer, I. and Phillips, N. (2000) Discourse as a strategic resource. *Human Relations*, **53**(9):1227–1248.

Hartfield, T. (1999) Competition and cooperation in an emerging industry. *Strategic Change*, **8**(4):227–234.

Hislop, D. (2005) *Knowledge Management in Organisations*. New York: Oxford University Press.

Hofstede, G. (2001) *Culture's Consequences: Comparing Values, Behaviours, Institutions, and Organisations Across Nations*. London, UK: Sage Publications.

Hofstede, G. (2003) *Cultures and Organizations: Intercultural Cooperation and its Importance for Survival-Software of the Mind*. London, UK: Profile Books Ltd.

Hotchkiss, L. and Borrow, H. (1990) Sociological perspectives on work and career. In: Brown, D. and Brooks, L. (eds) *Career Choice and Development, Applying Contemporary Theories to Practice* (2nd edn). San Francisco: Jossey-Bass.

Huddleston, J. (1999) *Standing up for Humanity*. New Delhi: Baha'i Publishing Trust, India.

Kaplan, H. and Gurven, M. (2005) The natural history of human food sharing and cooperation: A review and a new multi-individual approach to the negotiation of norms. In: Gintis, H., Bowels, S., Boyd, R. and Fehr, E. (eds) *Moral Sentiments and Material Interests: The Foundation of Cooperation in Economic Life*. Massachusetts: Massachusetts Institute of Technology, pp. 75–113.

Karlberg, M. (2004) *Beyond the Culture of Contest*. Oxford: George Roland Publisher.

Klein, J. (2002) Beyond competitive advantage. *Strategic Change*, **11**(6):317–327.

Latham, M. (1994) *Constructing the Team*. London: HMSO.

Lave, J. and Wenger, E.C. (1991) *Situated Learning: Legitimate Peripheral Participation*. Cambridge: Cambridge University Press.

Lewontin, R.C. (1991) *Biology as Ideology: The Doctrine of DNA*. New York: Harper Collins.

Lips-Wiersma, M. (2002) Analysing the career concerns of spiritually oriented people: Lessons for contemporary organizations. *Career Development International*, **7**(7):285–397.

Luthans, F. and Avolio, B.J. (2003) Authentic leadership: A positive developmental approach. In: Cameron, K.S., Dutton, J.E. and Quinn, R.E. (eds) *Positive Organizational Scholarship*. San Fransisco: Barret-Koehler, pp. 241–261.

Munshi, N., Oke, A., Stafylarakis, M. *et al.* (2005) *Leading for Innovation. AIM Executive Breifing*. London: Advanced Institute of Management Research (AIM).

O'Shaughnessy, N. (1996) Michael Porter's competitive advantage revisited. *Management Decision*, **34**(6):12–20.

Pajunen, K. (2006) Stakeholder influence in organisational survival. *Journal of Management Studies*, **43**(6):1261–1288.

Porter, M.E. (1980) Competitive Strategy – Techniques for Analysing Industries and Competitors. New York: Free Press.

Porter, M.E. (1985) *Competitive Advantage*. New York: Free Press.

Porter, M.E. (1990a) *The Competitive Advantage of Nations*. London: Macmillan.

Porter, M.E. (1990b) Competitiveness: Challenging the conventional wisdom. *Harvard Business Review*, **68**(3):190–193.

Porter, M.E. (1998) Clusters and the new economics of competition. *Harvard Business Review*, **76**(6):77–91.

Rocha, H. and Ghoshal, S. (2006) Beyond self-interest revisited. *Journal of Management Studies*, **43**(3):585–619.

Sherwood, D. (2002) Seeing the Forest for the Trees: A manager's Guide to Applying System Thinking. London: Nicholas Brealey.

Silk, J.B. (2005) The evolution of cooperation in primate groups. In: Gintis, H., Bowels, S., Boyd, R. and Fehr, E. (eds) *Moral Sentiments and Material Interests: The Foundation of Cooperation in Economic Life*. Massachusetts: Massachusetts Institute of Technology, pp. 43–74.

Smith, A. (1982 (1759)) *The Theory of Moral Sentiments*. Indianapolis: Liberty Fund.

Smith, A. (1991 (1778)) *The Wealth of Nations*. New York: Prometheus Books.

Starik, M. (1994) The Toronto conference: Reflections on stakeholder theory. *Business and Society*, **33**(1):89–95.

Strange, J.M. and Mumford, M.D. (2002) The origins of vision: Charismatic versus ideological leadership. *Leadership quarterly*, **13**(4):343–377.

Thomas, P. (2003) The recontextualisation of management: A discourse-based approach to analysing the development of management thinking. *Journal of Management Studies*, **40**(4):775–801.

Toor, S. and Ofori, G. (2007) Ethics and authenticity of leadership in the construction industry. In: *Proceedings of CIB World Building Congress*. 13–17 May 2007, Cape Town, South Africa.

Walker, G. (2007) *The Science of Morality*. London: Royal College of Physicians.

Wild, A. (2002) The unmanageability of construction and the theoretical psycho-social dynamics of projects. *Engineering, Construction and Architectural Management*, **9**(4):345–351.

12 Impact of Procurement on Stakeholder Management

Pantaleo D. Rwelamila

12.1 Introduction

According to Stewart (1982), project management is about controlling the demands of the project, and making choices, within constraints. The client's objectives (project parameters) are cost, schedule, quality and utility (Rwelamila and Savile, 1994). Cost generally refers to first cost; in this context choice means the allocation of budgets and contingencies. Time is better described as timing; choice means the adjustment of timing, time scales and float(s). Quality is the level of specifications and doing right first time and choice therefore means the adjustment of the specifications, that is to basic, medium or high. Utility includes such things as running costs, maintenance issues, constructability and flexibility for alterations or other uses. Choice, in this context, usually means making value judgements between higher initial costs and longer-term savings.

For the sake of completeness, it is important to recognise that construction projects do not occur in isolation, and that consequently they must be considered in relation to the prevailing environment. This background is specific to the country, year, location, project type, industry, etc. (Rwelamila and Savile, 1994). There is little choice for the project manager (PM); any changes that occur will probably be due to outside agencies/stakeholders and will cause a risk of serious disruption to the project. These changes may be in regulations, union policy, markets, technological innovation, etc. Thus an appropriate construction project procurement system is necessary for the PM in order to balance the project parameters, allocate risk appropriately and consequently meet stakeholders requirements. This chapter endeavours to explore the relationship between construction project procurement and stakeholder management – *specifically on the impact of procurement on stakeholder management*. Furthermore, the chapter takes a cross-section of typical construction project procurement systems (CPS) and how their characteristics relate to stakeholders management. The chapter also draws from the results of a number of research projects on how various construction project procurement systems have impacted on stakeholders management. Finally the chapter makes recommendations, with regard to understanding construction project procurement systems and the dynamics of managing stakeholders.

12.2 Theory and practice of construction procurement systems and stakeholders management

The theme of this chapter rests on three of the salient areas of project management: *construction procurement, stakeholders and stakeholders management*. A review and analysis of theory and practice of these is covered under this section.

12.2.1   Theory and practice of construction procurement system (CPS)

One of the principal arguments of this chapter rests on one fundamental aspect of the construction project process that requires early and particular attention if project success is to be achieved – *specifically in managing project stakeholders*. This is the selection of the most appropriate organisation for the design and construction of the project – here in referred to as the *construction procurement system (CPS)*.

12.2.2   Defining a construction procurement system

The term *construction procurement system* has become fashionable and more common in recent times with industry practitioners and researchers. In principle it determines the overall framework of responsibilities and authorities for participants within the construction process, and is a key factor contributing to project success and hence stakeholders satisfaction. The evolution of the term stems from the need for identification and communication of the organisation used to set up the construction process and bring the project to a successful conclusion. The definition described by Franks (1984) as a . . . *amalgam of activities undertaken by a Client to obtain a building* . . . has gained popularity in the industry but is somewhat narrow however, and thus many attempts of defining construction project procurement have been made aiming at the advancement of a more precise definition.

Root and Hancock (1996) refer to a project procurement system (PPS) as a strategy adopted by construction project clients to acquire construction projects. Mohsini and Davidson (1991) and Turner (1990) argue that it is not so much procurement but rather a process. Turner (1990) thus defines the CPS process as *the act of obtaining, acquiring, or securing something*. This is in keeping with McDermott and Jaggar's (1996) opinion that a CPS is the framework within which construction is brought about, acquired or obtained. Hindle (1994) states that a CPS is intended to describe the process by which industry customers procure new buildings. NEDO (1988) takes this view further and expresses a CPS as the manner in which clients buy specialist activities and resources from the construction industry needed to create a new structure. These definitions are somewhat lacking, however, in that the various stages of the project that the CPS affects are not included.

Latham (1994) argues that the choice of a CPS should precede the preparation of the project brief, since it affects who assists with the design brief as well. Govender and Watermeyer (2000) also state that the CPS may be defined as the process, which creates, manages and terminates contracts, and is therefore concerned with activities that both precede and follow the signing of a contract. Rwelamila and Ngowi (1996) propose the definition of a CPS as:

> *The project organisational structure, which is the collective action required to acquire the design, management, and installation inputs.*

The CIOB (1996) concurs, less specifically, that a CPS is the process by which the necessary contributions of the various participants in the design and construction phases of the project are secured. Masterman (1992) also agrees that a CPS is the organisational structure adopted by the client for the management of the design and construction of a building project. These definitions are by no means all inclusive since the relationships between those involved in the project and their individual responsibilities are not mentioned, but are merely implied.

Reporting on an examination of past research literature, Masterman (1992) refers to phrases such as *'building procurement method'*, *'procurement form'* and *'procurement path'*, which have been used by various authorities when referring to this concept. According to Franks (1984), CPS is *'the amalgam of activities undertaken by a client to obtain a building'*.

Franks's (1984) definition is revisited by Rwelamila (2000a) who states that a construction procurement system (CPS) is the organisation structure adopted by the client to manage all phases of the project from inception to completion and in certain situations including post-completion phase/s. According to Love *et al.* (1998), CPSs assign specific responsibilities and authorities to people and organisations, and define the relationships of the various elements in the construction of a project.

Ireland (1982) states that a CPS describes the roles of participants, the relationships between them, both formal and informal, the timing of events, and the practices and techniques of management that are used. Sanvido and Konchar (1999) concur that a PPS is defined as the relationships, roles and responsibilities of parties and the sequence of activities required providing a facility.

A general definition of a CPS has been formulated for this chapter, based on the foregoing descriptions but more specifically incorporating the definitions of Ireland (1982), Rwelamila and Ngowi (1996), Sanvido and Konchar (1999) and Rwelamila (2000a) definitions stated above – a PPS is hence suggested as:

An organisational structure that defines and describes the roles of stakeholders, the relationships between them – both formal and informal, their individual responsibilities, the sequence of activities and timing of events required to provide a facility, as well as the practices and techniques of management that are used.

This definition is supported by Newcombe (1999) who states that a procurement system is more than just the establishment of a contractual relationship or the allocation of roles, but also creates a unique set of social relationships and power structure within a coalition of competing or cooperating interest groups. The term CPS has therefore been adopted and used throughout this chapter.

12.2.3 *Towards selecting an appropriate CPS*

Once a client is satisfied about real need and feasibility of the construction project within overall budgetary constraints, the instinctive reaction is to retain a consultant to help in the assessment of risk towards devising an appropriate CPS. The client should decide how much risk to accept. No construction project is risk free. Risk can be managed, minimised, shared, transferred or accepted. It cannot be ignored. The client who wishes to accept little or no risk should take different CPSs from the client who places on detailed, hands-on control.

The basic decision on the appropriate CPS to take should precede the preparation of the outline (project) brief, since it necessarily affects who shall assist with the design brief as well. That choice of an appropriate CPS must be determined by the nature of the construction project and the clients' wishes over acceptance of risk. Such decisions are very difficult. There are a number of publications [Masterman (1992), Hughes (1990, 1992)], which give, detailed account of different CPSs, their risk distribution potential and merits and demerits of the same.

Once an appropriate CPS has been determined, the first stage of managing the project parameters (Q, C, T and U) has been reached. You now have a system with its associated project management structure, which will be partly, able to manage the whole spectrum of dynamics associated, with balancing the project parameters. *Why partly?* The two over-arching considerations will give the answer to this question.

For the selected CPS to lead to project success, it is important to have effective project teams (focus on teams and health and safety of every individual in those teams). The PM must have the skills required to build effective project teams. Team building is a process of getting a diverse mix of individuals to work together effectively as one unit. Team building involves integrating both project tasks and project management process. It requires leadership skills and a good understanding of the organisation, its culture, interfaces, authority and responsibility relationships, power structures among its stakeholders, and motivational factors. Project clients form one major arm of project stakeholders, hence the need for the PM to understand the composition of project stakeholders.

In fact there are many stakeholders for a construction project, for example a funding financial institution [external stakeholder (ES)], venture capitalist (ES), developer (ES), purchaser on completion (ES), the tenant (ES), and *a public client (ES) – using taxpayers money*, contractor [internal stakeholder (IS)], and subcontractors (IS) too, have different expectations of the project. Although profit may be a common desire, no one except the end user cares what is built. If the PM does not take it upon him/herself to assess the needs of all the project stakeholders, especially the ISs who range from highly skilled to site cleaners, it is unlikely that anyone else will, and meeting project goals could be a pipe dream.

This raises many hard questions. Who are we building for, apart from the external stakeholder (the client) who sets the brief? How can the PM take account of other stakeholder's (IS's) demands? With whom is the PM's construction project contract, and does this create a conflict of interests? The idea of the universal stakeholder (ES + IS) [Rwelamila and Savile (1994)] is neither new nor it is impossible that we should be able to satisfy the universal stakeholder's demand.

It is important to note that the presumption is that choice of an appropriate procurement system will lead to a successful project outcome; this makes an implicit assumption which forms the basis of this chapter that the primary objective of an appropriate procurement system is to provide a successful project and one of the variables of project success is appropriate stakeholders management (Rowlinson, 1999a). The philosophy adopted when dealing with good practice approaches in construction procurement is that of a contingency approach: by this, Rowlinson (1999a) argues that it is meant that there are a number of alternative procurement systems available from the beginning of the process to the end which are able to satisfy the objectives of the user (as a main stakeholder) and that a range of variables exist which, depending on their value, have an effect on the system as a whole.

Various established CPS currently available, forming the basic pool of information, are described primarily by analysing their common features. This is necessary in order to ascertain the way in which the risk is allocated and shared between the participating stakeholders in each of the CPSs, as well as the key project parameter/s that each of them favour, and hence are intended to argue about the importance of a suitable choice.

- Classification of CPSs

Any study involving CPSs leads to the adopting of a logical approach of classifying the currently available procurement systems. The CPSs are often closely related to one another making this classification difficult. However, in doing so, clarity can be achieved which will aid the analysis. Various attempts at classification have been made with the arrangement occurring according to certain factors. These include: *management styles depending on the degree of Client responsibility – NEDO (1983); the sophistication of the Client due to the extension of responsibilities to combine design and construction or design and management – Turner (1990); The level of Client involvement – Hamilton (1997); who leads the project depending on the extent of the integration of design, management, and construction – Franks (1990); the allocation and/or apportionment of associated risk in terms of complexity and uniqueness of the project, Client involvement, speed from inception to completion, and the degree of price certainty – Flanagan (1990) and Sawczuk (1996); on a cost basis in terms of the integration of design, construction, and finance – Ferry and Brandon (1991), Gordon (1994) and McDermott and Jaggar (1996); the responsibility for design and construction determined by the relationship between the two processes – Masterman (1992); Love et al. (1998); and the approximate circumstances under which each system should be used – Cox and Townsend (1998); and the occurrence of certain features in varying degrees namely, complete documentation before construction, competitive tendering, Contractor input in design process, Client responsibility, appropriate risk sharing, minimal variations, cost control, time management, and quality control – Ambrose and Tucker (1999) and Rwelamila (2000a).*

The latter two are not elaborate classifications into a readily understood system since Rwelamila (2000a), Cox and Townsend (1998) and Ambrose and Tucker (1999) believe that there does not appear to be a consistent approach to codifying the different CPSs and their derivatives.

In his review of the problems facing the construction industry, Emmerson (1962) stated that in no other important industry is the responsibility for design so far removed from the responsibility for production. This relationship between design and construction is perceived by Masterman (1992) to have the most bearing on the evolution of a consequent choice of CPSs. This view is supported by Hardcastle *et al.* (1999) and Edum-Fotwe *et al.* (1999), but it was not considered adequate on its own for the purposes of this study. This is simply because in most developing countries, the financing of construction projects carries approximately the same weight as the design and construction. It is for this reason that an amalgam of the various classifications are proposed for the purposes of the theme of this chapter, and this is illustrated diagrammatically in Figure 12.1.

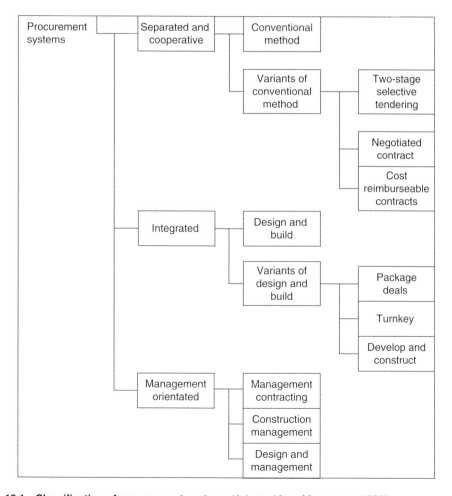

Figure 12.1   Classification of procurement systems (Adapted from Masterman, 1992).

As indicated in Figure 12.1, the classification describes the essence of CPS without compli-cation. Detailed information on various construction CPSs falling under each classification are provided elsewhere (Masterman, 1992; Sanvido and Konchar, 1999). Sufficient detail is provided with the basis being on the responsibility adopted by the various parties as well as the relationship between them, and it incorporates alternative methods of financing the project parallel to the integration of design and construction. The summary of characteristics for each of the project contract strategy briefed described above are shown in Table 12.1.

- The CPS selection process

As indicated above, every CPS falling under the three classifications has basic characteris-tics that define its framework. They only differ in their apportionment of authority, responsi-bilities and risk. When a particular CPS is selected for a specific project, these characteristics define the structure and interaction of stakeholders. There are no objective criteria for choos-ing a CPS since the choice is a strategic decision and should be made only by assessing and interpreting the stakeholders requirements. Therefore embracing a CPS has nothing to do with fashion and succumbing to the present trends in terms of choice. The construction industry exists to serve its stakeholders needs and these must be met by the industry. So the CPS must be matched with the stakeholders objectives. Therefore the selection of a CPS must be based on the scope and nature of the individual project, which will render it capable of balancing the project parameters.

Suffice to say the CPS is without exception designed to take cognisance of the project parameters of cost, time, quality, and utility, within health and safety, and environment requirements.

Cost generally refers to first cost and involves budget allocation and contingencies; *Time* is usually described as timing and includes timing adjustments, timescales and float; *Quality* is basically the level of specification and incorporates specification adjustment to basic, medium or high; and *Utility* encompasses running costs, maintenance issues, buildability and flex-ibility for alterations and requires making value judgements between higher initial costs and longer-term savings. All these four project parameters or constraints need to be balanced within *health and safety, and environment requirements*. It is through this combination, that it is possible to create value in project management. In order to have a successful project, the chosen CPS must have an appropriate framework to deal with these variables.

So an ineffectively chosen CPS can be one of the major causes of project success or poor project performance by way of serious problems with the quality of constructed work, which could result in extensive delays on planned schedules, cost overruns, and a general increase in claims and litigation. Rwelamila (2000b) argues that an appropriate CPS is necessary in order to balance the project parameters and allocate risk appropriately, hence the foundation for harmonious and symbiotic relationships towards project success.

12.2.4   Methods for CPS selection

It is clear from the above discussion that each type of CPS has evolved to suit a particular client/stakeholder or project needs and that each CPS has certain characteristics that can be said to offer different strengths and weaknesses. Love, Skitmore and Earl (1998) aptly point out that there is one CPS that is in some sense 'better' than all others for an individual project although no one CPS is likely to be better than others for any project. Suffice to say the per-formance achieved on one project may not be attained on another since the CPS may be valid in the one context but inappropriate in another. This cements the principle that CPSs exhibit quite distinct characteristics, which are entirely symptomatic of the circumstances and hence

Table 12.1 Summary of characteristics of procurement systems

	SEPARATED AND COOPERATIVE				INTEGRATED				MANAGEMENT-ORIENTATED		
	Variants of conventional method				Variants of design and build						
	Conventional method	2-Stage selective tendering	Negotiated contract	Cost-reim-bursable	Design and build	Package deals	Turnkey	Develop and construct	Management contracting	Construction management	Design & management
Complete documentation at tender stage	✓	X	X	X	X	X	X	X	X	X	X
Open competitive tendering	✓	✓	X	✓	X	✓	A	A	X	X	X
Contractor collaboration in design process	X	✓	✓	✓	✓	X	✓	✓	✓	✓	✓
Project/technical complexity	X	✓	✓	✓	X	X	✓	✓	✓	✓	✓
Appropriate risk shared to Client	✓	✓	A	A	✓	✓	✓	✓	✓	✓	✓
Client involvement	X	✓	✓	✓	✓	X	X	✓	✓	✓	✓
Facility for variations	✓	✓	✓		✓	X	X	X	✓	✓	✓
Extreme cost for variations	X	✓	✓	A	X	X	X	X	X	X	X
Time management	X	✓	✓	A	✓	✓	A	A	✓	✓	✓
Economic financial solution	✓	✓	X	X	✓	✓	A	A	✓	X	X
Control of expenditure	✓	✓	✓	A	✓	X	A	A	X	X	✓
Price certainty	✓	A	✓	✓	✓	X	A	X	X	X	X
Quality control	✓	✓	A	A	✓	A	✓	A	X	✓	✓
Desired quality level	✓	✓	A	✓	X	A	✓	A	X	✓	✓
Desired utility	✓	✓	A	A	X	X	✓	A	✓	A	A
Division of responsibility	✓	✓	✓	✓	X	✓	✓	X	A	A	A

LEGEND

Yes	No	Average
✓	X	A

Source: Adapted from Masterman (1992).

the individual project. Cox and Townsend (1998) state that there is currently a tendency within construction towards the discipline of benchmarking performance and practices, which extends to the manner in which CPS are selected. In fact, construction industries in most developing countries have already established a poor track record in choosing between either only partially understood CPSs or blindly copying those of others in the hope that they will lead to similar end results. It is fundamentally important therefore to acknowledge the importance of appropriate selection of CPSs as a good practice.

Such an approach can be traced back to the early reports in the UK on CPS performance by Wood (1975). The rationale behind this approach is that different projects will have varying degrees of complexity associated with them so the CPS must be fitted to this complexity and hence to the particular project. The current trend in procurement, however, is that the selection process is based upon biased past experience where default rather than appropriate systems are chosen. Rwelamila (1999) suggests that default systems in terms of management structure and contract arrangements, are used merely because project stakeholders do not seem to consider the issue of selecting suitable CPSs. Instead of going through the exercise of validating and checking the CPSs in order to establish the ones that are appropriate for the project, the *devil you know* syndrome is adopted. That is, tried-and-tested systems are relied on with sometimes even particular favourites being developed. In Rwelamila and Meyer's (1999) words:

> *This leads to projects going wrong because the actual tasks peculiar to each project were not identified.*

In 1989 one of the aims and objectives establishing the Construction Project Procurement Working Commission (W92) under the International Council for Research and Innovation in Building and Construction (CIB) was to formulate recommendations for the selection and effective implementation of CPS. The reason being, as pointed out by Rowlinson (1999b), that there are a number of alternative routes available with a range of variables which, depending on their value, have an effect on the project strategy as a whole. It is therefore essential that the CPS chosen should be appropriate to the circumstances in order to ensure that the required success measures as indicated above are achieved.

Project consultants are confronted with the same situation of selecting appropriate CPSs when they are involved in projects. There are strong indications to suggest that what has been described as a trap of using the same CPS for any project, regardless of its suitability with the scope and nature of the project, is a construction industry-wide problem facing most of the project consultants (Rwelamila, 2000a).

But the range of available CPSs has proliferated in recent years as a result of the increased complexity of designs, escalating costs, the magnitude of projects and increasing demands of Clients, which all stem from the need to provide more sophisticated commercial and industrial working environments. Despite this the majority of clients and project consultants continue to select CPSs from a small number of the wide range available. The popular perception is that the *known and established* CPSs will always satisfy his or her needs, provided every project client fulfils his or her responsibilities. Masterman (1992) recognises that one of the principal reasons for the industry's poor performance is due to this sort of mindset where CPSs have been chosen in this manner.

Murray *et al.* (1999) support Mohsini and Davidson's (1986) view that the tendency in the construction industry is to bring forward previous CPS approaches to present projects, even though the projects may differ significantly in *component* and *structure*. It is therefore evident that the appropriateness of the CPS with respect to how it may cope with changed conditions is not considered a criterion, neither is consideration given to alternate options. That which has been used for years is always simply fallen back on. At first, there may be a number of notable successes, as the CPS is used under conditions for which it was originally intended. Then, as word of its success gains momentum, the new approach may be used less and less appropriately. It then becomes a matter of time before it becomes discredited and there are calls for yet another new approach. Therefore each new form of a CPS has essentially been developed in response to practical limitations that appeared in practice in previously popular methods.

With the advent of more experienced and sophisticated clients as well as projects becoming technically more complex, there has been an increasing recognition that the established CPS selection approach is inadequate. McDermott (1999) emphasises that the process of CPS should assume a greater status than it is normally afforded because factors have combined to force the construction industry into the position where it has to change to survive. Consequently, there is a desperate need to move away from the approach of blindly implementing previous practices and copying those of others. There are strong indications to suggest that most construction industries – especially English speaking Commonwealth countries face significant challenges in CPS selection. The need to assess the impact CPS on stakeholder management is very opportune if CPS good practices are going to be understood by all construction project stakeholders.

12.3   Theory and practice of project stakeholders management

12.3.1   *Project stakeholders – defined*

In any project, and especially in construction projects, many different and sometimes discrepant interests must be considered (Olander and Landin, 2005). Representatives of these interests are referred to as the project stakeholders. There is a plethora of definitions of a *project stakeholder* (Gibson, 2000; Karlsen, 2002; Gardiner, 2005; Pinto, 2007). Gibson (2000) defines a stakeholder as a person or a group of people who have a vested interest in the success of a project and the environment within which the project operates. This definition is closer to that of Pinto (2007), Karlsen (2002) and significant number of other authors and researchers. Gardiner's (2005) definition adds more clarity to the definition and his definition is used throughout this chapter, where project stakeholders are defined as:

> . . . .*individuals, groups and organisations who are actively involved in the project, or whose interests may be positively or negatively affected as a result of the project.*

Stakeholders according to Gardiner (2005) can be internal or external to the project. Owners, funders, banks providing capital, senior management, government departments/ ministries, parastatals, individual citizens, lobbying organisations and society at large are all potential project stakeholders. Figure 12.2 shows an example set of stakeholders for a project.

12.3.2   *Project stakeholders within the project environment – context*

Past research (Karlsen, 2002) has shown that most construction projects are sensitive to changes in the environment. The environment as indicated in Figure 12.3 is complex with a number of forces affecting the project. On the other hand, Karlsen (2002) argues that many projects experience that clients, end users and public authorities make tougher demands

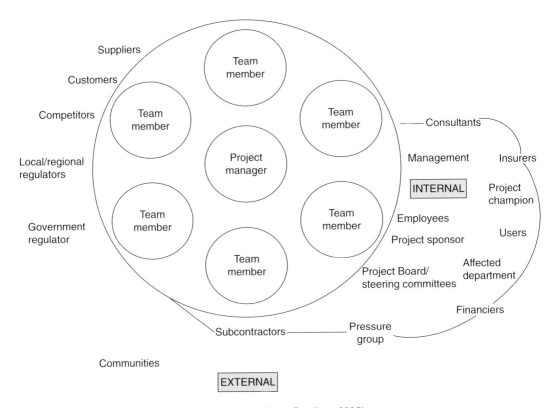

Figure 12.2   Stakeholders for a project (Adapted from Gardiner, 2005).

on project execution. Hence, he further argues that it is a mistake for project management (including procurement – *author's emphasis*) to ignore the stakeholders or attempt to impose a rigid control.

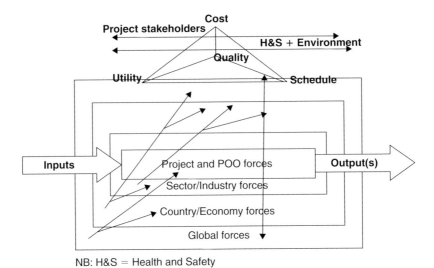

NB: H&S = Health and Safety

Figure 12.3   A 3-dimensional maze – *environment.*
(*Source*: Rwelamila, 2007)

According to Milosevic (1989), project stakeholders can be viewed as a network in which the actors interact with each other and exchange information, resources and results. He further argues that often the information and resources that are input to the project as indicated in Figure 12.3 are controlled by stakeholders. This control of information and resources gives the stakeholders a certain power. Changes in access to information or these externally controlled resources argues Milosevic (1989) can affect project *planning, organising, staffing and directing* of the project. It is the *project management process* which is primarily defined by the *PPS* that transforms inputs into outputs (see Figure 12.3), which implies the achievement of a unique change. Milosevic (1989) strongly argues that this alteration is beneficial for both a single client and a group of project stakeholders. It is the stakeholder who ultimately determines whether a project is a success, based on the project results (Jergeas *et al.*, 2000). From this discussion it is important to understand that some stakeholders have power because they control information and resources, while other stakeholders are important because they decide whether the project result is a success or not; therefore Karlsen's (2002) question – *Which stakeholder is the most important to the project?* need to be addressed.

Gilbert (1983) looks at stakeholders from the position of a project management model – *the 3-dimensional maze* (Figure 12.3). He argues that the project environment is complex and changing. If stakeholder management is not adequately addressed in the project, this can mean unexpected problems and uncertainty to the project caused by stakeholders. Addressing issues of project success and failure, he uses Meredith and Mantel's (2000) argument that a clear and comprehensive definition of project success and failure may not be determined, and consequently the PM may strive to meet goals that were never intended by the stakeholders. Additional problems and uncertainty caused by stakeholders that contribute to project failure include poor communication, inadequate resources assigned to the project, changes in the scope of work, unfavourable news about the project in the press, and negative community reactions to the project. According to Karlsen (2000) some stakeholders will cause high uncertainty during the project and it is not possible to predict which stakeholder is the most important to the project. The project set-up which include the project framework (defined by the PPS) and the way in which project stakeholders are managed will clearly influence the management and consequently the classification of stakeholders.

Earlier research results by Karlsen (2000) have identified that in many projects, management of stakeholders lacks strategies, plans and methods. He found that stakeholder management is often characterised by spontaneity and causal actions, which in some situations are not coordinated and discussed within the project team. The result of this practice is often an unpredictable outcome. To meet this challenge, several stakeholder management methods and guidelines have been introduced (Gilbert, 1983; Cleland, 1986; Savage *et al.*, 1991; Amabile, 1998; Newcombe, 2000; Jiang *et al.*, 2002; Gardiner, 2005; Pinto, 2007).

12.3.3 *Managing stakeholders expectations*

According to Gardiner (2005) there are often many stakeholders in a project and dissatisfied or disillusioned stakeholders can cause a project to fail. Managing project stakeholder perceptions and expectations is about generating agreement and harmony between the different views and beliefs held by the stakeholders. When all stakeholders are *'dancing to the same tune'*, the project moves towards a successful outcome.

Gardiner (2005) argues that in reality most projects have stakeholders that look at the project from different perspectives and hold different expectations. In a research project carried out by Newcombe (2000), it was found that project performance continues to be viewed in contrasting ways by different stakeholders. For example:

- The manager of a department that has requested a new information system may desire basic functionality and low cost, the system architect may emphasise technical excellence and the programming contractor may be interested in maximising profit.
- The owner of a property development project may be focused on timely performance, local government may wish to maximise tax revenue, an environmental group may wish to minimise adverse environmental impact and nearby residents may hope to relocate the project altogether.

These differences are not necessarily undesirable (Gardiner, 2005). Correctly managed, Amabile (1998) contends that different perspectives can promote creativity, which is often how the best solution eventually emerges. However, as a project progresses beyond the early planning stages, significant residual differences between the main stakeholders can be harmful and may interfere with the successful completion of the different phases of the project life cycle (Gardiner, 2005).

It is important to note that with congruence between stakeholders, the PM can be more confident that all effort is being directed towards the same project goals, and the project is less likely to suffer wasted time and effort, inefficiencies, scope changes, harmful conflict and unhappy customers/clients. One of the difficulties of managing this process, contends Gardiner (2005), is that many of these differences can remain hidden for a time. Through stakeholder analysis (as a technique) as discussed below, it can be used to help identify project stakeholders and discern their *values*, *beliefs*, *assumptions* and *expectations* at the start of a project. The stages involved are:

- identification and analysis
- mapping stakeholders according to their impact on the project
- response and resolution of issues

12.3.4 *Stakeholders identification and analysis*

According to Pinto (2007) and Gardiner (2005), identifying a few of the project stakeholders is easy, finding all of them can be fraught with difficulty. In addition to the more familiar technique of brainstorming, argues Newell (2002) other useful problem-solving techniques that can be used are brain-writing and the Crawford slip method, both adaptations of brainstorming.

Establishing what stakeholders really want is fundamental towards reducing uncertainty during the construction project development process. The process of establishing what stakeholders want is far from straightforward, but can be aided by asking them lots of questions. Gardiner (2005) argues that people are rarely clear in their first answers and a good PM (through an appropriate procurement system – *author's emphasis*) can build credibility by helping stakeholders to clarify their ideas. Furthermore, each individual stakeholder is embedded in some form of organisational culture that will fuel and/or influence their perception of and behaviour towards the project.

An appropriate PPS should be able to provide a project management framework when the PM is able to identify where one stakeholder's expectations clash with those of another and encourage the parties to talk to each other with a view to resolving areas of conflict (Rwelamila, 1999; Gardiner, 2005). This is a challenge in any project and the PM should be conscious of its dynamics. It should be said that this challenge is further complicated by the need to communicate, perhaps using different business language, with the various customer/client stakeholder groups (see Figure 12.2 above).

12.3.5   Mapping the impact of stakeholders onto project

Good practice stakeholders management approaches demand that once stakeholders and their expectations have been identified, they can be organised by mapping them in relation to their likely impact on the construction project. Gardiner (2005) insists by providing an example, that it is important to know:

- whether or not project stakeholders are likely to support or oppose the project
- the power and means available for project stakeholders to do so
- the predictability of project stakeholders behaviour and expected level of interest in the project

Although reporting on information systems development projects, Lederer and Sethi (1991) address stakeholders fundamentals which cut across industries and strongly suggest that these factors can be evaluated using matrices which aid the project sponsor and PM in understanding the various threats and appropriate management approaches to apply. For luck of space and brevity the mapping process, which is well covered by Nutt and Backoff (1992) and Bryson (1995) is not discussed here.

12.4   Impact of CPS on stakeholders management

In order to provide a clear analysis of CPSs impact on project stakeholders management it is important to acknowledge as described in detail above that project teams cannot operate in ways that ignore the external effects of their decisions. If project stakeholders are managed appropriately one of the requirements of project success will be met. In managing construction projects therefore, we are challenged to find ways to balance a host of demands and still maintain supportive and constructive relationships with each important stakeholder group. Knowledge about the impact of CPS on stakeholder management will go a long way to positively support this situation.

The primary areas to be considered when analysing the impact of CPS on stakeholders management are provided by Pinto (2007) and shown in Figure 12.4. These are:

1. Identification of project stakeholders – *both internal and external stakeholders should be identified for a fruitful stakeholder analysis.*
2. Assessment of the environment – *to assess if the project is low-key or if it is potentially so significant that it will likely excite a great deal of attention?*
3. Identification of the goals of the principal stakeholders – *as a first step in fashioning a strategy to defuse negative reaction, a PM will have a responsibility to attempt to paint an accurate portrait of stakeholders concerns (to be able to find an answer for the question: What, then, are the needs of each significant stakeholder group regarding the project?)*
4. Assessment of your own capabilities – *client organisation must consider what it does well. Likewise, what are its weaknesses? Do the PM and his or her team have political savvy and a sufficiently strong bargaining to gain support from each of the stakeholder groups?*

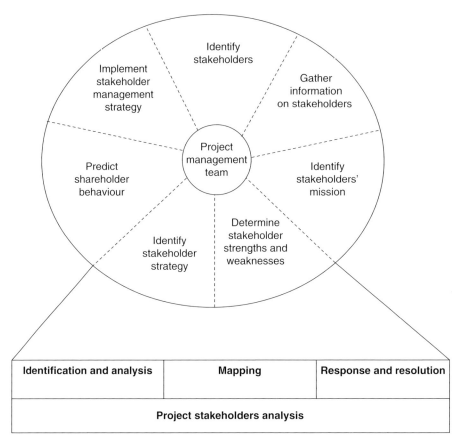

Figure 12.4 Project stakeholder management cycle.
(*Source*: Adapted from Pinto, 2007)

5. Developing solutions – *creating an action-plan to address, as much as possible, the needs of the various stakeholder groups in relation to the other stakeholder groups; and the need to have sufficient information before developing solutions (to avoid the trap of attempting to manage a process with only fragmentary or inadequate information).*
6. Testing and refining solutions – *the PM and team should realise that solution implementation is an iterative process-making a best guess, testing for stakeholder reactions, and reshaping strategies accordingly.*

It is important to note that the above six primary areas will bring different results to projects when analysed through the eyes of every procurement system. Since each procurement system has characteristics which make it different from the other, the characteristics will have an influence on the primary areas and consequently an impact on stakeholder management. In order to have a sound and robust argument on the impact of CPS to stakeholder management, Masterman's (1992) CPS classification and characterisation framework is used.

12.5   Impact of separated and cooperative procurement systems

12.5.1   The traditional method – traditional procurement system (TPS)

This is a typical dominant CPS under this classification known as the 'conventional method' or 'traditional method'/'traditional system' – where the unique characteristic of this system is the separation of the responsibility for the design of the project from that of its construction. Even where variants or hybrids of this basic (typical) system allow cooperation between the contractor and the client or his/her consultants, these two fundamental elements remain as two separate entities.

The fact that this CPS allows the process of project documentation to be completed before going to tender strongly suggests that there is sufficient time to identify internal and external stakeholders. This could be done parallel to the process of preparing project documents. Thus a PM with appropriate project management knowledge and skills will have a chance to manage the process of identifying project stakeholders provided by this CPS. Using the same argument, it will be possible to assess the project environment and establish its status in terms of its potential to excite a great deal of significant attention, which will need to be managed.

While this procurement method provides sufficient time to identify principal stakeholder goals and consequently their assessment and development of solutions, it is important to look at those stakeholder issues which it cannot help to address. If among stakeholders goals are issues like *contractor collaboration in design process and client involvement*, this procurement is very weak in dealing with these issues. Its framework is built on separating design from construction, hence the 'typical' TPS has no mechanisms to address these two issues. Without adjusting the TPS towards a responsive framework which could allow these issues to be accommodated, a negative impact on stakeholder management could be felt. There are two good practice options open to the PM towards a positive impact on stakeholder management – either to select a different procurement system (outside separated and cooperative procurement systems) which has inbuilt tools to deal with these issues or to use an adjusted TPS (hybrid TPS) – *TPS with additional mechanisms to deal with issues.*

12.5.2   Variants of the TPSs

- *Two stage selective tendering:* Under this variant the process is similar to the parent procurement system (the TPS) up to the time when tender documentation is produced (though most of these documents are approximate). If issues of high levels of variations are among the stakeholder's major concerns, then this procurement approach seem to generate a very low value of variations due to its inbuilt cooperative framework. Furthermore, this variant has a strong base for greater input from project stakeholders (especially the client and his/her advisors) and thus a great potential in strengthening teamwork among IS and consequently impacting positively on project stakeholder management. It is important to note that this procurement system can be an appropriate solution where it is desirable to secure the early involvement of a contractor (an IS) to provide form of expertise – where the client (the main stakeholder) wishes to commence work on site in advance of the detailed design of the project having been finalised. If minimising the overall project period is one of the main stakeholder's objectives, this system has a potential to provide a chance to overlap design and construction. All these possibilities will have a positive impact on stakeholder management.

- *Negotiated contracts:* Again this variant has a similar process to the TPS, but it is possible to appoint a contractor early in the design phase. The appointment of a contractor early could fulfil a number of stakeholder requirements – for example on *constructability, value engineering and construction methods* and thus impact positively on stakeholder management by strengthening the project satisfaction index of the project stakeholders. There is a negative side to this procurement system (during negotiations) which could have a negative impact on project stakeholder management – when the client has no option of being able to break off negotiations if the impediment to an equitable solution cannot be removed. Conflicts will become prominent and automatically affect the project stakeholders binding medium which is fundamental for stakeholder management.
- *Cost-reimbursable contracts:* This procurement system could fall under *'cost plus contracts'* or *'target-cost contracts' – described in detail elsewhere* (Masterman, 1992). The financial and practical risks associated with the construction project are shared more equitably by key stakeholders than when using most other procurement systems. This environment of sharing risk will have a positive impact on stakeholder management, because you will have project stakeholders who are committed to the project. Looking at this procurement system from the public sector projects angle, a very strong chance exist that public sector stakeholders could be affected negatively by the non-existence of contractual commitment – since there is no contractual commitment by the contractor to the final cost to be paid by the client or direct financial incentives for him/her to make the most efficient use of his/her resources means that the system cannot easily satisfy the concept of public accountability. Public accountability runs across the minds of most public sector stakeholders and the lack of it could have a strong effect on stakeholders cooperation and consequently this could have a negative impact on stakeholder management.

12.6   Impact of integrated procurement systems

12.6.1   Design and build (D&B)

Under this procurement system there is an arrangement where one contracting organisation takes sole responsibility, normally on a lump sum fixed price basis, for the bespoke design and construction of a client's project.

The process of identifying stakeholders is easy under D&B because the two central tasks of the project are implemented through a single point of responsibility – two of the key stakeholders are designers and constructors and these are operating under a single focus point. Furthermore, the process of assessing the environment around the project in terms of the impact of the project within its locality and beyond is easy to undertake since the two key stakeholders are working as a team. The same context will apply when identifying the goals of the principal stakeholders (designers and constructors under working as team) and when there is a need to develop solutions in order to address the needs of stakeholders.

The client responsibility of assessing his or her own capabilities is made easy under this procurement system because the client is not communicating with dispersed project experts. The majority of major experts under this system fall under designers and constructors who are primarily operating under one management umbrella creating an easy connection to the client towards assessing his or her capabilities.

It is important to note that if there are other stakeholders outside the project designers and constructors circle whose interests or expectations are on the management of *project/technical complexity; facilitation of high quality levels* and other related issues, this procurement system in its *'typical form'* does not have a mechanism to deal with these issues appropriately. If D&G

is selected in its *'typical form'* for a particular project where stakeholders are expecting to have a project framework to manage *project/technical complexity* and *facilitation of high quality work,* the results will be a clear dissatisfaction of the stakeholders, and thus a negative impact to stakeholders management. Looking at a *'typical'* D&G from the other side – say where project stakeholders are expecting *appropriate risk sharing; client involvement; facility for variations; time management; and close control of expenditure,* the results are likely to be positive and consequently this could have a positive impact on stakeholder management.

12.6.2   Package deals (PDs)

As the precursor and parent of D&B proper, the intention here is to use a proprietary structure system in order to produce a scheme which is unlikely to satisfy all of the client's needs. Provided that the client's requirements are flexible, this procurement system can be an attractive proposition particularly as the probable reduction in the design, approval and construction stages of the project can lead to savings in time and cost. For stakeholders who are not flexible in terms of their requirements this system will create conflicts and consequently affect issues of stakeholder management negatively.

   The majority of PD contractors, by their very nature, employ their own in-house designers and can thus be categorised as pure design builders and such be expected to perform well in terms of the speed and time criteria. If these two requirements are central to project stakeholders requirements, then smooth stakeholder management could be achieved. Some of the products of this system lack aesthetic appeal, but as the potential stakeholders are often able to see actual examples of the contractor's product before reaching a decision, this potential difficulty can often be avoided.

   In all other respects the PD replicates the characteristics of the D&B described above on stakeholder management.

12.6.3   Turnkey method (TM)

Since only one organisation (generally a contractor) is responsible for the total project from design through to the point where the 'key is inserted in the lock, turned and the facility is immediately operational', the responsibility of the contractor is extended to include the installation and commissioning of the client's process. It could also include installation and commissioning of equipment and sometimes the identification and purchase of the site, recruitment and training of management and operatives and the arranging of funding for the project. The fact that all processes are controlled through one point of responsibility (the contractor), strongly suggests that dealing with stakeholder management primary areas become very easy and if TM is applied with a good understanding of its characteristics its impact on stakeholder management could be more positive. Beside these approaches, TM generally echoes similar dynamics to its parent system (D&B).

12.6.4   Develop and construct (D&C)

Under this system one of the ISs – *the client's consultant* – prepares conceptual drawings/sketch designs, and a site layout often including the disposition of individual structures and their plan forms. Another IS – *the contractor* – develops the conceptual design, produces detailed drawings and chooses and specifies materials and submits these proposals with his/her bid in the same way as typical D&B. This system is most appropriate when key project stakeholders desire, or need to, determine the basic concept of a project before inviting competitive tenders and still requires a single organisation eventually to take responsibility for the detailed design and execution of the project. While a typical D&B system could affect stakeholder management negatively when key stakeholders are interested in using a consistently retained consultant with previous experience of similar types of projects, D&C facilitates this requirement. Beside this aspect, D&C impact on stakeholder management eco D&B, PDs and TM.

12.7   Impact of management oriented procurement systems

12.7.1   Management contracting (MC)

One of the project ISs – *the contractor (management contractor)* – is appointed on a professional basis as an equal member of the design team providing construction expertise, but the actual construction is carried out by other ISs – *the works or package contractors* employed, co-ordinated and administered by the management contractor. Almost all the primary areas of stakeholder management primary areas require a high degree of flexibility in order to enable delays to be overcome or reduced, changes to be absorbed (if any) and rescheduling of construction work to implemented. This system has an inbuilt framework of flexibility, hence a positive impact on stakeholder management equation. Furthermore, since one of the main concerns of key stakeholders is financial failures, the fragmented structure of the construction process (the use of works packages) reduces the effect of financial failures on the total project and consequently contributes positively to minimising stakeholders fears and thus brings a positive impact on stakeholder management.

   The need to develop solutions in order to create action plans to address the needs of various project stakeholders is at the heart of stakeholder management primary areas. One primary area of stakeholder management which affects both external and ISs is the maintenance of industrial relations fundamentals. The management contractor's knowledge and experience under MC ensures that industrial relations on the project are better than when using more conventional procurement systems, and thus strengthens one of the pillars of stakeholder management.

   The dynamics of risk distribution among stakeholders contribute significantly to 'messes' which affect the stakeholder management framework. Under MC a great deal of risk management connected with construction contractors, time overruns, remedying of work-package defects and even design seem to be the responsibility of the management contractor. This arrangement contributes significantly to the process of stakeholder management where the management contractor assumes the position of a single point of responsibility.

   It should be noted though that the MC system does not provide an effective framework to facilitate contractor(s) collaboration in design process, control of cost expenditure and higher levels of quality. When these issues are central to stakeholders expectations, other procurement systems with appropriate frameworks to deal with them should be used otherwise you could end up with dissatisfied stakeholders and thus a weak stakeholder management infrastructure.

12.7.2   Construction management (CM)

The construction manager as one of the ISs is appointed as a consultant during the initial stages of the project with equal status to other ISs responsible for design. The emphasis on management which is central to this system generally results in clients and designers making timely decision to match the needs of construction. It thus becomes easy for the majority of

ISs (as a team) to make a concerted effort to minimise the time and cost penalties that could be incurred by the client, and hence contributing positively to strengthening the relationship between the client and other stakeholders. When the relationship between the client and other key stakeholders is improved it translates to a conducive environment for stakeholder management.

It is important to note that all the six primary areas of project stakeholder management are predominantly connected to the dynamics of co-ordinating production of design information with the requirements of the construction programme. The MC has strong mechanisms for strict control to deal with co-ordination requirements, thus contributing significantly to a cohesive environment between design and production. This is a very positive feature of MC which makes project stakeholder management (especially when focusing on ISs) very smooth for the construction PM – the environment is characterised by a good sense of team work and a positive approach to problem-solving.

One of the major problems in managing projects is the problem of paying contractors suppliers late. Late payment affects these stakeholders negatively in terms of their cash flows and the consequence of this situation is that teamwork becomes an illusion and these stakeholders become difficult to manage. The fact that under MC the client is responsible for direct payment of the works contractor's accounts usually results in an improvement in the cash-flow position of the individual contractors as they are paid earlier than when operating under other procurement systems, on a *'pay when paid'* basis, when main or management contractors have this responsibility. This is one of MC characteristics which partly strengthens the stakeholder management equation.

Clients being among project key stakeholders need to be aware that the use of MC involves them in additional administrative duties and responsibilities to those accepted when using most other procurement systems and also increases the risks they carry, particularly those associated with cost overruns, delays and claims. If clients are not aware of these features of MC, their relationship with the construction PM (the construction manager) could be frustrated and consequently this could affect the whole framework of stakeholder management.

12.7.3   Design and manage (D&M)

Under this procurement system a single organisation is appointed to both design the project and manage the construction operations using package contractors to carry out the actual work. Whether contractor-led or consultant-led, this system has a common characteristic of enabling construction work to commence on site before the total design has been completed. With project stakeholders who are concerned with early commencement of construction work, the D&M framework provides the solution and thus contributing to the stakeholder satisfaction equation which consequently affects the stakeholder management environment positively.

A further common feature of D&M is the presence on site, generally on a semi-permanent basis, of the personnel responsible for the design of the project whose duties whilst resident include further detailed design, clarification of existing design details, liaison with works contractors to ensure constructability/buildability and working with the client's representatives so as to make certain that the project's functional requirements are maintained and achieved. This feature strengthens communication between ISs and contributes positively to stakeholder management because conflicts between designers and constructors are minimised.

When the impact of D&M on stakeholder management is observed by looking at a public project and private project the difference between the two environments need to be understood. When using D&M it is necessary that the client be prepared to commit to the project without any guarantee [unless the guarantee maximum price (GMP) approach is adopted], of final financial outcome. It is therefore likely that the private sector client will appreciate the value of the savings in time that can accrue, and be prepared to accept the need for this commitment, under conditions which would not normally be acceptable to the guardians of public accountability. Based on this argument, it is important to note that if a public client accepts D&M as the PPS without any guarantee (GMP), there are strong possibilities of conflicts in terms of issues of accountability which could affect the relationship between the client and the PM and consequently affect the stakeholder management environment negatively.

12.8   Conclusion

Project procurement systems have a significant impact on how construction projects are managed. It is through a PPS that the project management framework and project risks are shared between key stakeholders. The PPS framework provides the basis for determining the project management power structure. It thus creates an environment where project stakeholders are supposed to interact and the relationship between them. The management of stakeholders is conditioned by the type of procurement system.

Having said this, there is a remarkable lack of knowledge of construction procurement systems among construction experts and this situation has contributed significantly to a number of problems which have been part and parcel of the causes of construction project failures. The need to understand PPSs and their relationship to project management approaches cannot be over emphasised. It is through understanding different characteristics of PPSs that it will be possible to select appropriate PPSs for different projects. The selection of an appropriate PPS is a fundamental foundation of a balanced stakeholder management framework. The theme of this chapter is thus opportune in understanding the dynamics of PPSs and how a PPS influence project stakeholder interactions.

The impact of project procurements systems on stakeholder management has been covered in this chapter through a systems approach of looking at a project within an organic structure of 'an input–output model' (Figure 12.3). By reflecting on each procurement system, it has been established that PPS characteristics are the building blocks of the PPS effects on project stakeholder management. When a PPS is selected for a specific project and its characteristics are appropriate for that project, the implications are that there is a strong chance that stakeholder management issues will be dealt with smoothly. But when a PPS is selected for a particular project and its characteristics are not appropriate for that project, the likely result is to have perpetual conflicts during projects which will consequently affect stakeholder management negatively and frustrate the process of having a cohesive project tem.

It thus follows that in order to understand the impact of project procurements systems on stakeholder management you need to have a good knowledge of PPSs and their respective characteristics and fundamentals of stakeholder management. This chapter has managed to provide a good base necessary to understand from both theory and practice the relationship between PPSs and stakeholder management.

References

Amabile, T.M. (1998) How to kill creativity. *Harvard Business Review*, September–October:77–87.
Ambrose, M.D. and Tucker, S.N. (1999) Matching a procurement system to a client and project needs: a procurement system evaluator. In: Bowen, P. and Hindle, R. (eds) *Customer Satisfaction – A Focus*

for Research and Practice: Proceedings of CIB W55 & W65 Joint Triennial Symposium. Cape Town, South Africa: University of Cape Town, CD ROM, without page numbers.

Bryson, J.M. (1995) *Strategic Planning for Public and Nonprofit Organizations: A Guide to Strengthening and Sustaining Organizational Achievement* (rev. edn). San Francisco, USA: Jossey-Bass.

Chartered Institute of Building (CIOB) (1996) *Code of Practice for Project Management for Construction Development*. Singapore: Longman.

Cleland, D.I. (1986) Project stakeholder management. *Project Management Journal*, **17**(4):36–44.

Cox, A. and Townsend, M. (1998) *Strategic Procurement in Construction – Towards Better Practice in the Management of Construction Supply Chains*. London: Thomas Telford Publishing.

Edum-Fotwe, F.T, Thorpe, A. and McCaffer, R. (1999) Organisational relationships within the construction supply chain. In: Bowen, P. and R. Hindle (eds) *Customer Satisfaction – A Focus for Research and Practice: Proceedings of CIB W55 & W65 Joint Triennial Symposium*. Cape Town, South Africa, CD ROM, without page numbers.

Emmerson, Sir H. (1962) Survey of problems before the construction industries. *Ministry of Works, Controller of Her Majesty's Stationery Office*, London.

Ferry, J.F. and Brandon, P.S. (1991) The cost planning of buildings (6th edn). London: BSP Professional Books.

Flanagan, R. (1990) Making international comparisons in the global construction market. In *Building Economics and Construction Management: Proceedings of CIB 90*. University of Technology, Sydney, pp. 230–246.

Franks, J. (1984) *Building Delivery Systems – A Guide to Building Project Management*. Ascot: The Chartered Institute of Building.

Franks, J. (1990) *Building Procurement Systems – A Guide to Building Project Management* (2nd edn). Ascot: The Chartered Institute of Building.

Gardiner, P.D. (2005) *Project Management – A Strategic Planning Approach*. New York, USA: Palgrave Macmillan.

Gibson, K. (2000) The moral basis of stakeholder theory. *Journal of Business Ethics*, **26**(3):245–257.

Gilbert, G.P. (1983) The project environment. *International Journal of Project Management*, **1**(2):83–87.

Gordon, C.M. (1994) Choosing appropriate construction contracting method. *Journal of Construction Engineering*, **120**:196–209.

Govender, J.N. and Watermeyer, R.B. (2000) Potential procurement strategies for construction industry development in the SADC region. In: B. Ngowi and J. Ssegawa (eds) *Challenges Facing the Construction Industry in Developing Countries: Proceedings of CIB TG29A*. The Faculty of Engineering and Technology, University of Botswana, Late Paper.

Hamilton, A. (1997) *Management by Projects: Achieving Success in a changing World*. London: Thomas Telford.

Hardcastle, C., Tookey, J.E., Langford, D.A., Chair, B. and Murray, M.D. (1999) Re-engineering the building procurement decision-making process. *Customer Satisfaction – A Focus for Research and Practice: Proceedings of CIB W55 & W65 Joint Triennial Symposium*. Cape Town, South Africa, CD ROM, without page numbers.

Hindle, R.D. (1994) Construction education and the change from building procurement systems to construction marketing systems. *Direction – The UPE Construction Management Journal*, University of Port Elizabeth Construction Management Society, **2**(3):12–22.

Hughes, W.P. (1990) Designing flexible delivery systems. *Proceedings of CIB W92: Delivery Systems Symposium*, Zagreb, Yugoslavia.

Hughes, W.P. (1992) An Analysis of Traditional General Contracting. Construction chapter series No. 12, The Chartered Institute of Building, Ascot.

Ireland, V. (1982) Virtually meaningless distinctions between nominally different procurement methods. *Proceedings 4th International Symposium on Organisation and Management of Construction*, Waterloo, Canada.

Jergeas, G.F., Erin, W., Gregory, J.S. and Janice, T. (2000) Stakeholder management on construction projects. *AACE International Transaction*, P12.1–P12.6.

Jiang, J.J., Edward, C. and Gary, K. (2002) The importance of building a foundation for user involvement in information system projects. *Project Management Journal*, **33**(1):20–26.

Karlsen, J.T. (2002) Project stakeholder management. *Engineering Management Journal*, **14**(4):19–24.

Latham, Sir M. (1994) Constructing the team: final report of the government/industry review of procurement and contractual arrangements in the UK construction industry. *Controller of Her Majesty's Stationery Office*, London.

Lederer, A.L. and Sethi, V. (1991) Managing organizational issues in information systems development. *Journal of Business Strategy*, **12**(6):38–43.

Love, P.E.D., Skitmore, M. and Earl, G. (1998) Selecting a suitable procurement method for a building project. *Journal of Construction Management and Economics*, **16**(2):221–233.

Masterman, J.W.E. (1992) *An Introduction to Building Procurement Systems*. London: E & FN Spon.

McDermott, P. (1999) Strategic and emergent issues in construction procurement. In: S. Rowlinson and P. McDermott (eds) *Procurement Systems – A Guide to Best Practice in Construction*. New York: E & FN Spon, pp. 3–26.

McDermott, P. and Jaggar, D. (1996) A strategic exploration of procurement – seven years of CIB Working Commission W92 procurement systems. In: D.A. Langford and A. Retik (eds) *Shaping Theory and Practice – Managing the Construction Enterprise: Proceedings of CIB W65 Symposium*. Glasgow: University of Strathclyde.

Meredith, J.R. and Mantel, S.J. (2000) *Project Management – A Managerial Approach*. USA: John Wiley & Sons.

Milosevic, D.Z. (1989) Systems approach to strategic project management. *International Journal of Project Management*, **7**(3):173–179.

Mohsini, R.P. and Davidson, C.H. (1986) Procurement Organisational Design and Building Performance: A Study of Inter-firm Conflict. Translating Research into Practice 8.

Mohsini, R. and Davidson, C.H. (1991) Building procurement – key to improved performance. *Building Research and Information Journal*, **19**:106–113.

Murray, M., Langford, D., Hardcastle, C. and Tookey, J. (1999) Organisational design. In: S. Rowlinson and P. McDermott (eds) *Procurement Systems – A Guide to Best Practice in Construction*. New York: E & FN Spon, pp. 83–118.

National Economic Development Office (NEDO) (1983) Faster building for industry. *Controller of Her Majesty's Stationery Office*. London.

National Economic Development Office (NEDO) (1988) The changing building process. *Controller of Her Majesty's Stationery Office*. London.

Newcombe, R. (1999) Procurement as a learning process. In: S.O. Ogunlana (ed) *Profitable Partnering in Construction Procurement: Proceedings of CIB W92 & TG23 Joint Symposium*. Bangkok, Thailand: Asian Institute of Technology, pp. 285–294.

Newcombe, R. (2000) The anatomy of two projects: a comparative analysis approach. *International Journal of Project Management*, **18**(3):189–199.

Newell, M.W. (2002) *Preparing for the Project Management Professional (PMP) Certification Exam* (2nd edn). New York: American Management Association.

Nutt, P.C. and Backoff, R.W. (1992) *Strategic Management of Public and Third Sector Organizations: A Handbook for Leaders*. San-Francisco, USA: Jossey-Bass.

Olander, S. and Landin, A. (2005) Evaluation of stakeholder influence in the implementation of construction projects. *International Journal of Project Management*, **23**:321–328.

Pinto, J.K. (2007) *Project Management – Achieving Competitive Advantage*. New Jersey, USA: Pearson – Prentice Hall.

Root, D. and Hancock, M. (1996) Familiarity and procurement preference – putting the brake on the adoption of new procurement methods. *Economic Management of Innovation – Productivity and Quality in Construction: Proceedings of CIB W55*, Civil Engineering Institute of Croatia, Zagreb, Croatia, September, pp. 523–534.

Rowlinson, S. (1999a). A definition of procurement systems. In: S. Rowlinson and P. McDermott (eds) *Procurement systems – A Guide to Best Practice in Construction*. New York: E & FN Spon, pp. 27–35.

Rowlinson, S. (1999b) Selection criteria. In: S. Rowlinson and P. Mcdermott (eds) *Procurement Systems – A Guide to Best Practice in Construction*. E & FN Spon, pp. 276–299.

Rwelamila, P.D. and Savile, P.W. (1994) Hybrid value engineering: the challenge of construction project management in the 1990s. *International Journal of Project Management*, **12**(3):157–164.

Rwelamila, P.D. (1999) Harmonious and symbiotic relationships towards project success: myth or reality? In: S.O. Ogunlana (ed) *Profitable Partnering in Construction Procurement: Proceedings of CIB W92 & TG23 Joint Symposium*. Asian Institute of Technology, Bangkok, Thailand, pp. 93–106.

Rwelamila, P.D. (2000a) Selection of procurement systems – Why have we failed the test? *Unpublished report*, Department of Construction Economics and Management, University of Cape Town, Cape Town.

Rwelamila, P.D. (2000b) Project Delivery Systems and Contractual Arrangements. *Unpublished report*, Department of Construction Economics and Management, University of Cape Town, Cape Town.

Rwelamila, P.M.D. (2007). Project management competence in public sector infrastructure organisations. *Construction Management and Economics Journal*, **25**(1):55–66.

Rwelamila, P.D. and Meyer, C. (1999) Appropriate or default construction project procurement systems? *The International Journal of Cost Estimation, Cost and Schedule Control and Project Management*, **41**(9):40–44.

Rwelamila, P.D. and Ngowi, A. (1996) Allocation of project sources and the Inadequacy of the traditional procurement system. *Economic Management of Innovation – Productivity and Quality in Construction Proceedings of CIB W55*, Civil Engineering Institute of Croatia, Zagreb, Croatia, September, pp. 117–127

Rwelamila, P.D. and Savile, P.W. (1994) Hybrid value engineering: the challenge of construction project management in the 1990s. *International Journal of Project Management*, **12**(3):157–164.

Sanvido, V. and Konchar, M. (1999) Selecting project delivery systems: comparing design-build, design-bid-build and construction management at risk. *Project Delivery Institute*, Pennsylvania.

Savage, G.T., Timothy, W.N., Charlton, J.W. and John, D.B. (1991) Strategies for assessing and managing organizational stakeholders. *Academy of Management Executive*, **5**(2):61–75.

Sawczuk, B. (1996) *Risk Avoidance for the Building Team*. London: E & FN Spon.

Stewart, R. (1982) A model for understanding managerial jobs and behaviour. *Academy of Management Review*, **7**:7–14.

Turner, A.E. (1990) Building procurement. London: The Macmillan Press Ltd.

Wood, Sir K.B. (1975) The public client and the construction industries. *National Economic Development Office*, London.

13 Stakeholder Management in the Hong Kong Construction Industry

Steve Rowlinson, Tas Yong Koh and Martin Morgan Tuuli

13.1 Introduction

Hong Kong is world renowned for its impressive infrastructure, such as the airport at Chek Lap Kok, and famous buildings, such as the Hongkong Bank building and 2IFC, for the pace of the construction process and for the quality of the finished product. The construction industry is lauded for its 'can-do' attitude and the apparently high levels of integration and cooperation that enable its high level of performance. One might well imagine that an industry that can regularly complete 4-day floor cycles on high rise buildings over 40 storeys would be an innovative and relationship-based industry. However, this is not the case. For example, the predominant form of procurement in Hong Kong is still design-bid-build (the 'traditional' approach) and 'partnering' has been introduced into the industry but in a piecemeal fashion and in a manner which is hardly effective (Rowlinson and Cheung, 2008).

The historical context of the industry is important in understanding the current situation in Hong Kong. Hong Kong 'returned' to China after 150 years of British colonial rule in 1997. During these 150 years, a 'British' approach to construction was followed which focused strongly on the traditional approach and was regulated and administered by a strong civil service. This led to an industry which relied heavily on hierarchy, tradition and procedures in order to function effectively, but the industry was also heavily influenced by the Chinese culture in which it was situated. Hence, values such as face, harmony and conflict avoidance were also embedded in the industry culture. In such a situation, the issue of stakeholders and their management was paid scant regard; the government was used to making decisions on development rather than consulting widely and the other major players, the oligarchy of large property developers, adopted a simple, economic approach to their business plans and only over the past few years have issues such as corporate social responsibility (CSR) reached their boardrooms.

In this context, stakeholder management and relationship management can be said to be in their infancy and in some ways run counter to the ethos and philosophy of an industry where speed and money are king. However, Hong Kong people have become much more demanding of their government and institutions and have demanded that they be consulted and involved in all major and minor developments (e.g. the West Kowloon Cultural Hub, the Tamar Site Redevelopment, the demolition of the Star Ferry and Queen's Piers). Indeed, during the Handover period, Hong Kong people took to the streets demanding freedom and democracy and those demands continue to this day as political reform has come slowly to the colonial and post-colonial systems.

Having briefly set the scene, we present below two case studies, one a civil engineering project and the other a public housing project, which draw out a number of the themes

216

alluded to above by way of example. We then attempt to draw together some generalisations on how stakeholder and relationship managements are enacted in Hong Kong, identifying drivers and inhibitors to their successful implementation, and noting the impact of history, tradition and culture on how they are implemented and used in Hong Kong. We conclude with the assertion that stakeholder management and relationship management must be implemented in a context-specific manner in each instance and that a 'PMBOK' recipe style approach to these issues will not be effective nor efficient.

13.2 Stakeholder management initiatives

Project stakeholders are a person or group of people who have a vested interest in the success or failure of a project and the environment within which the project operates (Olander, 2007, p. 278). Vested interest, in turn, can be viewed as the actual or perceived benefits or risks/harms from the activities of construction project management (Donaldson and Preston, 1995). The project stakeholders may have a positive or a negative influence on the project. The challenge for the project team, hence, becomes one of implementing the project strategies such that positive stakeholder's influence is maximised and negative influence is minimised (Walker *et al.*, 2008). In analysing stakeholder management activities, it is useful to categorise stakeholders into two broad groups: primary and secondary stakeholders. Primary stakeholders are people or groups that have a legal contractual relationship to the project. Secondary stakeholders, on the other hand, are those who influence or are influenced by the project but are otherwise not regularly engaged in transactions with the project (Cleland and Ireland, 2007, p. 151). It is apparent that the client, the main and subcontractors, the quantity surveyor, suppliers and the like belong to the former group, while local communities and general public the latter.

13.2.1 Background

The Hong Kong Special Administrative Region (HKSAR) government has embraced the worldwide trend of sustainable development. Consequently, in the development front, the HKSAR government has emphasised sustainability and community development in procuring and implementing construction projects. Four sustainability dimensions have been adopted by the government when administering construction projects. These dimensions focus on economics, environment, society and resource utilisation. The client of the project in case study A implemented the four dimensions in all aspects of the procurement and the administration of the project. The thrust is based on sustainable construction, the aim of which is to progressively achieve sustainable development in public housing. The efforts are that of balancing the economic, social and environmental concerns of all the stakeholders in the project. To achieve these goals, various issues are embedded in the tendering and contracting procedures in the implementation of the project.

13.3 Project description – Case A

Bearing in mind the foregoing discussion, we now describe the context of project A. The economic dimension focuses on attainment of cost effectiveness of the project. Cost effectiveness is critical for economic sustainability because all aspects of the housing development, construction, through to operation and maintenance impact on the budget. Public funds are at stake.

The environmental dimension concerns the maintenance or betterment of the environment where the development is located. Construction activities have to be undertaken such that the impacts to the surrounding residents and community are kept to a minimum. Better construction methods and the use of more environmentally friendly construction materials are two strategies to achieve these objectives. Resource utilisation is related to the environmental dimension. The main thrust is to properly manage and reduce the consumption of resources in the construction processes. The production of waste and the use of energy are the two main areas of concerns. The social dimension is grounded in the client's belief that public housing and its development and construction have to promote social stability, economic prosperity and foster social cohesion. In the construction of the project, the client strives to provide a model working environment for those working on the site. As will become apparent, these dimensions are variously manifested in the procurement and stakeholder management of the project.

The project presented is one government project administered under such a backdrop. The project involves the construction of a public rental housing estate. Three 41-storey blocks are to be built. Each block measures approximately 50×34 m on plan. The blocks are approximately 117 m high from the ground floor to the main roof level. The three blocks consist of over 2300 rental domestic flat units of various types and sizes. Apart from these building works, there are also some civil engineering works. These works include excavation, filling, disposal, lateral support works for the raft foundations, and pilecap works for the three domestic blocks.

In addition, there are other structures that are incorporated in the project. The housing estate will be served by a neighbourhood elderly centre. The works involve the construction and fitting out of the centre. A bus terminus is to be built next to the estate. The works for the bus terminus include site formation, construction, backfilling, drainage works, street furniture and the associated services works. A two-storey lift tower with an attached footbridge connecting the estate to the adjacent residential areas and commercial centre are to be constructed. The pile foundations of these facilities are also included in the construction of these facilities. A double-deck walkway will be constructed to connect the current estate to the next estate. Finally, there is the construction of auxiliary structures. These include drainage and external works, slope improvement works, retaining walls, permanent protection to the existing gas offtake station, and road works within the estate.

The client has adopted innovative procurement initiatives for the project. Tagging along the works that are contracted out on a traditional design-bid-build approach are six Guaranteed Maximum Price (GMP) packages allowing design leverage and buildability scope on the part of the main and subcontractors. In addition, several contractual initiatives have also been pioneered in the project. It is under these innovative initiatives that various stakeholders of the project are engaged. We explore first the procurement arrangement in the next section.

13.3.1   Procurement arrangement

All works for the project were contracted via the traditional approach. The contract used for the project is the Government of Hong Kong General Conditions of Contract for Building Works (1993 Edition). Special conditions were added to the contract for the six GMP work packages. These six packages are: (1) the specialist external works (including the footbridge, lift tower, double-deck walkway, covered walkway and miscellaneous external works); (2) the enclosure to drainage reserve and the associated backfilling works; (3) the plumbing and drainage installation (including both the above and below ground drainage works); (4) the fire services and associated water pump installation; (5) electrical installation and (6) the superstructure of the domestic blocks other than the main structural frame. The majority of the packages involves design-and-build arrangement. The building services packages (i.e. the plumbing, electrical and fire services), however, involve only installation works with design provided by the client. These packages collectively represent some 30% of the project cost.

The procurement method is essentially a risk-reduced model developed from the private sector approach to target cost and GMP contracting. Risk reduction for the client is mainly realised through the contractor having to assume the risk associated with the design, development and the construction of the works, and the contractor's commitment on a price ceiling based on his design proposal at the start of the project. Two risk factors are particularly relevant for the project. These are the construction and excavation works at an extremely close proximity to the underground railway line and an existing live gas offtake station. The management of these risks is not only the responsibility of the main contractor, but is also vitally important because the contractor is also involved in the design of the works for these areas.

In addition to risks mitigation, the procurement approach enables the client to potentially reduce claims, integrate the diverse interests of a complex construction project, offers the contractor an incentive to provide value-added services by assimilating the contractor's expertise in the design and innovations in construction methods and materials to enhance buildability (Chan *et al.*, 2007). For the latter consideration, the contractor is rewarded for his creativity and improvement efforts on the design and construction of the works. The procurement route is depicted in Figure 13.1. The stages are briefly explained in the following paragraph.

In the design stage, the client identified work packages that require the integration of the contractor's input in design and buildability. The client then prepared a basic scheme design and performance specification for the work packages. These packages were subjected to subsequent detailed design development by the contractor. At the tender stage, due to the novel nature of the procurement method and the technical risks involved in the project, tenderers were invited from a pool of preferred contractors. These contractors had demonstrated their established track record in the areas of corporate strength, partnering commitments, capability to deliver a quality product, good safety and environmental performance, and experience with GMP arrangements. In so doing, the client hoped to achieve a balance between ensuring effective competition and selecting competent contractors. Tenderers submitted technical proposals for the modified GMP packages. Assessment was done on both technical and price evaluations. The ratio adopted was 70/30 on price to technical score. Tender interviews were conducted with the tenderers' teams. Criteria assessed for the technical category comprised of methodology and technical proposal for the work packages; the resources and expertise of the tenderer; relevant project experience; and safety, health, environment, and resources management of the tenderer. The exercise aims to exclude exceptionally low bids. Instead, the contractor was selected based on the best value offered considering the benefits of buildability, compliance to the specification, alternative design and construction proposal, future maintainability, and cost effectiveness. Because of the unique nature of the procurement arrangement that involves a design-and-build element, the GMP subcontract packages were let as domestic subcontracts. At the construction stage, the GMP packages operated under the 'open book' accounting arrangement where the client could 'see' the contractor's costs at both the tender and construction stage. It is said that the client can therefore understand the contractor's costs better. The improved understanding helped to facilitate mutual efforts in driving costs but not margins down. Given this context, we turn now to examine how various stakeholders are managed through a series of mechanisms.

Constructive engagement was implemented throughout the supply chain from the primary project stakeholders (the project team, the client, the subcontractors and suppliers) and secondary stakeholders (the community) in the project. These initiatives target each stakeholder's main concerns and attempt to match them. The avenues used in this aspect include shared saving among the main stakeholders of the client, main contractor and subcontractors;

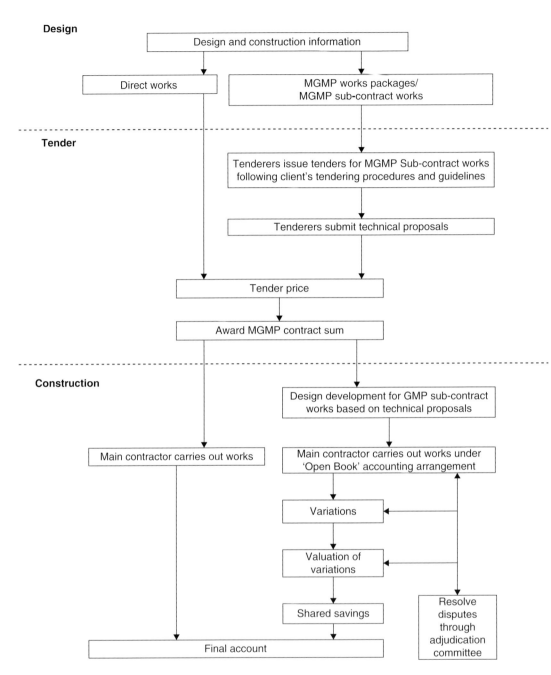

Figure 13.1   The contract procurement route (adapted with permission from HKHA (2005).

ensuring community benefits through various schemes administered by the main and sub-contractors endorsed by the client; ensuring workforce benefits and welfare; and project team members' human resource development. From the perspective of stakeholder management, the two salient thrusts in driving these initiatives are client proactiveness and farsightedness, and the main contractor 'coming-of-age' CSR awareness and its manifestations.

We discuss first the primary stakeholder management initiatives as administered by the client for the project. The impact of the procurement method on stakeholder management is manifested in the institution of supporting mechanisms that are crucial for the successful implementation of the procurement method. These mechanisms are: (1) 'gain share and pain share' arrangement, (2) project dispute resolution system and (3) promotion of a collaborative work environment.

Unlike the traditional GMP scheme whereby the sharing arrangement is only limited to the gain (Chan *et al.*, 2007), both the 'pain and gain' are shared for the project. The cost saving for GMP packages is shared equally between the client and the main contractor, i.e. 50%:50%. However, the contractor is only entitled to 15% of his portion of the saving. The remaining 85% is to be shared between the contractor and the GMP subcontractor(s) on a pro rata basis based on the contribution to net savings by both parties. This arrangement potentially motivates both the main and subcontractor(s). The project dispute resolution system is implemented to resolve disputes that might arise at source. The system laid down steps for amicable dispute resolution by inaugurating an adjudication committee. The committee comprised representatives from the client, the main contractor, subcontractors and the quantity surveyor. Under the system, the aggrieved parties first attempt to settle their dispute/s in good faith through the adjudication committee. If this step fails, the dispute will then be referred to the senior management of both parties. Arbitration is used as the last resort. An independent dispute resolution advisor is appointed to facilitate the resolution process. Clearly, a collaborative environment is fundamental in smoothening project works. As with most of the government projects, team spirit of the project is developed through a partnering approach. While minor disagreements were present, generally it was observed that better communication and understanding were achieved among the main contractor and client teams. Informal 'workshops' convened by the project architect especially at the beginning stage of the project were particularly effective in promoting cooperation among various parties. Although originally intended to solve technical problems, the constant contacts of participants throughout the workshop sessions had produced a 'side effect' of improving relationships due to close and frequent contacts. More problems were solved promptly on site. The parties became more proactive in working towards achieving common objectives. The cordial relationship between the client and main contractor's team was evident through an incident where both parties waived the design processing costs for the GMP packages. It appears that the procurement method with its associated supporting mechanisms facilitated positive stakeholder management.

There are other general initiatives adopted by the client which are not specific to the GMP procurement method. The main contractor's active participation also contributed to the smooth implementation of these initiatives.

The first initiative is the workers wage protection scheme. The scheme is a direct response to workers' grave concerns on the protection of their wages in the event of default by contractor or subcontractor as evident in recent cases (Lau, 2008). The institution of an on-demand bond in the contract can be used to secure payment of wages for the affected workers. Active monitoring of workers' wages is effected at operational level. A labour relations officer (LRO) is employed on site to check, verify and monitor workers' wage records. The LRO also receives, acknowledges and records complaints, and follows up those complaints on site. Active engagement and participation are required at the lower stream of the supply chain. Both the main and subcontractors are required to operate under the wages declaration system. Under the system, the subcontractors are required to pay their workers on time before applying to the main contractor for their monthly payment in conjunction with works done. On time payment to workers was stressed for the project. The computerised wage monitoring system employed on the project is equipped with a sophisticated mechanism to track wage payment such that if late payment to the workers is encountered, the system issues a warning and the subcontractor's payment will be delayed. As cash flow is vitally important to the subcontractors, it appears that information technology has been disguised as a powerful administrative instrument instituted to monitor workers' wages. The main contractor's responsibility is to implement and maintain the computerised system, and to manage the subcontractors' payments. The workers, then, are responsible to actively report work through the system with the use of their access passes.

The main contractor of the project has also adopted other primary stakeholder management initiatives. These initiatives concern mainly the onsite welfare provision for the workers and staff, and human resource development for the main contractor's site management team. The former involves a health promotion programme that includes basic health check and health counselling for workers with health conditions (e.g. hypertension), cash prizes for high performing workers, heat stress preventive programme in view of the high temperature summer working periods, the provision of mobile mist generating machines, installation of thermometers throughout the site, the provision of workers' quarters and laundry areas, etc. The latter initiatives address the project team members' concerns for their personal development and enrichment. The team members have been encouraged and sent to attend various personal development courses that include management skills, technical skills and leadership courses. However, what is more pertinent to the human resource development drive by the main contractor is the great length the main contractor went to in order to emphasise the shaping of a familial atmosphere among the site team. Chief of those initiatives administered by the main contractor, through the senior management staff on site (e.g. project manager and site agent), are the coaching programme, recognition, and the active seeking and provision of opportunities for site staff to try new things within their capability. A systematic recognition and promotion scheme (both financial and positional rewards) was administered in the project. The results observed were the promotions of some site staff and the re-joining of some junior engineers after the completion of their industrial training with the main contractor. In addition, various industrial awards and accolades won by the main contractor in recent times have attested to the efficacy of the approach adopted. The observation is in line with studies conducted elsewhere that indicate the clan type culture that emphasises people orientation is more conducive to successful project outcomes, albeit in the area of quality management (Thomas *et al.*, 2002).

The management of secondary stakeholders, in particular, on the part of the client has seen a saliency in the client's proactiveness in engaging stakeholders. The client identified principal stakeholders as those who are an integral part of their development and construction operations. Of importance is the recognition of the essential roles the stakeholders play in meeting their visions. The client has built into their planning and development processes community engagement initiatives. For the project, a series of activities designed to instill a greater sense of belonging and participation of the community in the project were implemented. A competition for mural painting was organised in the community in conjunction with the project. The winning design had been incorporated as a permanent mural feature for the estate. 'Action Seedling' was another activity implemented to promote community participation in the project. The community was encouraged to participate in the greening of the project during the construction stage. In this activity, local residents and school children from nearby schools participated in planting seedlings and nursing the plants for the estate under construction. In addition, the adoption of construction technology also contributed to the betterment of the community. Extensive use of prefabricated building elements and hard paved site areas are two technological measures aimed at reducing dust and noise.

It is apparent that all the activities cannot be smoothly rolled out without the active participation of the main contractor. In what appears to be the response to the client's push for active community engagement at the beginning, the notion of CSR has gradually evolved throughout the organisation of the main contractor over time. As a result of the increased awareness of the impact of their activities on the community, the contractor has been active in participating and responding to the client's drive for community engagement, at times, going beyond the requirements of the client. Two incidents exemplify the contractor's active involvement. The first involved the contractor's volunteer house improvement activities during a festival to help the elderly residents at the nearby estate. The main contractor dispatched two teams of personnel to help repair malfunctioning services within the flats of elderly residents. The second concerned the main contractor volunteering construction related information to the nearby residents in terms of prolonging construction activities beyond normal working hours (i.e., 7.00 pm). The improved communication between the project team (both the client and main contractor teams) and the community resulted in reduced complaints and a more positive impression from the residents.

The stakeholder management initiatives in terms of the identified stakeholders, their interests, impacts and the strategies adopted to address their concerns are summarised in Table 13.1 at the end of the next section.

13.3.2   Implications from case study A

Several implications can be drawn from the foregoing discussions of the project stakeholder management in this project.

Passive reaction among the subcontractors and junior staff members

The passive reaction refers mainly to the initiation and participation of the parties in the implementation of stakeholder management. The situation is particularly evident in the management of secondary stakeholders. For the project, initiation of stakeholder management was mainly driven by the management of the client and the main contractor. Little effort came from the lower echelon of the project organisation. The contribution from this hierarchy of members came mainly in the form of carrying out instructions from their supervisors/managers. It appears, therefore, that the members of the lower echelon are adopting a minimalist approach. For these members, engaging with the external stakeholders does not readily contribute to their immediate works. As both the main contractor and the client are fully committed to the stakeholder management paradigm, the issue is one of engaging the lower echelon of the project organisation so that a uniform and positive attitude can be inculcated.

The lack of a structured approach to project stakeholder management

The preceding observation is symptomatic of the present issue of the lack of a structured project stakeholder management system on the part of the main contractor. The deficiency is particularly acute with external stakeholder management. Despite considerable success in dealing with and tackling issues with the community, the main contractor admitted that their approach was one of trial-and-error and experimentation. Most of the stakeholder management initiatives rolled out in the project were implemented for the first time, at times without thorough deliberation. For the main contractor, while there are elaborate procedures and guidelines dealing with the internal stakeholders, the guidelines for managing external stakeholders, especially the communal stakeholders (e.g. surrounding residents, property and estate management agency, and district councillors), had not been established. Some of the initiatives appear *ad hoc*. In particular, there was no structured approach to identify external stakeholders, their impacts and the method of engaging them; yet methodologies currently

Table 13.1   Stakeholder management initiatives for the case project

Stakeholder	Stakeholder interest	Impact	Strategy
Procurement-specific initiatives			
Client	Embraced contractor expertise, improved buildability	Less buildable design	Design-and-build element in the GMP packages
Client	Cost certainty, risk reduction	Cost escalation	Introduce GMP scheme
Main contractor	Equitable cost and risk sharing	Cost escalation	Client administers pain and gain share scheme
Subcontractors	Enjoy the benefit of saving	Less motivated to suggest buildable design	Client administer pain and gain share scheme
Client, main contractor and subcontractors	Better disputes resolution	Cost escalation, delay and negative relations among parties	Client administers dispute resolution advisor system
Client, main contractor and subcontractors	More amicable working environment	Negative and adversarial working relationships	Developing team spirit through project partnering
General initiatives			
Workers	Prompt payment of wages	Low morale, work stoppage	Wage protection scheme
Workers	Welfare and safe working environment	Low morale, lost productivity due to incident/accident	Main contractor provides safe and comfortable working environment, health promotion
Project team members	Self-improvement and promotion	Low morale and productivity	Main contractor implements human resources development
Project team members	Familial working team	Low morale	Main contractor's project manager promotes team cohesion
Client, main contractor	Organisation and company image	Negative publicity	Active engagement with community and public to improve communication and impression
Client, main contractor	To be recognised as socially responsible corporate entity	Bad corporate image	Active implementation of corporate social responsible activities
Community and public	Participation in the development of estate	More complaints	Client and main contractor's engagement activities and communication sessions
Community and public	Less disruption of their living environment	More complaints	Noise and dust reduction construction methods
Elderly residents at adjacent estate	Malfunction within-unit services repaired at low or no cost	(NA)	Main contractor free attendance to the units

exist for their identification and management (see, for example, Walker *et al.*, 2008). The main contractor appears to be passive in that they were taking the cue from the client. However, the client, apart from stating an intention and commitment at the strategic level, did not provide an actionable guideline to the main contractor. While the efforts and achievement of the main contractor have to be commended, the situation reflects the somehow parochial mentality of the construction fraternity in terms of external stakeholder management.

Contracting firms have traditionally adopted the attitude that construction operations are confined within the boundary of the site. Site operations are therefore a closed system. This view overlooks both the direct (e.g. dust and noise) and indirect impact (e.g. bad impression resultant from direct impact) on the community. In terms of engaging external stakeholders and mitigating the impacts construction activities have caused, it is not in the interest of firms to do more than necessary as costs are incurred in extra efforts. Hence shareholder management and interest still overrides the stakeholder paradigm. The project is typically not described and hence not operated in terms of external stakeholders' interests. That is, the stakeholders' perspectives are not integrated into the project formulation processes despite the best intentions of both parties (cf. Cleland and Ireland, 2007).

No allowance for additional resources for stakeholder management

Despite the various external stakeholder management activities that had been carried out by the main contractor, there was no provision of additional resources available for the main contractor under the contract. The reward from the client comes in the form of recognition. Both the client and the main contractor are fully committed to making the project a success in most if not all aspects. In addition, given its status as a pilot project, the ensuing image issues and the high stakes involved especially for the two primary stakeholders of the client and main contractor (Mahesh *et al.*, 2007), the main contractor resorted to adsorbing the extra costs. However, while the costs involved in carrying out those activities are not considerably large, the lack of compensation from the client may lead to only token efforts from the main contractor. The situation may be more acute for the cost conscious contractor. It is therefore desirable to provide some financial support and introduce an appropriate disbursement mechanism to entice the main contractors to exert effort in managing external stakeholders.

Engagement of specialist subcontractors from the client's nominated list

The subcontractors for two GMP packages were 'novated' from the client's nominated list, but because of the nature and element of design-and-build inherent in the packages, these subcontractors were engaged as domestic subcontractors. The arrangement is seen as a move to improve the buildability thereby achieving a cost saving design. However, although the arrangement helps ensure quality control to some extent for the client, it can reduce the main contractor's capacity to stay within the GMP (Haley and Shaw, 2002). In addition, the level of cooperation between these novated subcontractors and the main contractor needs extra attention and promotion. For this project, it was observed that the client's intervention was invoked in the initial stage of the project to bring the parties together. In the long run, however, a more appropriate arrangement needs to be implemented.

13.4   Project description – Case study B

The project is an infrastructure project, comprising a 1.1 km elevated viaduct dual three-lane carriageway (average 65 m above ground) connecting a tunnel (under construction) on one end, to a cable-stayed bridge (under construction) at the other end. Together, they form an integral part of a 7.6 km long major highway. The project site is reclaimed land (to be handed over in phases) surrounded by industrial facilities, container terminals and an educational institution. The contract is a re-measurement type, traditional design-bid-build approach, with an initial contract period of 40 months. There is also a non-contractual partnering arrangement in place. The client is a major works department of the Government of Hong Kong and the contractor is a joint venture between a Hong Kong-based French company and a Chinese state-owned company. The consulting engineer is a Hong Kong-based international engineering consulting firm.

The peculiar features of this project, especially its size, location (vertically and laterally) and technical complexity, brought together a myriad of stakeholders, whose interests needed to be aligned at various phases to successfully deliver the project.

In the next section(s), the management of stakeholders on the project is analysed using data gathered through documentary records and interviews with key project participants. Five incidents, involving critical and contentious issues during the construction phase of the project, are used to illustrate how the stakeholders surrounding each incident were identified, managed or mismanaged individually and collectively in resolving the various issues, as in case A. The impact of the procurement arrangement on the configuration of the project stakeholders and the implications for their management are also discussed.

13.4.1 Incident analysis

Interface arrangement

The contractor proposed sometime after the commencement of the project to change the nature of the original arrangement regarding the temporal use of the deck of an adjoining bridge project (under construction), as a platform to station a launching girder in order to manoeuvre and launch viaduct segments. The proposed change was to position the launching girder beyond the point originally proposed in their technical proposal at tender and which was subsequently built into the contract as an interface arrangement. From the contractor's perspective, however, the change was necessary to make the launching operation simpler and safer. Yet, given the significant shift from the original plan, the new proposal had various implications for progress and risks. In particular, late resolution of the issue could jeopardise the achievement of the project Key Dates. To resolve this issue, however, the input and buy-in of a host of stakeholders were required. The stakeholders in this incident comprised the following, both internal and external to the viaduct project organisation:

1. The Client (same for both projects)
2. Viaduct Contractor (viaduct JV contractor)
3. Independent Checking Engineer (ICE)
4. Bridge Contractor (bridge JV contractor)
5. Engineer's Representative (ER) (viaduct project)
6. Engineer's Representative (bridge project)
7. The Engineer (viaduct project)
8. The Engineer (bridge project)
9. Project Board of Directors (Viaduct JV Contractor)
10. Project Board of Directors (Bridge JV Contractor)

A number of critical and contentious issues regarding the new proposal were apparent.

1. The structural stability of the bridge deck to withstand the imposed loads beyond the original point needed to be established.

2. Cast-in items were required on the pier and bridge deck to facilitate the positioning of the launching girder.
3. Partial removal of some of the temporary supports to the bridge deck was required to avoid collision with parts of the launching girder.
4. The works programme could be derailed if the issue was not resolved in a timely manner, jeopardising the achievement of Key Dates.
5. Responsibility for the risk and liability for any unforeseen circumstances regarding the proposed operations needed to be established.
6. Associated cost and time liability needed to be established.

It was therefore the contractor's responsibility to obtain buy-in of the various stakeholders identified above to resolve all of the above issues of contention. In doing so, the stakeholders were engaged both formally and informally. For example, the issues regarding the structural stability, partial removal of temporary supports and cast-in items, which were within the domain of the Bridge Contractor, were discussed in the first instance at their regular monthly interface meeting. At this meeting, the Bridge Contractor agreed in principle to check the feasibility of the issues raised and to give its response.

While the first three issues, which were technical in nature, were easier to resolve with the Bridge Contractor, the last three, which were contractual, were most problematic. In terms of risk and liability regarding damage to the bride deck works, this was covered under an Owner Controlled Insurance Programme (OCIP) taken by the client to cover all the projects within the 7.6 km highway. The contentious issue was however with potential claims from either contractor for extension of time or associated cost due to any unforeseen prolongation arising from the proposed arrangement. It became significantly more contentious when the client requested that The Engineers of both projects get undertakings from their respective contractors not to claim time or costs associated with the proposal if approval was granted. Apparently, a similar arrangement on one of the client's previous projects had resulted in huge prolongation claims from one of the contractors and thus reinforced the 'baggage' parties carry from one project to the other. The Client's suggestion was however at variance with the contract provisions in both contracts (the viaduct and the bridge) that allowed the contractor to claim extension of time and additional payment for interface issues if the issue requires the contractor to act in a manner which goes beyond his obligations under the interface provision. The ER on the viaduct project however consequently requested a full risk assessment on the issue from the contractor, emphasising as well that the client would only give consent for the proposal to go ahead if the Viaduct Contractor was willing to accept full liability for any eventuality. A contingency plan was also requested from the contractor, in the event that the proposal was not approved by the client.

To obtain buy-in of all parties regarding the viability of their proposal, the Viaduct Contractor organised and delivered a presentation on the sequences involved in their new proposal regarding the use of the bridge deck. Yet, this did little to persuade the parties to shift their positions. The Client maintained his position of no approval without waiver of rights to claim time and associated cost by the contractors. The contractors also maintained that they could not waive that right. While this was generally a contractual matter, it also highlights the cultural disposition of uncertainty avoidance in a Chinese work context and the tendency to work strictly according to the rules (or contract in this case). Not even the double assurance provided by the use of the ICE to provide an independent assessment of the safety and structural soundness of the proposal could persuade the parties to reach an agreement.

At this time it had been about 5 months since the proposal was put forward and there was still no end in sight, and the launching girder was within weeks from the point where access to the bridge deck was required. The Client continued to emphasise the need for an early resolution, yet was not prepared to compromise on its stand. All parties then agreed that this interface issue be resolved as soon as possible and that a *drop dead* date for a conclusive decision on the proposals be established. As the issue dragged on the contractor's revised works programme could not also be approved, partly because it was contingent on the new proposal and thus failed to meet the established Key Dates. At this stage, the remaining contentious issue unresolved was still that of waiver of right to claim. The Engineer for the viaduct project then wrote to request the contractor to confirm their acceptance of all direct and consequential costs if the launching girder was in an accident or incident involving the bridge works resulting in prolongation. The contractor however indicated in three related letters to the ER that they could not accept additional liabilities as that constituted *additional constraint* under the contract. They further indicated that they were studying alternative temporary works to enable the undertaking of the launching of the segments without using the proposed bridge deck beyond what was originally proposed, but that these had cost implications. In a reply, the ER reminded the contractor of their contractual obligation to indemnify the client irrespective of which proposal they chose to go with. To put an end to the *ping-pong letters* that were becoming the main mode of communication regarding this issue, a meeting was then scheduled to specifically deal with the issue. However, as the issue could not be resolved, the contractor was requested to revert to the original sequence of segment erection in the technical proposal at tender or submit alternative proposals for consideration. Out of options and running out of time as well, the contractor agreed to revert to the original proposal and thus prepared and submitted a proposal to the ER accordingly. This proposal included a method statement, risk assessment, detailed interface arrangement and various ICE certificates as required. This was approved by the ER. As this was also the outstanding issue making it impossible to have the revised programme of works approved, the contractor also prepared and submitted the programme in line with the original arrangement.

Evidently, about 6 months was spent needlessly, only to revert to the original proposal. Ironically, the segment launching operation which was the subject of about 6 months back and forth discussion and 'ping-pong letters' actually took less than 3 weeks to complete after reverting to the original plan. It is interesting also that the various stakeholders in this, especially The Client, took positions that appeared at variance with the spirit of the non-contractual partnering that was in place on the project and that was continuously reinforced through various workshops. Indeed, an attempt to use the partnering process to resolve this issue was met with silence from all parties, reinforcing the sceptics' belief that many parties who sign up to such non-contractual partnering arrangements have little commitment to working in 'real' partnership. One of The Client's team members was particularly unequivocal when he put it rather bluntly in an interview that:

> *Under the partnering spirit, we organize . . . workshops and . . . discussions with facilitator where we can express our opinion, . . . but still the roles of the engineer, the employer and the contractor are still clear under the contract. . . . partnering . . . [is] there to facilitate any exchange of opinions, but not as a forum for making decisions. . . . It is not a forum for making decisions. Of course, we have our own decision, whether to proceed with a certain idea, but that's not contractual. So you have to slightly distinguish this . . .*

While there appear to have been genuine efforts by the contractor (may be because the contractor stood to benefit most if the proposal was approved) to engage and obtain buy-in, it is doubtful whether any alternative mode of engaging, especially The Client, could have yielded a different outcome. Public project settings are particularly replete with risk averse and fear of blame attitudes. This, rather than the means of engagement of the parties, may be why a proposal such as this was predisposed to failure.

In a related incident a couple of months later, the Bridge Contractor needed to erect cranes on the same bridge deck area. Given the elevated nature of the bridge (about 70 m above ground), the only viable access point to the deck was through the viaduct, as the other end of the bridge under construction was still hanging over water (sea). The Bridge Contractor therefore requested to use the viaduct in order to move the cranes and associated equipment to the bridge deck. This proposal was accepted, apparently because the trucks and crane parts that were to be delivered were not expected to impose any loads beyond what the viaduct has been designed to withstand. Given that the element of uncertainty was greatly reduced in this case, a resolution was a straightforward matter. But more importantly, this suggests that when two projects share boundaries like this, interdependence is inevitable. Thus, parties must always remember that often, 'what goes around, comes around'. If the Bridge Deck Contractor had refused to consider the proposal put forth by the Viaduct Contractor to extend use of the bridge deck, then the Viaduct Contractor could also have used that precedent to refuse access to the viaduct or unnecessarily delay the granting of such access.

Temporary Traffic Arrangement

To facilitate the works and safeguard the public, it was necessary from time to time during the project to temporarily divert traffic passing through the site. This normally involved full or partial closure of some or all roads. These changes to the normal movement of traffic are handled under what is called 'Temporary Traffic Arrangement' (TTA), and is governed by the 'Code of Practice for the Lighting, Signing and Guarding of Road Works' 4th issue which came into force on the 1 July 2007. This incident revolves around the TTA schemes on the project and how they were managed to reduce non-compliance (NC) and inconvenience to the public and the engagement of various stakeholders.

The stakeholders in the TTA schemes included the following:

1. Road users (General Public)
2. Client's Audit Team (Research and Development section)
3. Traffic Management Liaison Group (TMLG) (which comprised the Client, Contractor (and his Transport Consultant), ER, Police (Road Management Office and Traffic Management Bureau), Transport Department (Engineering and Operations section), Representatives of Adjoining Businesses, Lands Department, Local Council representative)

The key stakeholder was the TMLG, whose decisions supersede the contract provisions regarding the TTAs. The key players in the TMLG were the police and the Transport Department, with the other members tending to go with whatever these two decided.

The contractor and his Transport Consultant were responsible for the design of the TTAs. The proposed arrangements are then presented to the TMLG for deliberation and approval. Once approved, the TTA becomes the standard against which NC is determined. Given the importance The Client attaches to the TTAs, an Audit Team (the Research and Development section) carries out on average about 10 audits every month regarding the performance of TTAs and issues NC for breaches.

For several consecutive months in the course of the project, the Audit Team continually issued NCs for various breaches. The Client's project team expressed their unhappiness about the situation and asked the ER to step up their own inspections to forestall any future breaches. It was then agreed that representatives of the ER and the contractor would check on a daily basis to ensure that the TTAs were implemented to the required standard.

The ER together with the contractor then instituted various measures to prevent contraventions of the TTA arrangements. Central to this was increased joint inspections. Three inspections of the TTAs were carried out daily. The first inspection often took place in the morning where breaches were identified for rectification. In the afternoon the ER carried out another inspection to ascertain that the breaches that were identified in the morning had been rectified. The last inspection was then undertaken jointly by the ER and contractor at about 4 pm to make sure all TTAs are still in order. Following the success of these measures in reducing the NCs to zero for the following months, the Client suggested that the ER circulate such measures to the Client's other projects for possible use, since they were apparently having similar problems. In a show of support for the contractor's efforts in upholding the standards regarding the TTAs, the Client's project team personally appealed on one occasion on behalf of the contractor when the Audit Team issued an NC which the Client's team thought was unfair. The inspection team subsequently withdrew the NC.

The management of public expectations was also central to the success of the TTAs. To help reduce the inconvenience, the TTAs sometimes caused to the public, the Client was usually given 3 days advance notice by e-mail or phone of all future TTA leading to diversions or road closures to allow their prior notification to concerned members of the public.

Feedback from the public was also a central element of the TTA implementation, since it was often not possible to envisage and cater for the expectations of all road users. Although not exclusive to this project, the Government of Hong Kong has in place various channels through which the public can send in enquiries and complaints on a wide range of issues including issues relating to roadwork activities such as TTAs. The most commonly used channel is 1823 Citizens Easy Link (CEL). This is an integrated contact centre operated by the Efficiency Unit of the Government of Hong Kong on behalf of about 20 Government Departments including all works departments. Once a complaint is received, it is processed and passed on to the department of concern. In addition to this, the client also runs a 24-hour hotline, enquiry and complaint e-mail addresses. When the client receives a complaint regarding, for example, the TTAs on the project, the complainant is usually contacted for further details or clarifications. These are then passed on to the ER or contractor for appropriate actions. A range of actions are possible depending on the content of the complaint, but could include inviting the complainant on a joint site visit to better understand the problem for a more appropriate resolution. Indeed, the Client has made a pledge to resolve all complaints within 7 days. If this is not possible, complainants are normally sent a preliminary response on progress of resolving their concern with information on when they expect to completely resolve the situation. On this project in particular, they have instituted what they call the *complaint walk* where the client goes on site to walk through with the ER and contractor to ensure that the complaints from the public are being addressed. In one episode, a lorry driver launched a complaint when he was affected by the closure of one of the roads passing through the project site. The issue was resolved to his satisfaction by the installation of additional traffic signs. The Client then advised that the planning and implementation of the TTAs should take into account the perceptions of the road users in addition to meeting the minimum standards. In another episode, a passenger had to pay an additional HK$30 as taxi fare due to diversions resulting from a TTA. He launched a complaint regarding this and was taken on a site visit to explain the situation. However, he was dissatisfied with the explanation and demanded a refund of his taxi fare. This was rejected by the contractor as it was considered unjustified as all the necessary signage was complied with. The passenger was considering taking the issue up with the ombudsman.

TTAs are an important feature in roadwork projects and are considered one of the most challenging tasks on most road projects (Chan, 2003). Yet, as shown here, the project team, especially the contractor, is keen on ensuring that inconvenience to the public is reduced as much as possible by engaging all stakeholders for successful implementation of all TTAs.

Community Planting Exercise

In December 2003, the Works Bureau of the Hong Kong Government issued a circular (No. 34/2003) on 'community involvement in greening works'. This directive was to show the government's commitment to the promotion of greening activities to enhance the quality of the living environment, and to promote community involvement and a sense of ownership among local residents. It required that all capital works contract with the estimated value of the landscape works in excess of HK$3 million should involve consultations with the respective district councils with regards to the greening works prior to tender and that the community be invited to participate in the planting works near to or after the completion of the project.

Since the value of the landscape works on the project was less than HK$3 million, the adjoining bridge project whose value for landscape was also less than HK$3million was invited to join the community planting exercise. Thus, both the contractor and consultant confirmed that the community planting exercise was not part of the original contract but only in their opinion a public relations exercise by the client. Yet, the ER was quite supportive asserting that:

> . . . I think this is, one of the reasons is to let the public know that [the Client] . . . is very keen in greening the environment or [the Client's] projects are not just a concrete bridge, concrete 'spaghettii' built in urban or rural areas but [the Client] at the same time, [the Client] thinks about the aesthetics of the bridge work and [the Client] thinks about what has been affected in terms of the planning so [the Client] tries to compensate the area by putting more plants at the same time [the Client] enhances more greening works.

The details of the onsite community planting on the project were however discussed at one of their monthly progress meetings. The client advised that the exercise be arranged earlier to avoid the hot weather and typhoon season if possible. Following this, the ER (viaduct project) was requested to attend a similar community planting activity being organised by an adjoining project for first-hand information on how it is done, so that he would be in a position to advise the contractor on how to organise for this project.

The key participants for the community planting project were pupils from two selected primary schools in the neighbourhood and some district council members. The contractor however had some concerns about the composition of the volunteers for the planting exercise and expressed some reservations:

> . . . there is some hidden risk in this, because for us at the moment, this is still a construction site, so under the law anybody who comes into the site will require a Green Card. If he is a worker, he needs to have a registration card, . . . the kids who will be doing the planting, they are actually doing [the contractor's] work. Technically they are doing our permanent works because they are planting the area where [the contractor] is supposed to plant themselves, so they don't have a green cards, they don't have workers' registration cards and they are all underage (Rep. Contractor)

He added regarding insurance that ' . . . technically the insurance people will say, if something happens such as claims issue who is responsible?'.

On the question of whether these issues had been raised with the client, the response was that the volunteers should be classified as visitors. Since the circular on the community planting mandates the Client to take responsibility among others for insurance and safety matters, it was the contractor's view that this has been given due consideration.

Taken together, however, the community planting exercise appears to be well received by the volunteers and attracting public enthusiasm. This can be attributed to the fact that it presents them with the opportunity to get closer to projects than they normally would, and in the process learn more about what the tax payer's money pays for. Government and community representatives are also keen to show up at such exercises as it gives them the opportunity to closely engage with and interact with their constituents.

Construction Noise Permit

Following a proposal to change from the use of two launching girders as proposed in the technical proposal at tender to one launching girder and a crawler crane, the contractor further proposed a 24-hour cycle for the erection of the viaduct segments in order to achieve an equivalent productivity level. While there was no issue with this part of the proposal, as that was within their contractual right, the continuous supply of precast segments to the launching girder beyond 11 pm to ensure the 24-hour cycle was achieved was problematic. This was because the proposed storage area for the precast segments was directly beneath a student hall of residence and the carrier that supplied the segments to this area produced noise beyond the acceptable Environmental Protection Department's (EPD) limits. The stakeholders in this case included:

1. Authorities of Educational Institution
2. The residents of school hall of residents (warden)
3. Environmental Protection Department
4. The Client
5. Viaduct Contractor (viaduct project)

To mitigate the situation, the contractor proposed some modifications to the segment carrier to reduce the noise. A noise enclosure was specifically designed and installed to the engine part of the carrier. A trial was then run and the noise levels at various times and from various points were recoded and the results presented to EPD. Given the importance of this permit to the progress of work, the Client played a key role in facilitating the approval process as testified by the contractor:

> . . . [the Client] *was involved in some of the discussions, so, everyone was involved trying to satisfy EPD, even* [the Client] *went with* [the contractor] *to discuss with EPD, about what can be done, what is acceptable to* [EPD] *in terms of noise level from the point of view of EPD for them to issue a permit* (Rep. Contractor)

In separate discussions with school authorities whose hall of residence was close by, the contractor suggested that the windows of the hall facing the site be closed at all times to reduce the noise. The school agreed to do so but also requested that the contractor replace their old air-conditioner with a much quieter new one.

While the approval process for this issue took an unusually long time to resolve, it is interesting to note that throughout the 24-hour segment erection operations, only one person (the warden) actually lived in the hall of residence, and thus shows the desire to fully adhere to the law no matter the circumstances.

Miscast segments

An estimated 67 number precast viaduct segments were miscast by the precast subcontractor due to wrong setting-out information provided. This resulted in the incorporation of crossfalls in the wrong direction. The ER subsequently issued a non-conformity notice which required that the segments be scrapped and recast. In view of the significant and unrecoverable delay to the work that this error could cause, there was the urgent need to review the procedures relating to the production of the precast segments in the precast yard in Mainland China, by strengthening supervision. There was therefore an immediate review of the setting out and checking procedures for the production of the precast segments.

The stakeholders in this case included:

1. Precast Subcontractor (in Mainland China)
2. Contractor (Viaduct)
3. Independent Engineering Consultant
4. The Client
5. Client's Maintenance Unit
6. Clients Audit Team
7. Government Department (in charge of waste disposal site)
8 The Engineer
9. The ER

When the error was detected, some of the wrongly cast segments were already erected. The consequence of the errors in the already erected segments was that the alignment of the finished road surface was unlikely to meet the requirements in the specifications. This therefore required that the approval of the Client's maintenance unit and the Transport Department be sought for the acceptance of those works. Given the implications of the lost production time had for the progress of the works, the contractor further proposed incorporating as many of the miscast segments as possible into the works since the errors had no implications for the structural capacity of the viaduct. In line with this, a full report on the segment errors was prepared and submitted to the ER so that the feasibility of further incorporating as many of the miscast segments (without rectification) into the works could be evaluated. The miscast errors were also picked up by the client's technical audit team following their prevention of substandard works audit and called for rectification.

The contractor engaged the services of an engineering consultant to undertake an independent review of the miscast segment situation. The independent review was then submitted to the Client for his comments. Queries were raised by the Client and replied to by the contractor. Following a 'no further objection' from the Client's project team, the contractor submitted the final report to the maintenance unit for approval of the incorporation of as many miscast segments as possible into the works.

While some segments were redeemed and incorporated into the works, about 35 miscast segments became redundant and needed to be discarded. However, the mode of disposal became another issue. The Client proposed that the contractor could consider sinking the miscast segments to the seabed to form an artificial reef. The Client however left it to the contractor to decide on his preferred method of disposing of the miscast segments and with a promise to assist as required. The contractor eventually decided to have them demolished. To facilitate their gaining consent to demolish the miscast segments at a waste disposal site from the Government Department in charge, the contractor requested the Client to provide them with a support letter. The Client agreed and provided them a letter supporting their proposal. The Government Department in charge however rejected the contractor's proposal to demolish the miscast segments at the waste disposal site and noted that the contractor can have them demolished in China where they are still stored in the precast yard.

It is clear here that the consequence of the miscast error for all stakeholders was an incentive to work together for a fruitful resolution of the issue. This demonstrates the power of joint interest or joint risk in motivating stakeholders to work for the common good of the project. Yet, the inability to agree on how to dispose of the remaining precast segments also shows how lack of alignment of interests forestalls consensus building.

13.4.2 Impact of procurement arrangement

This project was procured under a traditional design-bid-build approach. As the most common procurement arrangement in Hong Kong, it presupposes that the parties were generally familiar with the procurement route. Yet, it is apparent from the discussion so far that the arms-length mindset associated with this approach contributed to how some of the incidents played out. It is however commendable that the interface arrangements were built into the contract. This approach clearly defined the interdependence between the two projects from the onset as an issue to be managed during the project. However, the interface arrangement appears to have been structured without consideration for the uncertainties that can arise in a project of this size and complexity. This was further exacerbated by the inflexibility of the various parties. Ironically, there was a non-contractual partnering arrangement in place, in which the parties promised to work in partnership. Yet, when it mattered most all the stakeholders held on to their contractual rights.

The structuring of the project organisation also had implications for the number of stakeholders on any issue and thus their management. First, the Client organisation was a plural one. On many issues, three or more different departments of the Client organisation needed to be satisfied, and this became more problematic when they disagreed. The fact that the contractors on the two adjoining projects were joint ventures also had implications for engaging them. In this case, the board of directors of the JVs appear to have played only a passive role, as most of the issues were considered site matters, which were within the domain of the site teams. Some contractual provisions also had implications for the number of stakeholders who needed to be engaged, e.g. the ER as a separate entity from the Engineer; the use of an ICE, whose role was to independently check all the contractors' designs and the TMLG.

13.4.3 Implications

Five incidents have been analysed above to show how stakeholder management on a Hong Kong infrastructure projects manifested itself. The different incidents showed management of relationships among stakeholders internal to the project organisation as well as relationships among stakeholders external to the project. In both cases it was clear that when the stake of all stakeholders on an issue of contention was high, there was a tendency to reach an agreement easily (Table 13.2). Culture-specific dynamics also manifested themselves in the positions different stakeholders took on issues and there was a general tendency to follow or adhere strictly to the contract. This may be attributable to the fear of blame culture pervasive in public project settings and the conflict avoiding view inherent in the Confucian value system.

Taken together, however, this case study demonstrates an element of progress towards public engagement on projects in Hong Kong, an element which was unheard of a decade ago. Yet, the arms-length mindset, perpetuated by decades of use of the traditional procurement arrangement, is still prevalent. Indeed, when collaborative initiatives such as partnering are bolted onto the traditional procurement system, little evidence of real partnership is manifested. Thus, a shift in culture, both in terms of the way stakeholders are engaged and projects are procured, appears a viable option for project delivery in Hong Kong.

13.5 Conclusion: Lessons learned from the case studies

It is apparent from the case studies above that tradition, custom and practice, politics and culture have a major influence on how stakeholder management is undertaken in the Hong Kong construction industry. Without a strong tradition of democracy, it is not surprising that

Table 13.2   Impact of stakes on project developments

Stakeholder	Stakeholder interest	Impact	Strategy
Interface arrangement			
Viaduct Contractor	Safer work environment; simpler site operations	Escalation of risks, non-achievement of Key Dates	Buy-in of key stakeholders; formal and informal engagement; interface meetings; ping-pong letters
Bridge Contractor	Structural stability of bridge	Risk and liability	
Client	Limit liability and claims; structural stability of bridge	Blame/reprimand from superiors; escalation of risk	
ER	*Projecting Client's interests*; enforcement of contract	Loss of Client's trust	
The Engineers	*Projecting Client's interests;* enforcement of contract	Loss of Client's trust	
ICE	Neutral assessment	Neutral	
Project Board of Directors	Safer and simpler site operations	Passive observer	
Temporary Traffic Arrangement			
Viaduct Contractor	NC, least inconvenience to road users	Inconvenience to road users; loss of reputation of key project participants; public complaints	Management of public expectations; three-cycle daily joint inspections; feedback from road users; complaint walk; Government's central complaints unit (1823 Citizens Easy Link (CEL))
Road users (general public)	Least inconvenience		
Client	Reduction in NC, least inconvenience to road users		
Client's Audit Team	Enforcement of TTA		
ER	Reduction in NC, least inconvenience to road users		
TMLG	Faster resolution of TTA issues,		

(Continued)

Table 13.2   Impact of stakes on project developments   *(Continued)*

Stakeholder	Stakeholder interest	Impact	Strategy
Community Planting Exercise			
Client	Community involvement; PR, promotion sense of ownership, public enthusiasm	Public agitation; negative publicity	Invitation to participate; community out-reach; onsite community planting
Contractors	Liability and safety issues; insurance; composition of volunteers	Lack of commitment	
ER	*Projecting Client's interests;* enforcement of contract	Loss of client's trust	
Public (school children)	Participation		
Construction Noise Permit (CNP)			
Contractor	24-hour cycle; constant supply of segments; storage area	Delays to works	Mitigation measures; meetings; Government's central complaints unit (1823 Citizens Easy Link (CEL))
Client	Noise level; public complaints	Delays to works; public complaints	
School (hall of residence)	Noise level	Inconvenience; public complaints	
EPD	Enforcement of noise regulation		
Miscast segments			
Contractor (precast subcontractor)	Significant and unrecoverable delay and loss of resources	Delays to works; waste of resources	Review of precast procedures; strengthening supervision; mitigation measures
Client Departments/Units	Build as designed, easy maintenance	Maintenance difficulties	
Engineer's Representation	Enforcement of contract	Damaged reputation	Ping-pong letters
ICE	Neutral assessment	Neutral	

the move to draw the public, green groups and other parties into the development process has moved forward slowly; there is no evidence of resistance to change, rather an inertia grounded in the traditional values of society and the structure of Government Departments and institutions which puts a brake on change. This is not totally surprising: if one studies the position of Hong Kong on Hofstede's dimensions of culture, it is obvious that nations such as United Kingdom and United States have a value infrastructure which is more open to

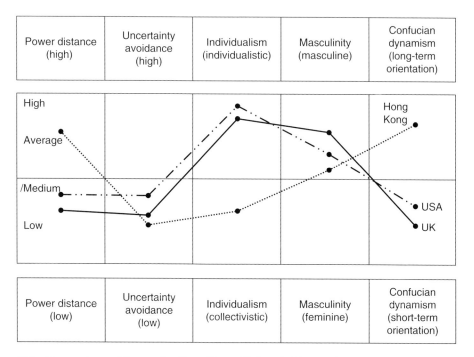

Figure 13.2   Comparison of Scores for Hong Kong, USA and UK on Hofstede's Cultural Dimensions.

stakeholder involvement and empowerment (Figure 13.2). The Confucian values of harmony and conflict avoidance are often an opposing force to the drive for stakeholder empowerment.

This having been said, there is evidence from the case studies that change is taking place and that the post-colonial administration is becoming more attuned to the legitimate demands of its stakeholders and a 're-education' process is taking place (Table 13.3). This cannot be described as a cultural revolution but a culture change is taking place. A move away from traditional procurement forms is now underway with the Hong Kong Housing Authority leading the way and the Works Bureau departments commencing a range of 'experiments' with more open procurement forms. Indeed, the incorporation of partnering type agreements into many projects has contributed to a change in culture and led to more open attitude to cooperation and collaboration in construction projects (see, for example, Anvuur, 2008). In line with this, there needs to be a recognition that performance measures need to be refocused to take into account medium- and long-term objectives in line with the arguments put forward by Walker *et al.* (2008).

In recent years, employees and stakeholders have become much more aware of the need for firms and government to show a commitment to CSR (see Rowlinson, 2009) and this has raised awareness in all sectors. Indeed, major infrastructure and property developers have taken on board stakeholder management as part of their CSR commitment; however, time will tell whether this is a marketing fad or a genuine culture change in the industry. With the establishment of the Construction Industry Council in 2008, there is now an industry-wide body dedicated to improving performance in the real estate and construction industries. One of its first tasks has been to improve construction site safety and this has involved an attempt to engage workers, managers and directors in a framework that provides a basis for joint problem-solving and initiative development. Such approaches augur well for the future development of stakeholder management and empowerment in Hong Kong.

Table 13.3 Stakeholder management issues and strategies

Issue	Project	Strategy	Example
Procurement systems reformation	Case A	Greater contractor participation, equitable sharing of costs and benefits	MGMP packages, dispute resolution system, pain and gain share scheme
	Cases A and B	Partnering	Non-contractual partnering; interface management
Improved collaboration	Cases A and B	Relationship management	Partnering, promote dialogue sessions
Life cycle value consideration	Case A	Life cycle costing	Design with maintainability in mind
	Case B	Emphasis on what is best for the project in the long run	Build with maintainability in mind (incorporation of miscast segments); Owner Controlled Insurance Programme (OCIP)
Community involvement	Case A	Proactive engagement, greater community participation	Community planting, mural wall design, dialogue sessions, volunteering information
	Case B	Buy-in of key stakeholders; formal and informal engagement; management of public expectations; community out-reach	Onsite community planting; Government's central complaints unit (1823 Citizens Easy Link (CEL))
Community benefits	Case A	Provision of direct and indirect benefits	Low dust and noise generating construction methods, free house improvement services, improved greenery around construction site
	Case B	Buy-in of key stakeholders; formal and informal engagement; management of public expectations; community out-reach	Onsite community planting

For further progress to be made in stakeholder management, the Hong Kong real estate and construction industry needs to address:

- The reform of existing procurement systems to allow for more innovative and collaborative approaches to the development process.
- A change in culture throughout the industry to allow participants to focus on cooperation and collaboration rather than defensive reactions – the establishment of relationship management approaches throughout the industry.
- A focus on the real meaning of value in the project context rather than a decision-making process based on lowest initial costs.
- A commitment to community involvement and a full implementation of the principles of CSR in both public and private sectors.
- A refocusing on community benefit as a mechanism for assessing the performance of projects.

Acknowledgement

The authors would like to acknowledge the contribution from the Research Grants Council, Hong Kong (Grant No. HKU712204E and HKU715807E) for the funding of this research.

References

Anvuur, A.M. (2008) *Cooperation in Construction Projects: Concept, Antecedents and Strategies*, PhD Thesis, The University of Hong Kong, Hong Kong.

Chan, D.W.M., Chan, A.P.C., Lam, P.T.I., Lam, E.W.M. and Wong, J.M.W. (2007) Evaluating guaranteed maximum price and target cost contracting strategies in Hong Kong construction industry. *Journal of Financial Management of Property and Construction*, **12**(3):139–149.

Chan, K.Y.J. (2003) A challenging task – temporary traffic arrangement. *Highways Department Newsletter*, 54, p. 15e, available athttp://www.hyd.gov.hk/eng/public/publications/newsletter/Issue54/eng/e15.pdf (Accessed on 21 June 2008).

Cleland, D.I. and Ireland, L.R. (2007) *Project Management: Strategic Design and Implementation* (5th edn). New York: McGraw-Hill.

Donaldson, T. and Preston, L. (1995) The stakeholder theory of the corporation: concepts, evidence, and implications. *Academy of Management Review*, **20**(1):65–91.

Haley, G. & Shaw, G. (2002) Is 'guaranteed maximum price' the way to go? Hong Kong Engineer (Online Journal), January, available at http://www.hkengineer.org.hk/program/home/articlelist.php?cat=business&volid=19, accessed 10th September 2008.

Hong Kong Housing Authority (HKHA) (2005), Modified Guaranteed Maximum Price (MGMP) Contracting for the Public Rental Housing Development at Eastern Harbour Crossing Site Phase 4, available at http://news.apmhk.org.hk/_attachments/1277971/Partnering%20SIG%20-%20MGMP_EHC%20SITE%20PHASE_4.ppt (Accessed on 2 September 2008).

Lau, N. (2008) Backpay row brings traffic to standstill. *The Standard*. 10 January 2008 (Online newspaper), available at http://www.thestandard.com.hk/news_detail.asp?pp_cat=11&art_id=59794&sid=17036540&con_type=1&d_str=20080110&sear_year=2008, (Accessed 2 September 2008).

Mahesh, G., Kumaraswamy, M., Anvuur, A. and Coffey, V. (2007) Contracting for community development: a case study based perspective of a public sector client initiative in Hong Kong. *Fourth International Conference on Construction in the 21st Century (CITC-IV) – Accelerating Innovation in Engineering, Management and Technology*. Gold Coast, Australia, 11–13 July 2007.

Olander, S. (2007) Stakeholder impact analysis in construction project management. *Construction Management and Economics*, **25**(3):277–287.

Rowlinson, S. (2009) Corporate social responsibility in the Hong Kong and Asia Pacific construction industry. In: Murray, M. and Dainty, A. (eds) *Corporate Social Responsibility in the Construction Industry*. Abingdon: Taylor and Francis, pp. 327–350.

Rowlinson, S. and Cheung, Y.K.F. (2008) Stakeholder management through empowerment: modelling project success. *Construction Management and Economics*, **26**(6):611–623.

Thomas, R., Marosszeky, M., Karim, K., Davis, S. and McGeorge, D. (2002) The importance of project culture in achieving quality outcomes in construction. In: Formoso, C.T. and Ballard, G. (eds). *Proceedings of the 10th Annual Conference on Lean Construction*. Gramado, Brazil, August 2002.

Walker, D.H.T., Bourne, L. and Rowlinson, S. (2008) Stakeholders and the supply chain. In: Walker, D.H.T. and Rowlinson, S. (eds). *Procurement System: A Cross-Industry Management Perspective*. Abington: Taylor and Francis, pp. 70–100.

14 A Mental Modeling Approach to Study Decision-Making in Dynamic Task Environments

Matthew Wood, Amlan Mukherjee, Todd Bridges and Igor Linkov[1]

14.1 Introduction

If the environment in which decisions are made, changes as a function of the sequence of decisions, independently of them, or both, then such an environment is referred to as a dynamic task environment (DTE) (Edwards, 1962). Such environments also tend to be complex because of inter-relationships between different parts of them, which tend to evolve in a nonlinear fashion. Expertise in managing complex dynamic environments is dependent on the ability to make critical decisions, and select appropriate strategies under constraints of time and resources. In addition, planning for contingencies, and apprehension of risk and uncertainty are crucial for the successful management of such environments.

There is anecdotal evidence that expertise in decision-making in a DTE is built through years of experience. However, experiential knowledge is often considered 'intuitive.' As a result it is difficult to formalize, even by expert decision-makers who often cannot explain their own actions in ways that can lead to the development of education and training of stakeholders, novices and laypersons. When expert decision-makers retire, they take with them significant knowledge, and abilities to communicate and analyze risk, leaving voids in the domain that become difficult to fill. The ability to study and formalize the way expert decision-makers analyze risk in DTEs and cognitively structure and organize their domain knowledge is critical to educating novice decision-makers as well as increasing awareness of the nature of crisis scenarios posed by DTEs to relevant stakeholders.

In this chapter we consider decision-making in two different DTEs in the area of civil and environmental engineering: (i) Construction management (CM) and (ii) Natural hazard management (NHM) specifically with respect to floods. Within the context of the problems posed by these domains, we discuss a mental modeling framework of expert decision—making. Within this framework the specific questions that we address are: can we develop mental models of decision-making that reflect the cognition of decision-makers, expert or novice, specifically with respect to the apprehending uncertainty, and responding to rapidly unfolding crisis scenarios? Can we quantitatively characterize the anatomy of an effective decision and the cognition of an effective decision-maker? Can such mental models support decision-making aids that can complement human judgment? In the process, can we capture implicit human knowledge that informs 'intuition' and drives expertise?

CM and NHM provide very good examples for studying decision-making in DTEs, because even though they both share similarities, they are inherently different. Interestingly, the differences between the two domains lie along a fault line that spans some of the fundamental challenges in studying decision-making in DTEs. In the following section we discuss the similarities and differences between these domains and the need to characterize expert decision-making in them using mental-modeling approaches.

14.1.1 Similarities and differences between CM and NHM

Construction management scenarios are examples of high stakes DTEs. Delay resulting from material delivery, the need for rework, lowered labor productivity, weather, or some other circumstance, can result in cascading delays that impact the final cost and schedule of the project, with impacts running into millions of dollars of litigation and liquidated damages. This point is well illustrated in the construction of the New York Times Co.'s fifty-two-story office tower in mid-town Manhattan (Post, 2007). The owner (The Times and its development partner Forest City Ratner Cos.) invested 3 years during the planning phase by collaborating with the architect to develop a variety of studies, simulations, building mock-ups and expert panel meetings to compare options and produce a complete set of bid drawings. Besides significant cost savings, this gave them the ability to bid a competitive guaranteed maximum price (GMP) contract transferring the risk onto the construction execution phase. AMEC Construction Management Inc. won the contract for $350-million, but due to various problems (discussed later), they barely managed to complete the steel construction on time, lost their entire budgeted contingency, and incurred losses due to unforeseen construction complexities.

Management of natural hazards such as hurricanes Katrina and Rita also present the challenges of a high stakes DTE. In the fall of 2005, hurricanes Katrina and Rita revealed inadequacies in severe storm and flood protection plans for the US Gulf Coast. Initial criticism centered around engineering design and management issues (e.g., impact and loss projections, flood protection infrastructure) and degradation over time of the region's wetland defenses by various industries (e.g., energy, transportation; Johnson, 2005; Cigler, 2007). In recent analyses, the importance of human factors in disaster prevention planning is a more prominent theme. For example, Gheytanchi *et al.* states:

> An interdisciplinary approach to the field of disaster management that views psychology as a central element—rather than second to engineering or information science—will lead to stronger, more resilient communities, [and] result in better decisions on the part of government . . . (2007, p. 129)

Even before hurricane Katrina, similar recommendations were made by the Institution of Civil Engineers panel in the United Kingdom as part of their assessment of the state of flood management and their proposed improvements in response to severe flooding there in 1998 and 2000. Two key recommendations of their report were to 'learn to live with rivers' by accommodating waterway expansion from rainfall and provide greater weight to human and social factors when assessing flood risk. For example, anticipated victim distress should be considered when designing flood mitigation strategies (Fleming, 2002a,b).

The first major difference is in the different types of stakeholders that define risk and loss in each of these domains. Complex CM projects – especially ones that are in the private sector – rarely tend to involve the public in a significant way. While public projects (such as highway construction) in the United States are openly bid and paid for directly by tax-payer monies, the public is only indirectly impacted and often barely involved or aware of the losses incurred. In general user costs and traffic delays are considered by public agencies, and of late they often involve communities in the planning and construction – however, most impacts due to delays and cost over-runs during the construction process are best classified as inconveniences. On the contrary, NHM involve the public as a very significant stakeholder due to the risk posed by natural hazards to life and property. In reality, though, the public tends to take scant responsibility for apprehension and planning for natural hazards, and they generally expect public agencies to perform during an emergency, irrespective of their ability or adequacy to provide such services.

This difference highlights the ultimate beneficiaries of this research. Mental models of decision-making in uncertain CM scenarios will be used to educate novice construction managers, and prepare them to fill in the shoes of experienced decision-makers. In the domain of NHM, the more pressing need is to efficiently communicate risk to public stakeholders, and prepare them to cooperate in the preparation for managing natural hazards. In addition, mental models of decision-making in the face of rapidly unfolding events can play a significant role in educating decision-makers – especially in scenarios created by infrastructure failures resulting from hurricanes. In the long run mental modeling will serve a dual purpose. As public agencies increasingly involve communities in the planning of public infrastructure projects, with increasing focus on sustainable design and construction, mental modeling will improve the ability of construction managers and engineers to communicate with community stakeholders and deliver systems that indeed cater to the social bottom-line.

The second major difference lies in the temporal and spatial scales over which these environments evolve. CM environments span over short time periods (usually between a few months to years) that are well structured using schedules. Spatially, they are also well defined by the span of the project site. As the project evolves, decisions and uncontrollable external events (such as weather, labor strikes, delayed material deliveries) play a significant role in delays, which in turn can create feedbacks and cascading impacts throughout the schedule that can quickly require adept and timely rescheduling in order to mitigate losses. NHM, on the other hand, involves evolutions over varying timescales with very quick damage occurring in very short periods of time followed by impacts, and secondary emergent scenarios over prolonged periods of time. This makes the domain prone to disproportionate temporal distribution of damage. Spatially, they often play out over very large geographical areas with diverse contexts.

These different temporal scales allow the investigation of the following significant challenges encountered by decision-makers in DTEs:

- Ability to take into account rate of change of the environment and relate their own sense of timing of their actions to the system evolution rate (De Keyser, 1990; Kerstholt, 1994, 1995).
- Capacity for selecting strategies by appropriately updating uncertainties and identifying risks when the system is changing (moving targets) (Ford et al., 1989; Payne et al., 1993), resulting in sub-optimal performance.
- Sub-optimal decisions that can result from misperceptions of feedback (delayed rather than immediate) (Diehl and Sterman, 1995).
- Failure to adapt to changes in the environment with a tendency to continue working on models of the environment that have ceased to exist (e.g., Lusk and Hammond's (1991) work with weather forecasters).

The CM and NHM domains each provide different angles to studying decision-making with respect to the perception of temporal scales, the differing rates at which the domain evolves, and the nature of feedbacks that are delocalized in time and space and the impact of decisions made in such environments.

Kerstholt and Raaijmakers (Raynard *et al.*, 1997) have questioned if it is at all possible to make optimal decisions in a dynamic environment as they conjecture that human cognitive ability might be limited to dealing with sequential tasks. They also note that there are nonetheless many expert decision-makers in DTEs who 'are able to maintain an overview of the system under control, whereas others tend to fixate too much on single local diagnosis problems.' This is one of the main differences (Chi *et al.*, 1982; Bransford, 1999) identified between expert and novice approaches to problem solving. Mental models can be very appropriately used to establish the differences in risk perception and analysis of problems between experts and novices/lay stakeholders.

14.1.2   Why use mental models?

Before furthering the discussion on using mental models, it is important to establish that they can be appropriately used to measure knowledge organization of expert decision-making in DTEs. The evidence lies in the learning and cognitive science research where the development of mental models has been considered to be an indicator of learning or improved organization of knowledge of a particular domain. Dreyfus and Dreyfus (1986), argued that expertise develops in proportion to a students' sensitivity to the environment in which problems arise, and their increasing willingness to 'break the rules' to solve such contextualized problems. This indicates that there is a connection between a learner's internal knowledge organization and expertise within a specific domain. Indeed, the ability to 'break the rules' indicates that students have a deeper understanding of the domain structure, and are willing to explore alternative problem-solving approaches. In addition, research has shown that explicit instruction in how to build mental models improves learning (Mayer, 1989; Seel, 2003). Research on mental representation, expertise, and conceptual change, from Posner *et al.* (1982) to Duit and Treagust (2003), have shown that learning takes place when existing mental representations change to accommodate new experiences that occur as students interact with the environment.

The importance of accommodating social and human dimensions in disaster preparedness has recently been recognized, but specific tools for integrating knowledge, interests, and values of stakeholders (defined as individual or organization with a direct or indirect investment in flood preparedness and response for the purpose of this paper; United States Environmental Protection Agency, 1997) with those of United States Army Corps of Engineers (USACE) groups are still underdeveloped. Mental modeling has been recognized as a useful method and framework for better understanding and addressing deeply held risk and value beliefs that can enhance stakeholder involvement in strategic planning. Some work has been done on representing layperson perceptions of floods (Lave and Lave, 1991; Kolkman *et al.*, 2007; Wagner, 2007). However, the techniques used in these studies do not directly inform the decision-making process for emergency plan alternatives. Similarly recent work in CM (Rojas and Mukherjee, 2005, 2006) strongly makes the case for understanding construction decision-making not only as a combination of resource interactions, but also as a combination of human-resource combinations. The importance of human decision-making in managing construction crisis scenarios and improving performance on complex projects strongly justifies the use of mental models.

Generally speaking it is critical to use mental models to study decision-making across the many different and similar problems that are encountered diverse DTEs such as CM and NHM. Section 14.2 presents a thorough survey of mental models literature.

14.2   Mental models

In this section, we will review mental model theory and four cognitive tools for representing mental models of floods and associated risks.

14.2.1   Mental model theory

The mental model concept is a long-recognized one in cognitive science. First formulated by Craik (1943), a mental model is a psychological term referring to the internal representation of a state of affairs in the external world. He suggests three processes essential for the human as information processor to function (cited in Johnson-Laird, 1983):

- A process that represents external stimuli as an internal code of words, numbers, or other symbols.
- A process that can convert one type of internal representation into another.
- A process that can turn internal representations back into an external representation.

Craik further claims that internal representations are used as a way for humans to understand the world around them. Johnson-Laird (1983) points out that this claim has implications on two levels. First, mental models are a simplification of real-world events and may leave out some aspects of the external stimuli, which the mental model is trying to emulate. Second, if cognitive science wishes to understand the human mind, it must construct a model of it. Since this model would exist to explain how the mind works, it need only provide information related to its functional organization to be an adequate tool. Information at a more minute level of analysis would not help us to explain how the mind operates.

Work in the 1980s by Johnson-Laird (1983), Norman (1983), and others (see Stevens and Gentner, 1983) sought to provide a clear theory of mental models. However, there is little consensus between mental model theorists beyond principles outlined earlier (Craik, 1943; Johnson-Laird, 1983). For example, while Norman (1983) proposes that mental models are difficult to manipulate because of cognitive capacity, Johnson-Laird (1983) provides no such constraints. Adding to the confusion, authors use many different ways to explicitly represent the content of mental models, using methods that range from mathematic and logic descriptions, to syllogisms and conceptual networks (Stevens and Gentner, 1983). The field of system dynamics has recently provided a mental model working definition that reflects compatible elements of most theories (Doyle and Ford, 1998, 1999; Lane, 1999). This definition states:

> [A mental model] is a relatively enduring and accessible, but limited, internal conceptual representation of an external system (*historical, existing*, or *projected*) [italics in original] whose structure *is analogous to* [italics in original] the perceived structure of that system. (Doyle and Ford, 1999)

This definition highlights the perseverance, brevity, and structure of mental models. Representations of mental models have been used to measure knowledge of a domain, and can be useful as a means for communicating how individuals and groups think about that domain. Mental model representations can be compared between parties to identify commonalities and inform how differences in mental models may be reconciled. Mental models can be represented propositionally in mathematical formulae, syllogisms, and logic statements (Gentner and Stevens, 1983), or diagrammatically via influence diagrams (Morgan *et al.*, 2002; Mendoza and Prabhu, 2006), concept maps (Mannes, 1989), or semantic webs (Novak and Gowin, 1984).

Johnson-Laird (2006) has found that use of mental model diagrammatic representations can assist individuals when reasoning and making inferences. Others (Hoffmann, 2005; Kolkman *et al.*, 2005) have provided evidence for the usefulness of diagram-based methods when attempting to understand group perspectives. Given their benefits, only diagram-based methods of mental model representation will be reviewed here. This review will help to identify a method for synthesizing views of flood and natural disaster between stakeholder and USACE groups.

14.2.2 Diagram-based representation of mental models

Four methods for diagram-based representation are introduced and compared below. A summary of each method is provided (*Method*) followed by a discussion of quantitative analysis available for the method (*Metrics*), and the final product(s) of the representation process (*Outcomes*). In addition, the *Strengths* and *Concerns* of each of the methods have been summarized in Table 14.1 while a comparison chart of approaches can be found in Table 14.2.

Table 14.1   Strengths and concerns related to model integration using representation methods

Name	Strengths	Concerns
Risk Communication Influence Diagram (RCID)	• compares models of stakeholder groups to 'real world' phenomena • can identify misconceptions and gaps in mental model for clarification • may help locate differences in Concepts that expert groups identify as important	• assumes layperson mental model has same structure as expert mental model, but is less developed • method cannot identify structural differences in mental models across expert groups • no explicit procedure for identifying causal links in models • conventions for drawing causal diagrams are somewhat inconsistent
Concept Map	• procedure is highly standardized • more reliable than other measures • little method-specific training is required • mapping procedure is transparent; process is clearly participant-driven	• suggests relatedness of concepts, but not causality • high degree of statistical expertise is required • lack of stakeholder statistical knowledge may make method seem more obscure than it is
Semantic Web	• diagram is simplest and most Intuitive of summarized approaches • propositional relationships can be converted into meaningful sentences quickly • available scoring systems are easy to calculate	• diagram creation is highly subjective • assumes layperson mental model has same structure as expert mental model, but is less developed • may not be appropriate for comparisons between expert groups
System Dynamics Influence Diagram (SDID)	• creates a 'real world' reference model from expert input • diagrams can handle a great deal of complexity • can execute simulations with quantitative inputs, allowing users to improve their understanding of the model through trail-and-error learning • handles diverse inputs well	• algorithms for determining relationships between concepts are not well-specified • level of client involvement is not disclosed

Table 14.2   Comparison chart: diagram-based mental model representations

NAME	FOCUS	DATA COLLECTION	METRICS	OUTCOME	CITATIONS
Risk Communication Influence Diagram (RCID)	• influence of Factor X on Factor Y • probability or magnitude of influence • expert vs. layperson knowledge	• semistructured expert interviews • unstructured/ semistructured layperson interviews • confirmatory layperson questionnaires	• probability/ magnitude estimates • completeness • specificity	• expert influence diagram • layperson influence diagram • expert/layperson mental model comparisons	Bostrom, Fischhoff, & Morgan (1992) Morgan, Fischhoff, Bostrom, & Atman (2002) Fischhoff (1995) Atman, Bostrom, Fischhoff, & Morgan (1994) Bostrom, Atman, Fischhoff, & Morgan (1994) Bostrom, Morgan, Fischhoff, & Read (1994) Read, Bostrom, Morgan, Fischhoff, & Smuts (1994) Vislosky & Fischbeck (2000)
Concept Map	• semantic similarity of statements & concepts • semantic difference of statements & concepts	• brainstorming • card sort • statement rating • group discussion	• binary symmetry similarity matrix • multidimensional scaling • cluster analysis • closeness, contrast, combined	• statement list • cluster list • cluster map • point map • point-and-cluster map • cluster-rating map • point-rating map	Trochim (1989a) Trochim (1989b) Mannes (1989) Galvin (1989) Ifenthaler (2008) Kearney & Kaplan (1997) Pirnay-Dummer (2007) Pirnay-Dummer, Ifenthaler, Johnson & Al-Diban (2008)
Semantic Web	• noun-level concepts of varying specificity • qualitative descriptions of concept relationships	• Piagetan structured interviews • analyst consultation	• relevant concepts • propositional linkages • misconceptions • total score • total concepts • total connections • complexity	• semantic web with concept nodes and connections with relationship descriptions	Novak & Gowin (1984) Novak & Musonda (1991) Freeman & Jessup (2004)
System Dynamics Influence Diagram (SDID)	• cyclic interaction between factors • rates of change between factors	• group discussion • business operations/ accounting information	• stock (variable level) • flow (rate of change)	• influence diagram • improved dynamic intuition • improved complex systems learning ability	Karkkainen & Hallikas (2006) Lane (2000) Sterman (2001)

Influence diagrams for risk communication

Method. Risk Communication Influence Diagrams (RCIDs) were developed as a means to detect differences between layperson and expert knowledge of a domain for use in communication of environmental risks to the public (Bostrom *et al.*, 1992; Morgan *et al.*, 2002). Fischhoff (1995) describes risk communication as fulfilling an implicit social agreement between those that create risk (e.g., government planners, industry, natural resource managers) and those that bear risk (e.g., laypersons, plan implementers). This method has been used to create brochures for laypersons to learn more about risks associated with radon exposure (Atman *et al.*, 1994; Bostrom *et al.*, 1994), climate change (Bostrom *et al.*, 1994; Read *et al.*, 1994), and other hazards (Morgan *et al.*, 2002). Some have even used this method in the development of business research and development plans (Vislosky and Fischbeck, 2000).

Bostrom *et al.* (1992) describes RCID creation as a four-step process. First an expert diagram (called an expert model) is created (Figure 14.1). Experts participate in semi-structured interviews where the interviewer is guided by a written protocol intended to elicit what the expert knows to identify important concepts for the domain of interest and how they may be causally related to each other. Next, layperson beliefs are elicited often through one-on-one interviews using similarly structured or semi-structured protocols that typically begin with general questions to identify what individuals know about the topic, and then move systematically to increasingly focused questions to identify what laypersons know about specific expert concepts. Layperson beliefs elicited in this way are then mapped onto the expert diagram or expert model. Finally, alignments as well as misconceptions held by laypersons and gaps between expert and lay knowledge are described using dissociations identified in the previous step. The severity of misconceptions can be measured by administering questionnaires to new lay participants where interviewee responses are typically given on a five-point Likert scale, ranging from 'definitely false' to 'definitely true.'

Metrics. Several quantitative metrics can be used to compare between-group mental model structure (Bostrom *et al.*, 1992; Vislosky and Fischbeck, 2000). *Completeness* is a measure of the layperson model identifying how much of the expert reference model is covered by a layperson's mental model. It is computed as a ratio of the number of expert concepts identified by a layperson divided by the total number of expert concepts. *Specificity* assesses the level of detail in a layperson model. A ratio of the number of specific concepts to general concepts is calculated for both layperson and expert models. The layperson ratio is then divided by the expert ratio. As with completeness, specificity is only calculated using concepts that were included in the expert model.

Outcomes. This method produces an expert mental model influence diagram that encompasses all the concepts important to a particular domain. It also provides an influence diagram of the layperson mental model whose structure reflects that of the expert influence diagram. Finally, comparison of expert and lay diagrams identifies a set of issues that laypeople should be taught to achieve understanding of a domain more consistent with expert knowledge of how a set of phenomena actually work.

Strengths. The RCID method typically compares an aggregated mental model diagram from laypersons to an aggregated expert model. The expert model can represent an estimate of the 'true' state of the world from those who are best able to report its structure and content. For us, the strength of this method comes from the comparisons that can be made between laypersons, who presumably have little knowledge of a technical domain, and experts who possess all important knowledge about the domain. Comparisons against the expert consensus model can illustrate areas where informing laypersons may be helpful. RCID can also serve to identify points of emphasis between experts of one specialty (e.g., meteorologists) and all experts in general. Quantitative measures for comparisons across diagrams are intuitive and easy to understand with limited technical knowledge.

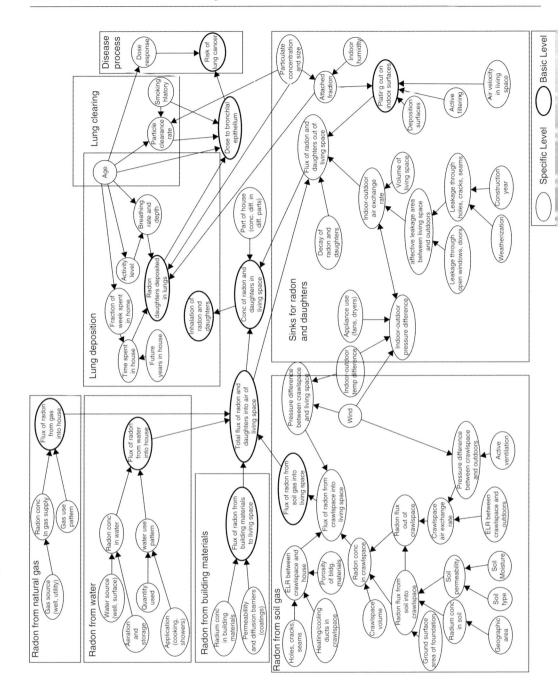

Concerns. A primary concern for use of this approach in mental model integration is that RCIDs assume the novice mental model is a simplified expert model in its structure and content. There may be circumstances where mental model structure is different across groups, especially as it concerns causality, which this method does not always detect. For example, meteorologists may know that low atmospheric pressure and high humidity together causes rain, while laypersons may believe that changes in atmospheric pressure cause changes in humidity which in turn cause rain. This difference in causal attribution between laypersons and experts would not always be detected by the RCID method; there may also be places where the structure of expert subgroup models differs considerably from the mean expert model.

Another problem related to causality with RCIDs is that procedures for determining concept connections and causality concepts are poorly defined in the literature. There are two possibilities for this lack of procedural disclosure. First, it may that formal procedures or computational algorithms do not exist yet for converting interview and questionnaire data into mental model diagrams. In this case, the reliability of the method may be suspect; depending on the analyst, aggregate expert models of the same phenomenon might look quite different in structure and content. Given the same inputs and the assumption that the aggregate model represents the 'real world' as it exists today, it is difficult to explain how aggregate models may differ between analysts. Certainly differences between models cannot be explained by saying Analyst A (along with his/her expert informants) lives in a world that works in one way, while Analyst B (along with the same informants Analyst A has) lives in a world that works in a different way. The opportunity for experimenter bias makes the reliability of analyses fair at best.

A second possibility why procedures are not elaborated may be proprietary. In this case, the merits of RCID cannot be compared to those of others diagram-based methods if its procedures are formalized but deliberately undisclosed. Even the best elaboration of the RCID method (Morgan *et al.*, 2002) provides little on how frequency counts of concepts mentioned in semi-structured interviews (a primary data source for RCID) are used to inform the diagram's structure or the degree to which the process is standardized.

Lastly, a casual review of available diagram examples from the literature cited in this section suggests sizable variability in how mental models are represented. Some identify superordinate-level concepts as large sections of the influence diagram; others identify the most general concepts using the same node representation as more specific sub-category concepts. All other things (mental model structure, emphasis, etc.) being equal, even small differences in how mental models are visually represented may make it difficult to identify differences in the mental models of different groups.

Overview of influence diagrams for risk communication

The interviewing process of RCID is its primary strength. The method attempts to reduce analyst bias by letting participants explain everything they know about a topic without interruption, and then asks them to discuss specific topics not covered in the earlier unstructured portion. Though current techniques for the analysis of interview content are a potentially large source of methodological error, quantitative analyses of transcripts may help to reduce this error to acceptable levels.

Concept maps

Method. Trochim (1989a, 1989b) developed concept mapping as a method to synthesize mental models of various groups, usually within a single organization, to inform program evaluation and decision planning. This method has been used in several domains, including the implementation of family-based intensive in-home services for the elderly (Mannes, 1989) and evaluation of a Big Brothers/Big Sisters program (Galvin, 1989). Computer programs for mental model diagramming based on this method are also available (Kearney and Kaplan, 1997; Pirnay-Dummer, 2007; Ifenthaler, 2008; Pirnay-Dummer *et al.*, 2008).

Concept mapping uses a six-step process as described by Trochim (1989b). Preparation (Step 1) begins by selecting individuals or groups to participate in the concept mapping exercise. A focus statement for later brainstorming and assessment criteria for brainstorm statements is also developed. The Generation of Statements (Step 2) uses the brainstorming focus statement from the Preparation step to elicit statements from participating individuals and groups related to the focus statement. These statements should describe one cohesive thought that addresses the focus statement in some way. After statements are generated via brainstorming, the Structuring of Statements (Step 3) asks participants to group statements using a variation of the Q-sort technique. Each statement is written on an index card, and a facilitator asks each individual to sort these cards into as many self-defined categories as the participant would like, with the exception that there are fewer total categories than the number of statements and that not all statements are grouped into one category.

Data from the Q-sort exercise is aggregated into a matrix representing the likelihood that any two statements were grouped into the same category. This matrix is used in the Representation of Statements (Step 4) to create a point-and-cluster map via 2D nonmetric multidimensional scaling and hierarchical cluster analysis. This graphically represents how closely statements are related to each other in visual space. A point represents each statement, and the distance between two points is used to qualitatively assess how related the two statements are to one another. The point-and-cluster map also proposes categories that statements may be grouped into based on cluster analysis of Q-sort results. These concept maps are provided to the group of participants so concept map categories can be labeled, and themes from these statement categories can be defined (Interpretation of Maps; Step 5; Figure 14.2). Finally, these concept maps are used (Utilization of Maps; Step 6) to identify key concerns for a plan, or to evaluate a program's success and outcomes (Trochim, 1989a,b).

Metrics. Computational techniques are the foundation of this method. In the concept map creation process, binary symmetry similarity matrices, 2-D nonmetric matrix multidimensional scaling, and hierarchical cluster analysis are all utilized. All these methods give outputs that provide a measure of concept relatedness, though some interpretations may need to be inferred qualitatively. Computer programs based on this concept mapping technique (Ifenthaler, 2008; Pirnay-Dummer, unknown) produce a few additional relationship metrics, including *closeness* (how related two concepts are rated by participants), *contrast* (participant ratings of how different two concepts are from each other), and *combined* (uses both closeness and contrast; measures overall model quality).

Outcomes. This method creates a host of products in the process of developing or evaluating the plan of focus. These include a statement list, a cluster list, a cluster map, a point map, and both point-rating and cluster-rating maps (using assessment focus data).

Strengths. The concept mapping approach's standardization is its premier attraction for synthesizing perspectives. The formalized procedure makes measurement more reliable than the other methods summarized. If two analysts are given input from the same brainstorming and Q-sort sessions, their results should look very similar to each other; sources of error can then be attributed exclusively to the method, and not to the method's administrator. Special training in concept mapping is not required to use this method.

Algorithms for determining relationships between statements and concepts are largely computational and have been publicly disclosed; a good statistician need only read a few articles to create a functional concept map. The high degree of structure and public disclosure of the process demonstrates that concept mapping is more participant-driven than analyst-driven, which works to build the method's transparency.

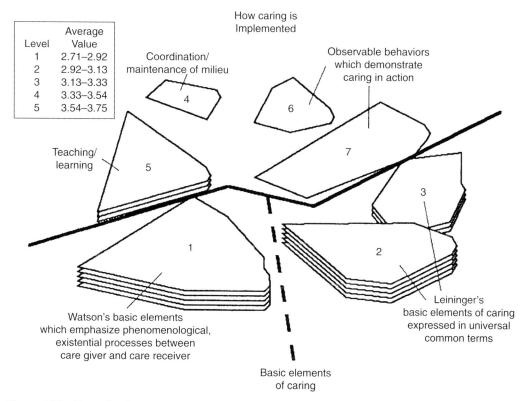

Level	Average Value
1	2.71–2.92
2	2.92–3.13
3	3.13–3.33
4	3.33–3.54
5	3.54–3.75

How caring is Implemented

Coordination/ maintenance of milieu

Observable behaviors which demonstrate caring in action

Teaching/ learning

Watson's basic elements which emphasize phenomenological, existential processes between care giver and care receiver

Leininger's basic elements of caring expressed in universal common terms

Basic elements of caring

Figure 14.2   Example of concept map.
(*Source*: Mannes, 1989)

Concerns. Though the concept mapping method is good at quantifying relatedness of themes and propositions within a mental model, it only suggests 'how related' statements are to each other. It provides no information on how statements or categories may influence one another. A fairly high degree of technical expertise is required for this method, even though expertise requirements are in a general domain (advanced statistics) rather than a specific one (concept mapping). The concept mapping process is transparent to those with a background in statistics or the social sciences. Some explanation of the procedure may be necessary to get laypersons and experts in other domains onboard, as the method's transparency may not be self-evident without some technical sophistication to understand how participant input influences concept map outcomes.

Overview of concept maps
Concept mapping gives a good standardized framework for creating mental model representations, but identifies only relationship strength and not the nature of the relationship between concepts; participants are left to their own devices to determine how highly related concepts may interact with each other. Though binary symmetric similarity matrices might be integrated into text analysis tools for extracting information from interview transcripts, the data collected with concept mapping is typically aggregated for all analyses. By comparison, RCID allows for a more suitable parsing of analyses and permits multiple permutations of participant-group assignments without the need for gathering additional information.

Semantic web

Method. The semantic web method of mental model representation was first developed by Joseph Novak and colleagues (Novak and Gowin, 1984; Novak and Musonda, 1991), to detect changes in children's understanding of science concepts and make outcome comparisons from the use of different pedagogical methods. This method provides a more qualitative assessment of knowledge than other diagram-based methods. Participant interviews are used to produce a diagram with a set of noun concepts represented as nodes in a network, while directional arrows labeled with relationship terms (mostly verbs) show relatedness between concept nodes (Figure 14.3). Though developed with primary and secondary school children, this method has also been used at the university level (Freeman and Jessup, 2004).

Data is collected through a Piagetan clinical interview technique. Participants are asked to think about something related to the topic of interest, and then report what would happen if the scenario were changed in some way. Statements are drawn from interview transcripts and recordings that illustrate the participant's perception of a relationship between concepts (Novak and Musonda, 1991). A similar framework can also be used in dyads, with the help of a properly trained analyst, to help elicit these maps directly from participants without coding transcripts (Freeman and Jessup, 2004).

Metrics. Some comparative statistics are available for this method based on assimilation learning theory, though the precise coding scheme is admittedly somewhat arbitrary (Novak and Musonda, 1991). *Relevant concepts* are scored by counting the number of nodes in the semantic web that are related to the topic of inquiry, with greater emphasis placed on superordinate-level concepts (10 points each) compared to basic- (5 points) or subordinate-level (2 points) concepts. *Propositional linkages* are assessed in similar fashion, with greater weight given to connections between superordinate connections (20 points each) relative to basic (10 points) or subordinate (5 points) ones. *Misconceptions* are assessed on a negative, with major misconceptions (−10 points) weighted heavier than minor misconceptions (−3 points). A *total score* is derived by adding *relevant concept*, *propositional linkages*, and *misconceptions* scores. A higher total score indicates the participant has a more robust mental model for the topic under investigation.

Freeman and Jessup (2004) consider alternative metrics for semantic web analysis. They simply add the total number of concepts and connections present, without weighting for concept generality. They also include a measure of *complexity*, estimated by subtracting the total number of links by the number of links required to connect the nodes in a linear fashion (or *complexity = total links − [total nodes − 1]*).

Outcomes. The semantic web method of mental model diagramming creates a network of noun concepts connected by a description of the relationships between them. While it does provide a few quantitative metrics, this approach is much more qualitative in nature.

Strengths. The clear benefit of the semantic web method lies in its simplicity. Participants are provided with explicit detail of *how* concepts are related to one another. Little in the way of technical knowledge is necessary to understand semantic web diagrams; relationships between concepts are provided propositionally and can be quickly converted into a few meaningful statements. Available scoring systems described here for semantic webs are also easy to calculate.

Concerns. A key concern for semantic webs is that they are as subjective as they are simple. Novak & Musonda (1991) explicitly state their algorithm is arbitrary. Also, scoring with their method introduces the same comparative problems as the RCID method when assessing layperson/novice mental models based on expert information; should we assume that the structure and content of these mental models is similar, although layperson mental models may be less robust? Also, since the developmental goal of mental model semantic webs was to compare differences in mental models of the same domain to expert competence as experience

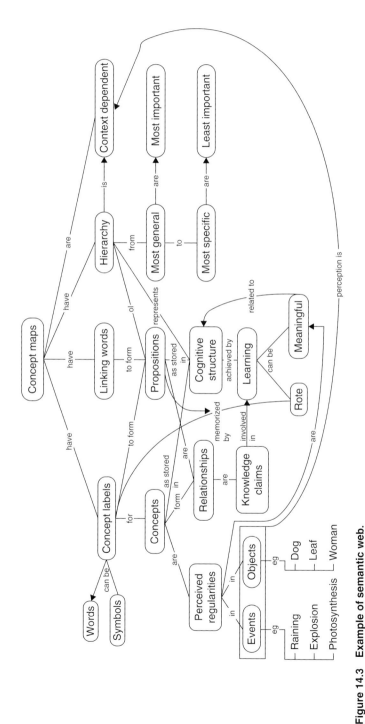

Figure 14.3 Example of semantic web.
(*Source*: Novak and Musonda, 1991)

increases, it may not be appropriate for comparisons between groups of experts where experience is equal, but the domain of information is different.

Reliability is another concern with this method. The construction of semantic web mental model representations is highly subjective. The chance of achieving consistency in diagrams across analysts is therefore small. Novak & Musonda (1991) state the reliability of representations across analysts is good in their work, but use anecdotal evidence to back their case. Also, as the diagramming method is highly subjective, differences in model structure across groups are as likely due to artifacts of the method as to actual differences in internal representations. An analyst could easily use transcript information to create diagrams that unwittingly provide confirmatory evidence of mental model structure and do not provide new information about stakeholders' mental models.

Overview of semantic web

Semantic webs provide complex information on relationships between concepts, and are intuitive to participants both in their development and dissemination; only a brief introduction to the method is required to be able to competently read these diagrams. That said, analyst bias in model development is more problematic here than in other methods, as the analyst has considerable influence in model development. Also, the practitioners of the method to date have not specified procedures for model integration. While a possible option for individual-level or single-group case studies of mental models, semantic webs are not well suited for mental model integration of groups with disparate views.

System dynamics influence diagrams

Method. The System Dynamics Influence Diagram (SDID) approach visually represents the behavior of complex systems (usually in business and industry) through use of feedback loops,

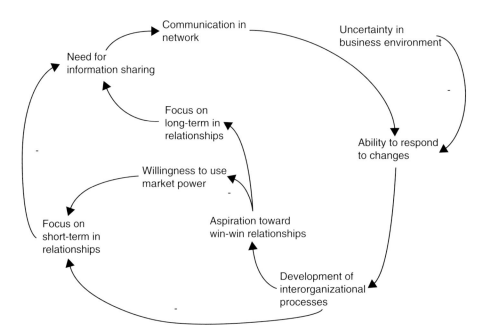

Figure 14.4   Example of system dynamics influence diagram.

(*Source*: Karkkainen and Hallikas, 2006) NB: Figure used with permission; Inderscience Enterprises Limited retains the copyright to this Figure.)

computer simulation, and the problem owners' mental model of the problem with related concepts (Figure 14.4). It uses a Gestalt approach in assuming a problem must be viewed as a whole of inter-related parts that together produce behavior that cannot be accounted for by summing individual contributions alone. This method focuses on identifying a problem's structure by defining the relationships between parts, and presumes that causalities between parts are often looped together (or bidirectional; Karkkainen and Hallikas, 2006).

Though assumptions of systems theory are included in most reviews of system dynamic approaches provided here (including the method's foundation in nonlinear dynamics theory; Sterman, 2001), the process by which concept maps are drawn to reflect system behavior is not well specified (Lane, 2000; Karkkainen and Hallikas, 2006). Causal loop diagrams are a first step, where processes that antagonize one another are identified; stock and flow structures are later added. Though differential and integral calculus are used to compute time delays and flow rates, specific methods or process steps are unclear (Sterman, 2001). Lack of specification may have provided the primary impetus for Doyle and Ford's (1998) unitary definition of mental models in the system dynamics field, to sharpen a concept that was important to the field but previously ill defined. Hazy descriptions of system dynamics processes may be an artifact of the field's 'atheoretical style' (Karkkainen and Hallikas, 2006).

Karkkainen and Hallikas (2006) identify three basic steps in SDID creation. First, during *Conceptualization*, the problem is defined and described, causal factors are identified, and policies that turn information into action are formulated. Next, *Dynamic Hypotheses* are generated that make inferences on how elements of the system interact, and are derived from externalized mental model representations of participants in the process. The emphasis when creating dynamic hypotheses is on how components of the system are 'looped,' or how they causally act with each other. Finally, during *Model Building*, relationships in the model are formulated as levels (stocks) and rates (flows), and the consequences of different policies are tested. A model can be constructed with binary descriptions of stocks and flows, or made in conjunction with fuzzy dynamic logic to create a more refined model of the system. If fuzzy dynamic logic is used, simulations are run assuming a likelihood of node activation and a magnitude estimate of the strength (based on a 0 to 1 scale) and valence (positive or negative) of the relationship between variables (Mendoza and Prabhu, 2006).

Metrics. SDID produces two metrics: stock or variable level (e.g., gross income, employee satisfaction) and rate of flow or influence (e.g., what is the proportion of change on gross income from employee satisfaction?). Often the mental models of systems that are represented with system dynamics frameworks are very complex, and computer simulation is often required to efficiently map this complexity.

Outcomes. The key outcome of the system dynamics influence diagram process is the development of a shared mental model based on input from consulted stakeholders, managers, and other interested groups. Other outcomes include an improvement in dynamic intuition for participants, as well as an improvement in learning ability for complex systems (Lane, 2000).

Strengths. Like the RCID approach, SDIDs gain their power from the use of expert input to infer key properties of the domain of interest. System dynamics representations may also be useful for their ability to tolerate complexity, especially in the relationships between concepts and groups of concepts. In addition, model representations typically accept quantitative inputs once mapped to demonstrate differences in process and output given a specific input. The ability to provide the model with inputs allows users to manipulate the model to identify optimal configurations. Since system dynamics is interested in illuminating underlying processes, it may be better at handling model-building input from different expert domains.

Concerns. As with the other summarized methods (except for concept mapping), the algorithms used to generate representations of mental models are underspecified. As with RCID, this may result from a lack of formalized algorithms, or from concerns around intellectual property rights. From what can be inferred with available information, relationships between concepts are determined mostly by the analyst's discretion. Little indication is given as to the level and manner of involvement the client has in the concept identification and structuring process.

Overview of system dynamics influence diagrams

SDID gives a strict quantitative assessment of mental models, but available literature does not specify how relationships between variables are first identified. RCID, in contrast, has explicit information available on how to conduct data-collection interviews in a way that reduces analyst bias. While current methods of translating interview information to mental model representations are crude and a source of error in the process, a quantitative framework for transcript analysis may serve to address this concern.

Mental models and expert versus novice cognition

Research conducted to understand cognitive structure has hinted at differences between expert and novice approaches to problem solving. Experts notice features and meaningful patterns of information, which cannot be reduced to isolated facts and propositions but are instead 'conditionalized' to specific circumstances (Bransford *et al.*, 1999). The process of conditionalizing allows experts to develop the 'expertise' that guide their decision-making processes. Experts also have the ability to retrieve information on a selective basis befitting the context of the problem at hand. Carey and Wiser (1983) also concluded that the shift from novice to expert is a shift from one system of beliefs about the world, one set of concepts and one set of problem-solving capabilities to another.

Finally, Lee (2005) studied differences between novice and expert oceanographers using the same simulation. She found that experts looked for and found general principles that they could connect systemically to explain ocean processes. Novices learned the information that was present, but were less successful at connecting the pieces to make a 'big picture.' Experts were also quick to grasp the limitations of the model on which the simulation was built, and easily filled in the conceptual gaps. Novices did not do this. This suggests that the metrics developed above can be useful for determining expertise in decision-makers. The differences in the mental models and/or indicative metrics can be used to study critical differences in risk perception and analysis of experts and novices/laypersons.

14.3   Mental models in NHM and CM

In this section we investigate the different kinds of mental models that can be attributed to expert and novice/laypersons in the context of CM and NHM decision-making.

14.3.1   Differences in risk perception in NHM

This section reviews perceptions of flood response and recovery in detail by laypersons and experts with the intent of identifying multiple factors reflecting risks. Data gaps and needs identified in this review call for the strategic and tactical use of mental models. Experts in this field are professional engineers and emergency managers working at major flood risk management agencies such as the U. S. Army Corps of Engineers (USACE). They have views of flood processes and consequences that differ from those of layperson stakeholders. In fact it has been noticed that even expert groups have trouble communicating with each other to take appropriate action in the face of a flood event. For example, professional emergency managers working with the public have difficulty understanding forecasting uncertainties without help in translating the science (Faulkner *et al.*, 2007). This section will outline layperson and expert views of the flooding process, flood prediction, the consequences of floods, and how to manage flood risk with both preventative (pre-event) and reactive (post-event) methods. Such views can inform mental models research.

Laypersons tend to have many misunderstandings related to the flooding process. Research by Lave and Lave (1991) indicates that those who live in flood-prone areas generally know little about their risk of flooding and are uninterested in learning about that risk. Furthermore, an individual's prototypical example of a flood event may be the most destructive flood they experienced, even if that event was years ago and if they have experienced less severe events since. Lave and Lave also found that individual resistance to evacuation orders may also come from denial about the possibility of another major flooding event.

Wagner (2007) found that personal experience and visibility of process are two main influencing factors that explain the accuracy of layperson flash flood knowledge. Laypersons were better informed about floods than landslides, in part because the physical process of flooding is easier for laypersons to understand and recognize through direct observation. For flash flooding, those with more experience with flash floods had a more complete understanding of the flooding process. Layperson knowledge about flash flooding changed very little over a two-year period, demonstrating that public perception of flood events can be stable over time (Wagner, 2007). Presumably processes here that lead to a better understanding of flash floods are the same with other types of flooding and natural disasters.

In comparison, experts tend to have structured approaches. For example, they consider three factors when devising and evaluating flooding preparation options. The first of these, exposure, refers to the frequency, water level, flow velocity, and flow duration in a particular area. Sensitivity, or how problematic exposure may be to a region, includes factors like population density, buildings, and economic values of that region. Adaptation is the ability to avoid damage through adjustments in ecological, social, or economic systems in response to floods and their impacts. These adaptations can happen at the public (government) or private (individuals, business) level, and can also be precautionary (pre-flood action) or reactive (in-flood or post-flood reaction; Grothmann and Reusswig, 2006).

Expert perception of risk is also significantly different from that of laypersons. For example, with respect to structural flood mitigation methods such as dams and levees, laypersons tend to believe (incorrectly) that these structures 'prevent' damages (Pielke, 1999). Lave and Lave may state the truth best, 'No engineering structure can guarantee protection for people living in a flood plain (1991, p. 257).' This mindset highlights the discrepancy in risk mitigation between laypersons that demand certainty in risk (Is this safe or unsafe?), while the whole story is more nuanced (How safe is this?). Advocates of dams who make claims about the safety they provide may cause people to believe a dam reduces flooding exposure to negligible levels. Laypersons then become more likely to build in the flood plain. Paradoxically then, the dam leads indirectly to increased sensitivity to floods seen in property damage rates and loss of life (Lave and Lave, 1991).

14.3.2   *Differences in expert-novice decision-making in CM*

Recent research in construction decision-making has investigated using mental model to understand strategic decision-making of construction managers. Dissanayake and AbouRizk (2007) applied methods that aim to quantify the associations in concept maps of construction managers who are given the opportunity to reason with the causal knowledge of key performance factors and indicators (Dissanayake and AbouRizk, 2007). A subjective method for modeling construction performance was presented using cognitive maps to represent mental models or internal knowledge representation of construction managers. Fuzzy cognitive maps model the cause-and-effect relationship between concepts that present themselves in a construction project. In this method, the concepts are represented as nodes in a graph, and the links between the nodes represent the cause-and-effect relationship between concepts.

Mukherjee *et al.* (2005) analyzed the differences in problem-solving approaches of novice and expert construction managers. This involved testing human subjects and quantitatively studying the results using ConProFac, a program that creates descriptions of how people organize their ideas and qualitatively analyze their responses.

The human subject testing involved providing construction managers, with different levels of experience, a construction scenario. The scenario described the construction of a $104 million project of a state-of-the art library facility, to be built over a period of 24 months. The scenario was based on real life construction projects, and was developed in collaboration with a Seattle-based general contractor. Four distinct areas of concern to construction decision-making: Space Management, Schedule Management, Labor Management and Materials Management. For each of these areas, each participant was required to indicate briefly his/her plan of action, using available information. For example a decision to 'hire-skilled laborers' would be a plan of action to 'manage labor,' while a decision to 'crash activity X' would be a plan of action to 'manage schedule.' Participants were also required to rank, on a scale of 1–10, how they believed their 'plans of action' would affect the project schedule, project cost and the reputation of the company, for each of the four areas of concern listed above. A value lower than 5 would indicate an adverse impact, while a value higher than 5 would indicate a positive impact. A value of 5 would indicate maintaining the 'as-planned' schedule. This provided a structured approach to collecting data that could be investigated for knowledge organization.

In addition, ConProFac a software tool was used to calculate an index i, indicative of the structuredness of the numerical responses. Qualitative plans of action from the participants were also analyzed. The ConProFac program creates descriptions of how people organize their ideas about a content area at different levels of generality: CONcepts, PROpositions or FACets. It can be used with any information about how collections of concepts are connected by predicates to form propositions. The truth of propositions about concepts can either be binary, that is either true or false, or weighted to express the likelihood that the proposition is true, or that a participant believes it to be true. ConProFac can calculate the *structuredness index i* that can be used to compare differences among the ways subjects organize their ideas and how these change over time, using standard statistical procedures.

Qualitative analysis of the participants 'plans of action' shed more light on the differences in the mental models of experts and novices. Experts tended to focus more on developing and enumerating different contingency plans, with specific focus on assessing the time it would take for a decision to take effect and the impact it would have on the schedule, showing the ability to deal with complex inter-relationships as well as recognize time lags between decision and impact. In comparison, participants with fewer years of experience tend to emphasize more on what needs to be done immediately to solve the problem at hand.

The study was very preliminary in scope and limited by the very small number of participants it involved. Most relevant to this study is its contribution in developing an experimental set-up that can be used to collect and investigate human-subject data to find knowledge organization patterns in decision-making. It also highlighted the need to develop data-collection environments that are interactive and can truly allow construction managers to explore 'what-if' scenarios. This idea is further discussed in the next section.

With experience, decision-making mental models of construction managers tend to be based on contingent plans and apprehensions of different possible outcomes of current situations. They also tend to support their decisions by rough calculations using thumb rules, even in situations with incomplete information. The thumb rules that they use are usually based on cost patterns that they tend to recall from previous project experiences. In contrast, managers with fewer years of experience tend to concentrate more on immediate actions without considering long-term impacts.

14.3.3   Discussion: operations of mental models in different scenarios

The discussion in this chapter highlights tools and methods that can be used to develop mental models of decision-making in DTEs such as CM and NHM (specifically flood risk management) to understand differences between how expert and novice/laypersons analyze and perceive risk.

One characteristic point worth discussing is that the perception of risk and uncertainty of experts in both domains, while different in specifics, shows similar cognitive organization. In CM, experts tend to apprehend future outcomes of decisions and think in terms of alternative plans. In managing flood risk, experts take a structured approach to risk and rank design alternatives in terms of how much risk they pose/mitigate. Both these approaches are driven by a need to account for different outcomes that can result from uncertainty at a particular point in time. The critical difference is that in the CM domain, managers think in terms of contingency plans, where as in NHM, experts think in terms of varying levels of mitigated risk. It can be argued, that this difference arises due to the different nature of temporal and spatial scales over which these two domains evolve. The structured and constraint/schedule driven time span over which CM domains evolve makes it possible for experts to 'simulate' possibilities in their minds and develop alternate plans, while the unstructured time spans (as discussed earlier) associated with flood risk management limits experts to think in terms of ranking outcomes by how much risk they mitigate, rather than specifically 'simulate' each outcome. This also reflects yet another difference between the domains: CM scenarios are limited in space and occur within well defined construction site, while NHM scenarios often span across large geographical spaces, involving diverse stakeholders with diverse socio-economic experiences and education backgrounds.

While, CM and NHM are fundamentally different domains they provide a platform to understand how the differing time spans over which they evolve, and the different nature of the stakeholders, influence the mental models of decision-makers. It can also be seen from the research that the mental models of expert and novice/laypersons in both these fields show similar cognitive structure and reflect results from research in the cognitive sciences.

14.4   Conclusion: the future of mental models

The discussion in the previous sections show that mental models of risk among expert decision-makers in complex DTEs tend to be informed by mental 'simulations' of the future – be it in the form of alternative plans within structured environments, or in the form of ranked outcomes in domains in which damage is temporally unstructured, disproportionate and spatially delocalized. Further studies need to be conducted in order to collect reliable human decision-making data from within crisis scenarios in these domains to provide more clarity to these results and highlight similarities and differences between mental models of risk.

The challenge lies in analyzing large volumes of decision-making data to find the anatomy of effective decisions made within the urgency of crisis scenarios. Case studies can provide important guidelines to classify such scenarios and to investigate patterns in effective decision-making. However, they can be misleading as they narrate limited scenario specific responses, instead of providing general statistically significant trends. In addition, it is not possible to explore 'what-if' scenarios to test varying impacts of different responses in a case study (Pennell *et al.*, 1997). Decision-making data from crisis scenarios as they unfold can be difficult to collect. Direct observation, while not impossible, is time consuming and costly. Besides, collecting data is a complex operation, if the level of detail at which data is being collected, and the level of abstraction at which it is to be modeled, are not known.

Existing construction databases have been studied and can often be used to construct project histories. Construction databases have been analyzed using data mining methods to investigate delays in construction projects. Soibelman and Kim (2002) analyzed the US Army Corps of Engineers construction database. Their research focuses primarily on developing a framework for identifying how project variables such as weather can be related to the occurrence of delays in the project. It does not, however, include the influence of the contexts in which construction managers on projects make decisions. They also do not provide information regarding decisions that shape the course of the project, and the factors that influence the anatomy of effective decisions.

The first step is to address appropriate levels of detail at which decision-making data from CM and NHM domains need to be collected, and the level of abstraction at which such data need to be analyzed. The second step is to develop situational simulations of crisis scenarios in DTEs that can be used as experimental test beds for collecting human decision-making data. In addition, they can be used to educate novice/laypersons about the complexities of DTEs. Mayer (2004) illustrates that interactive simulation environments can be effectively used to achieve changes in mental models through guided discovery learning.

Situational simulations provide an interactive simulation platform that can be used to explore 'what-if' construction scenarios, estimate risks and contingencies, test alternative plans during construction, and facilitate the capture and analysis of decision-making data. They create temporally dynamic clinical exercises of construction project scenarios that expose users to rapidly unfolding events and the pressures of decision-making. The design, development, and use of general-purpose situational simulations can be found in previously completed research (Rojas and Mukherjee, 2003, 2005, 2006). The Interactive Construction Decision Making Aid (ICDMA) is a specific implementation of a general-purpose situational simulation framework and its description can be found in previous work by one of the authors (Anderson *et al.*, 2007).

Situational simulations allow large quantities of human-subject data to be quickly and easily obtained within a digital environment. It is also easy to duplicate scenarios for multiple human subjects, providing the ability to conduct controlled experiments by exposing human subjects to similar scenarios. Watkins *et al.* (2008), discuss the role of situational simulations in studying human decision-making in construction management and present a formal method to capture decision-making data from a simulated construction environment. Preliminary analysis of human decision-making data collected using ICDMA show that the proposed method can be used to analyze complex decision-making behavior, specifically with respect to tracking how managers dynamically assign resources in complex crisis scenarios. In addition, project metrics indicating the health of the system can be tracked and correlated with direct resource allocation decisions as the project evolves in time.

The agenda for future research includes the development of simulation test beds, such as the ICDMA that can be used to collect significant human decision-making data, which can in turn be modeled using one, or many of the methods described in section 2, to develop mental models of decision-making in DTE. Understanding from such models of decision-making can be used to enhance the planning and management of complex DTEs.

ACKNOWLEDGMENTS

We would like to thank Drs. Jack Davis and Edmond Russo of the US Army Corps for advice and discussions. Gordon Butte and Alexander Tkachuk helped in manuscript preparation and review. Information described and the resulting data presented herein, unless otherwise noted, were obtained from research conducted under the sponsorship of the Civil Works Basic Research Program by the U.S. Army Engineer Research and Development Center. Permission was granted by the USACE Chief of Engineers to publish this material. This work was also supported by the NSF grant SES 0624118 to Amlan Mukherjee.

Endnote

1. Corresponding author.

References

Anderson, G. R., Onder, N. and Mukherjee, A. (2007) Expecting the unexpected: Representing and reasoning about construction crisis scenarios. *Proceedings of the 2007 Winter Simulation Conference*, Washington, DC, 9–12 December 2007.

Atman, C.J. Bostrom, A. Fischhoff, B. and Morgan, M.G. (1994) Designing risk communications – completing and correcting mental models of harzadous processes 1. *Risk Analysis*, **14**(5): 779–788.

Bostrom, A., Atman, C.J., Fischhoff, B. and Morgan, M.G. (1994) Evaluating risk communications – completing and correcting mental models of harzadous processes 2. *Risk Analysis*, **14**(5): 789–798.

Bostrom, A., Fischhoff, B. and Morgan, G. (1992) Characterizing mental models of hazardous processes: a methodology and an application to radon. *Journal of Social Issues*, **4**( ): 85–100.

Bostrom, A., Morgan, M.G., Fischhoff, B. and Read, D. (1994) What do people know about global climate-change 1. Mental Models. *Risk Analysis*, **14**(6): 959–970.

Bransford, J.D., Brown, A.L. and Cocking, R.R. (1999) *How People Learn: Brain, Mind, Experience, and School* (Expanded edn). Washington DC: National Academy Press.

Carey, S. and Wiser, M. (1983) When heat and temperature were one. In: Gentner, D. and Stevens, A.L. (eds) *Mental Models*. Hillsdale New Jersey: Lawrence Earlbaum Associates Publishers, pp. 267–298.

Chi, M.T.H., Glaser, R. and Rees, E. (1982) Expertise in problem solving. In: Sternberg, R.J. (ed.) *Advances in the Psychology of Human Intelligence* (Vol. 1). Hilldale, NJ: Erlbaum, pp. 7–76.

Cigler, B.A. (2007) The 'Big Questions' of Katrina and the 2005 great flood of New Orleans. *Public Administration Review*, **67**(1): 64–76.

Craik, K.J.W. (1943) *The Nature of Explanation*. Cambridge, UK: University Press.

De Keyser, V. (1990) Temporal decision making in complex environments. In Broadbent, D.E., Reason, J. and Baddeley, A. (eds) *Human Factors in Hazardous Situations*. Oxford: Clarendon Press.

Diehl, E. and Sterman J.D. (1995) Effects of feedback complexity on dynamic decision-making. *Organizational Behavior and Human Decision Processes*, **62**(2):198–215.

Dissanayake, M. and AbouRizk, S. (2007) Qualitative simulation of construction performance using fuzzy cognitive maps. *Proceedings of the 2007 Winter Simulation Conference*, Washington, DC, 9–12 December 2007.

Doyle, J.K. and Ford, D.N. (1998) Mental models concepts for system dynamics research. *System Dynamics Review*, **14**(1):3–29.

Doyle, J.K. and Ford, D.N. (1999) Mental models concepts revisited: some clarifications and a reply to Lane. *System Dynamics Review*, **15**(4): 411–415.

Dreyfus, H.L. and Dreyfus, S.E. (1986) *Mind Over Machine*. New York: The Free Press.

Duit, R. and Treagust, D.F. (2003) Conceptual change: a powerful framework for improving science teaching and learning. *International Journal of Science Education*, **25**(6):671–688.

Edwards, W. (1962) Dynamic decision theory and probabilistic information processing. *Human Factors*, **4**(2):59–73.

Faulkner, H., Parker, D., Green, C. and Beven, K. (2007) Developing a translational discourse to communicate uncertainty in flood risk between science and the practitioner. *Ambio*, **36**(8):692–703.

Fischhoff, B. (1995) Risk perception and communication unplugged – 20 years of process. *Risk Analysis*, **15**(2):137–145.

Fleming, G. (2002a) How can we learn to live with rivers? The findings of the Institution of Civil Engineers Presidential Commission on flood-risk management. *Philosophical Transactions of the Royal Society of London Series a-Mathematical Physical and Engineering Sciences*, **360**(1796):1527–1530.

Fleming, G. (2002b) Learning to live with rivers – the ICE's report to government. *Proceedings of the Institution of Civil Engineers-Civil Engineering*, **150**:15–21.

Ford, J.K., Schmitt, N., Schechtman, S.L., Hults, B.M. and Doherty M.L. (1989) Process tracing methods: contributions, problems and neglected research questions. *Organizational Behavior and Human Decision Processes*, **43**(1):75–117.

Freeman, L.A. and Jessup, L.M. (2004) The power and benefits of concept mapping: measuring use, usefulness, ease of use, and satisfaction. *International Journal of Science Education*, **26**(2):151–169.

Galvin, P.F. (1989) Concept mapping for planning and evaluation of a Big Brother/Big Sister program. *Evaluation and Program Planning*, **12**(1):53–57.

Gentner, D. and Stevens, A.L. (1983) *Mental Models*. Hillsdale, NJ: Lawrence Erlbaum Associates.

Gheytanchi, A., Joseph, L., Gierlach, E., Kimpara, S., Housley, J., Franco, Z.E., *et al.* (2007) The dirty dozen – Twelve failures of the Hurricane Katrina response and how psychology can help. *American Psychologist*, **62**(2):118–130.

Graham, W.J. (2000) Should dams be modified for the probable maximum flood? *Journal of the American Water Resources Association*, **36**(5):953–963.

Grothmann, T. and Reusswig, F. (2006) People at risk of flooding: Why some residents take precautionary action while others do not. *Natural Hazards*, **38**(1–2):101–120.

Hecker, E.J., Zepp, L.J. and Olsen, J.R. (2008) Improving public safety in the United States – from Federal protection to shared flood risk reduction. Paper presented at *The European Conference on Flood Risk Management*, Keble College, Oxford, UK; 30 September to 2 October 2008.

Hoffmann, M.H.G. (2005) Logical argument mapping: a method for overcoming cognitive problems of conflict management. *International Journal of Conflict Management*, **16**(4):304–334.

Ifenthaler, D. (2008) Relational, structural, and semantic analysis of graphical representations and concept maps. *Education Technology Research and Development*. [Online] Available from: <http://www.springerlink.com/content/x20657853q70552p/?p=a44229a029244249a68f8e455a006afaπ=27.> [Cited 15 June 2008].

Johnson-Laird, P.N. (1983) *Mental Models*. Cambridge: Harvard University Press.

Johnson-Laird, P.N. (2006) Models and heterogeneous reasoning. *Journal of Experimental and Theoretical Artificial Intelligence*, **18**(2):121–148.

Johnson, B.L. (2005) Hurricane Katrina and that vexing 'What If?' question. *Human and Ecological Risk Assessment*, **11**(6):1081–1082.

Karkkainen, H. and Hallikas, J. (2006) Decision making in inter-organisational relationships: implications from systems thinking. *International Journal of Technology Management*, **33**(2–3): 144–159.

Kearney, A.R. and Kaplan, S. (1997) Toward a methodology for the measurement of knowledge structures of ordinary people – The Conceptual Content Cognitive Map (3CM). *Environment and Behavior*, **29**(5):579–617.

Kerstholt, J.H. (1994) The effect of time pressure on decision making behavior in a dynamic task environment. *Acta Psychologica*, **86**(1):89–104.

Kerstholt, J.H. (1995) Decision making in dynamic situations: the effect of false alarms and time pressure. *Journal of Behavioral Decision Making*, **8**(3):181–200.

Kerstholt, J.H. and Raaijmakers, J.G.W. (1997) Decision making in dynamic task environments. In: Raynard, R., Crozier, R.W. and Svenson, O. (eds) *Decision Making: Cognitive Models and Explanations*, pp: 205–217.

Kolkman, M.J., Kok, M. and van der Veen, A. (2005) Mental model mapping as a new tool to analyse the use of information in decision-making in integrated water management. *Physics and Chemistry of the Earth*, **30**(4–5):317–332.

Kolkman, M.J., van der Veen, A. and Geurts, P. (2007) Controversies in water management: frames and mental models. *Environmental Impact Assessment Review*, **27**(7):685–706.

Lane, D.C. (1999) Friendly amendment: a commentary on Doyle and Ford's proposed re-definition of 'mental model'. *System Dynamics Review*, **15**(2):185–194.

Lane, D.C. (2000) Should system dynamics be described as a 'hard' or 'deterministic' systems approach? *Systems Research and Behavioral Science*, **17**(1):3–22.

Lave, T.R. and Lave, L.B. (1991) Public perception of the risks of floods – Implications for communication. *Risk Analysis*, **11**(2):255–267.

Lee, Y.-L. (2005) A comparison of learning in immersive virtual reality and print learning environments about oceanography. Paper presented in April at the Annual Meeting of the American Educational Research Association, Montreal, Quebec.

Lusk, C.M. and Hammond K.R. (1991) Judgment in a dynamic task: microburst forecasting. *Journal of Decision Making*, **4**(1):55–73.

Mannes, M. (1989) Using concept mapping for planning the implementation of a social technology. *Evaluation and Program Planning*, **12**(1):67–74.

Matthews, R. (2007) Modelling the sustainability of rural systems: concepts, approaches and tools. In: Crompton, L.A. and Wheeler, T.R. (eds) *Proceedings of the Thirty-Ninth Meeting of the Agricultural Research Modellers' group. Journal of Agricultural Science*, **145**(6): 636–641.

Mayer, R.E. (1989) Models for understanding. *Review of Educational Research*, **59**(1):43–64.

Mayer, R.E. (2004) Should there be a three-strikes rule against pure discovery learning? *American Psychologist*, **59**(1):14–19.

Mendoza, G.A. and Prabhu, R. (2006) Participatory modeling and analysis for sustainable forest management: overview of soft system dynamics models and applications. *Forest Policy and Economics*, **9**(2):179–196.

Morgan, M.G.; Fischhoff, B., Bostrom, A. and Atman, C. (2002) *Risk Communication: A Mental Models Approach*. New York: Cambridge University Press.

Mukherjee, A., Rojas, E. and Winn, W. (2005a) Understanding cognitive and meta-cognitive processes in construction management: the system dynamics perspective. *Proceedings of the Construction Congress 2005*. ASCE, pp.

Norman, D.A. (1983) Some observations on mental models. In: Gentner, D. and Stevens, A.L. (eds) *Mental Models*. Hillsdale, NJ: Lawrence Earlbaum Associates, pp. 7–14.

Novak, J.D. and Gowin, D.B. (1984) *Learning How to Learn*. New York: Cambridge University Press.

Novak, J.D. and Musonda, D. (1991) A twelve-year longitudinal study of science concept learning. *American Educational Review Journal*, **28**(1):117–153.

Payne, J.W., Bettman, J.R. and Johnson, E.J. (1993) *The Adaptive Decision Maker*. New York: Cambridge University Press.

Pennell, R., Durham, M., Ozog, C. and Spark, A. (1997) Writing in context: situated learning on the web. In Kevill R., Oliver R., and Phillips R., (eds), *What Works and Why: Proceedings of the 14th Annual Conference of the Australian Society for Computers in Tertiary Education*. Held at Curtin University, Perth, WA: Academic Computing Services, pp. 463–469.

Pielke, R.A. (1999) Nine fallacies of floods. *Climatic Change*, **42**(2):413–438.

Pirnay-Dummer, P. (2007). Model inspection trace of concepts and relations: a heuristic approach to language-oriented model assessment. Paper presented at the *Annual Meeting of American Educational Research Association*, Chicago, IL.

Pirnay-Dummer, P., Ifenthaler, D., Johnson, T. and Al-Diban, S. (2008) Reading with the guide of automated graphical representations: How model based text visualizations facilitate learning in reading comprehension tasks. Paper presented at *The Annual Meeting of the American Educational Research Association*, pp. 25–27, March 2008, New York.

Posner, G.J., Strike, K.A., Hewson, P.W. and Gertzog, W.A. (1982) Accommodation of scientific conception: toward a theory of conceptual change. *Science Education*, **66**(2):211–227.

Post, N.M. (2007) High-profile Manhattan Tower is Veteran Builder's Last Hurrah. *Engineering News Record*, (Cover Story, March 26). (Available online at) http://enr.construction.com/features/_Covers/2007/archives/070326.asp (Retrieved on 31 August 2008).

Read, D., Bostrom, A., Morgan, M.G., Fischhoff, B. and Smuts, T. (1994) What do people know about global climate-change 2: survey studies of educated laypeople. *Risk Analysis*, **14**(6):971–982.

Rojas, E. and Mukherjee, A. (2003) Modeling the construction management process to support situational simulations. *Journal of Computing in Civil Engineering*, ASCE **17**(4):273–280.

Rojas, E. and Mukherjee, A. (2005) Interval temporal logic in general purpose situational simulations. *Journal of Computing in Civil Engineering*, ASCE **19**(1):83–93.

Rojas, E. and Mukherjee, A. (2006) A multi-agent framework for general-purpose situational simulations in the construction management domain. *Journal of Computing in Civil Engineering*, ASCE **20**(6):1–12.

Seel, N.M. (2003) Model-centered learning and instruction. *Technology, Instruction, Cognition and Learning*, **1**(1):59–86.

Soibelman, L. and Kim, H. (2002) Data preparation process for construction knowledge generation through knowledge discovery in databases. *Journal of Computing in Civil Engineering*, **16**(1): pp. 39–48.

Sterman, J.D. (2001) System dynamics modeling: tools for learning in a complex world. *California Management Review*, **43**(4):8–25.

Stevens, A.L. and Gentner, D. (1983) Introduction. In: Gentner, D. and Stevens, A.L. (eds) *Mental Models*. Hillsdale, NJ: Lawrence Erlbaum Associates, pp. 1–6.

Trochim, W.M.K. (1989a) Concept mapping: soft science or hard art? *Evaluation and Program Planning*, **12**(1):87–110.

Trochim, W.M.K. (1989b) An introduction to concept mapping for planning and evaluation. *Evaluation and Program Planning*, **12**(1):1–16.

United States Environmental Protection Agency (1997) *Terms of Environment: Glossary, Abbreviations, and Acronyms*. Washington, DC: United States Environmental Protection Agency.

U.S. Army Corps of Engineers, New Orleans District. (2008) *Louisiana Coastal Protection and Restoration Technical Report Draft*. [Retrieved July 7, 2008] from http://lacpr.usace.army.mil/%5CReport%5CDraft%20LACPR%20Technical%20Report.pdf United States Environmental Protection Agency (1997). *Terms of Environment: Glossary, abbreviations, and acronyms*. Washington, D.C.: United States Environmental Protection Agency.

Vislosky, D.M. and Fischbeck, P.S. (2000) A mental model approach applied to R&D decision-making. *International Journal of Technology Management*, **19**(3–5):453–471.

Wagner, K. (2007) Mental models of flash flood and landslides. *Risk Analysis*, **27**(3):671–681.

Watkins, M.T., Mukherjee, A. and Onder, N. (2008) Using situational simulations to collect and analyze dynamic construction management decision-making data. In: Mason, S.J., Hill, R.R., Moench, L. and Rose, O. (eds) *Proceedings of the 2008 Winter Simulation Conference*, Miami, FL. December 2008.

Working Group for Post-Hurricane Planning for the Louisiana Coast. (2006) *A New Framework for Planning the Future of Coastal Louisiana After the Hurricanes of 2005*. Cambridge, MD: University of Maryland Center for Environmental Science.

15 Stakeholder Management – The Gains and Pains

Stefan Olander and Brian L. Atkin

15.1   Introduction

The effective management of stakeholders on a project is an important key to success (Jergas *et al.*, 2000). A construction project affects many different stakeholders before, during and after construction work on site and a negative perception by stakeholders can severely obstruct the implementation of a project. Inadequate management of the concerns of stakeholders may lead to controversy and conflict about the implementation of the project. Community attitudes are one example that has been shown to be an important factor when planning for and locating a construction project (Rogers, 1998). The demands of different stakeholder groups are likewise different. A construction project can benefit one stakeholder group and have a negative impact on another. Understanding the viewpoints of different stakeholders helps the construction project manager build relationships and thus avoid preconceived ideas and assumptions (Watson *et al.*, 2002). In this chapter, we examine, *inter alia*, the prevalence of multiple stakeholder perspectives.

The difficulty of addressing the concerns of stakeholders is due to a failure to recognise and avoid the technical rationale versus the cultural rationale, lack of success in defining and presenting the benefits and costs to all affected stakeholders, and failure to reach equitable and fair agreements on the redistribution of these benefits and costs (Dorshimer, 1996). Improper and arbitrary decision-making often becomes an issue when engineers make decisions on issues they believe to be purely technical and professional in nature, but which those affected see as questions of political power (Connor, 1998). Many of the sources of disagreement in such disputes involve value-trade offs rather than technical issues (McAvoy, 1999). Furthermore, engineers tend to explain problems in technical and economic terms, which may not be sufficient to address the concerns of stakeholders. The premise is that decisions are best based on data; the best decisions are based on the clearest, least ambiguous data. Those subscribing to this view believe that when the technical facts are clearly communicated, all reasonable hearers will arrive at similar conclusions. Yet, engineers must sometimes present data to audiences that do not share the values of the technical culture they represent (Hynds and Martin, 1995). The role of a construction project manager should involve not simply an understanding of the technical realities at hand, but also of the links between technology, the environment, the community and the people in it. Thus, a stakeholder management process should, if conducted properly, be seen as representing an opportunity for improving the project.

When handled adequately, controversies can have a positive side. Conflicting stakeholder interests can result in valuable improvement to the construction project in question (Dear, 1992). The responsibility for sufficiently managing stakeholder concerns should lie with the project manager who should make every effort to conduct planning, authorisation and evaluation in such a way that problems are negotiated and eliminated instead of being allowed to resurface as delays and cost overruns during implementation (Flyvbjerg *et al.*, 2004). The early planning stages of a construction project should have the aim of avoiding unnecessary expenditure and eliminate any steps that ultimately do not contribute to its success and thus be geared towards maximising stakeholder participation and acceptance in the decision-making process of the project (Beekman, 2002).

The purpose of the chapter is to review through the medium of case studies the stakeholder management process and, in particular, to show where and how the prospects for success can be enhanced. The case study approach is appropriate for exploring in-depth issues of organisational and human behaviour (Yin, 2003).

15.2   Stakeholder management process

In the construction industry stakeholders include a wide range of entities that directly or indirectly can provide support or resistance to the accomplishment of project objectives (Walker, 2000). Karlsen (2002) argues that there are at least four reasons for employing a stakeholder management process. First, to be acquainted with the project's stakeholders; second, to ensure the balance between contribution and reward in the relationship with stakeholders; third, to plan and define how to manage stakeholder concerns; and last, to set a basis for deciding which stakeholders are to be involved in determining the project goals and the measurement of success. A stakeholder management process is necessary in order to determine the reactions various stakeholders have to project decisions (Cleland, 1986); for example, what are the reasons for a specific reaction and how will different stakeholder groups interact with each other to affect the outcome of a proposed project strategy? Full alignment of all stakeholders is not possible because some will be fundamentally opposed to the project (Jergas *et al.*, 2000). Even so, a stakeholder management process will identify these stakeholders to the project manager. An awareness of their concerns will help to manage them sufficiently in order to prevent them from adversely affecting the project.

Freeman (1984) relates the stakeholder concept to different views of the firm or the project. In the production view, the major concern is input versus output, which means that the stakeholders considered in this view are the supplier and the customer. A more complicated model is the managerial view, where besides the supplier and customer, managers must also pay attention to owners and employees. These four stakeholder groups represent the internal change agents with the firm. However, Freeman (1984) argues that the more difficult task is to understand external changes that originate from the environment of a firm or a project and which affect its ability to cope with internal changes. External change produces uncertainty, which cannot be readily assimilated into the relatively more comfortable relationship with suppliers, owners, customers and employees.

External change may have a particularly strong effect on a project because of the environment within which a project operates changes from one project to the next. This is especially true of construction where the project is located on a specific site and where the environment of the external change that occurs can shift between projects and during project implementation (Olander and Landin, 2005). There is no strong tradition in the project management literature for discussing problems of the external environment (Engwall, 1995; Crawford *et al.*, 2006). There are, however some examples in construction (Winch and Bonke, 2002; Newcombe, 2003; Bourne and Walker, 2005; Olander, 2007). Even so, the emphasis is still on the internal processes involved. There is a need to emphasise the importance of the external environment both for projects in general and for construction projects in particular.

Internal stakeholders, representing the internal exchange, can be defined as those who are formally connected with the project (e.g. owners, customers and employees), whereas external stakeholders, representing the external exchange, are those affected by the project in some way (Gibson, 2000). Figure 15.1 shows a schematic picture of the potential stakeholders in a construction project, divided into internal and external stakeholders. Construction project managers need to be capable of managing internal *and* external stakeholder exchange. In a way, it can be argued that more effort and resources should be committed to the management of external stakeholders because there are more of them and relationships with them are more diffuse than in the case of internal stakeholders. For the internal exchange, there is often an agreement or a contract to set the framework of the relationship and the resolution of potential conflicts.

The stakeholder environment has a strong effect on project implementation. The environment will change in its nature depending on actions taken by different stakeholders and will affect project implementation in a certain way. The stakeholder management process can thus be defined as having the aim of maintaining the desired implementation of the project and avoiding unnecessary conflict and controversy with stakeholders. The Hallandsås project

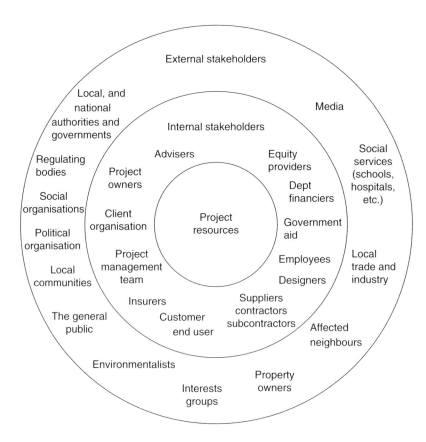

Figure 15.1   Potential stakeholders for construction projects *(Adapted from Cleland (1999) and Leiringer (2003)).*

(the construction of an 8.8-kilometre twin-bore railway tunnel in southern Sweden) is one example of a project in which controversy and conflict have had an impact on project implementation. The project was started in 1991, with basically all permits that were needed being approved, and it was scheduled to be finished in 1996. As of late 2007, the project was still unfinished and the budget has been exceeded by a factor of ten, due mainly to bad management choices that have affected the surrounding environment and the people in it. There have been several studies that directly or indirectly use the Hallandsås project as a research case (Hydén and Baier, 1998; Boholm *et al.*, 1998; Boholm, 2000; Danielsson and Holmberg, 2002; Baier, 2003). Experiences from the Hallandsås project can be summarised as follows:

- The local community felt frustration at not being able to influence the planning and implementation of the project.
- The information given to external stakeholders was not correct, timely or appropriate.
- The local community was inclined to form action groups outside the parliamentary system in efforts to exert an influence over decisions made within the project.
- The project's management did not fully address the concerns of stakeholders.

The Hallandsås project is an example of a failed stakeholder management process, due mainly to an inadequate analysis of how project decisions would affect stakeholders and how these would, in turn, affect those decisions.

  If the potential impact of a construction project on various stakeholders is not adequately communicated in the early stages of a project (as was the case with Hallandsås) this may lead to controversy and conflict over the project's location, size and design. Experience gained from the construction of the tunnel under the English Channel also highlighted the need to address the interests of stakeholders and that the management of them should be considered as being an essential cost element in the implementation of any construction project.

  Poor public perception can damage or stop a project as surely as can bad ground or shortage of labour . . . The Channel Tunnel project is a classic example: for much of its formative period it existed in an often destructive climate of adverse public opinion. Most of this was avoidable, but it resulted in the project team spending much of its time fighting a rearguard action rather than simply getting on with the job (Lemley, 1996).

The Channel Tunnel and Hallandsås are examples of major projects. The problems connected with stakeholder influence are not, however, limited to projects of such size; the influence of stakeholders is an important topic to consider for just about any construction project. Indeed, construction projects, independent of their size, can become embroiled in a process of controversy and conflict with external stakeholders and without warning if adequate steps are not take to engage them; even then there is the unpredictability factor. The image issue is also a relevant aspect of many construction projects. The UK's Considerate Constructors Scheme (Barthorpe, 2002, 2003; Olander 2004) is one example where external stakeholders and the effects they suffer 'at the hands of construction' merit proper consideration.

  The arguments and reasoning about the stakeholder management process for construction projects that are discussed here are based on the in-depth study of five case projects: two housing projects and three civil engineering projects, which were undertaken with varying degrees of success from the perspective of stakeholder management. In addition to related literature, the study revealed two important aspects of the stakeholder management process. By acknowledging the concerns of all stakeholders there is a base for communication with regard to relevant issues addressed by stakeholders. The knowledge obtained from this communicative process can then be used in the implementation of a construction project to perform a stakeholder management process that will maintain or increase the level of acceptance.

15.3 Acknowledging concerns from all stakeholders

The acknowledgement of stakeholder concerns is an integral part of achieving a communicative process of conflict resolution (Susskind and Cruikshank, 1987; Susskind and Field, 1996). The concerns of stakeholders can be described by the variables of perceived benefits or perceived negative impacts (Rogers, 1998). These variables are essential in understanding the dimension of support and opposition to a project (Grimes, 2000a) and will influence the decision by a stakeholder to accept or refuse the implementation of a construction project (Davy, 1997). The perceived negative impacts and the determination of its acceptability are influenced by many factors that are not related to the actual level of risk concerning these consequences (Young, 1990). Construction projects are rejected or accepted due to the trade-offs between risk and the amount of gain (Sokolowska and Pohorille, 2000).

The empirical data from the projects studied clearly indicated that the acceptance level sets the stakeholders position towards the project and defines the extent and direction of stakeholder influence. The level of acceptance depends on two basic considerations: the concerns of stakeholders and the stakeholder management process, i.e. what are they worried about and how they have been treated. The acceptance level towards the project is based on the ability of the project manager to acknowledge the concerns of stakeholders and maintain or increase the received acceptance level through an effective stakeholder management process. The challenge for the project manager is that of communicating and implementing the construction project in such a way that the perceived benefits and the negative impacts are presented realistically and minimise the effects of negative impacts and, to the extent possible, maximise the benefits for all affected stakeholders.

In order to analyse the influence and concerns of stakeholders adequately, it does not suffice to simply identify them, since the dynamics of the environment and the power of the stakeholder in relation the organisation (or project) need to be assessed (Mendelow, 1981). Mitchell *et al.* (1997) propose a set of stakeholder attributes for assessing potential stakeholder influence of power, legitimacy and urgency. A stakeholder can have the power to impose its will on the relationship. The power of stakeholders may arise from their ability to mobilise social and political forces, as well as from their ability to withdraw resources from the project organisation (Post *et al.*, 2002). Legitimacy can be defined in terms of stakeholders who bear some sort of risk in relation to the organisation, be it beneficial or harmful. The dynamic character of stakeholder influence is covered by the term urgency, which is defined as the degree to which claims (or stakes) call for immediate attention. At any given time, some stakeholders will be more important than others (Jawahar and McLaughlin, 2001). Newcombe (2003) argues that the value of stakeholder analysis and mapping is in addressing the following.

- Whether the political/cultural situation is likely to undermine the adoption of a particular project strategy.
- Whether to pursue project strategies to reposition certain stakeholders.
- The extent to which it is necessary to assist or encourage stakeholders to maintain their level of predictability, interest and power in order to ensure the successful implementation of project strategies.

Concerns and priorities change over time, new classes and configurations of stakeholders appearing in response to changing circumstances. From a purely project perspective, it can be argued that stakeholders possessing the attribute of power are the most relevant to consider in the stakeholder management process. However, stakeholders who possess the attribute of legitimacy are in a sense more important, because they are the risk bearers in the project. Thus, it is important from a moral standpoint, to address the needs of the legitimate stakeholders fully. If not, they may try to achieve a powerbase by themselves or by forming an alliance with more powerful stakeholders. In either case, the project manager loses control over the stakeholder management process.

From the projects studied it was evident that the main factor for opposing or accepting a construction project was the perception of benefits. If there were no perceived benefits, stakeholders would oppose the project even if the perceived negative impacts could be viewed as low or reasonable. Thus, there were insufficient trade-offs between the negative impacts and the perceived benefits. If no benefits can be presented, the stakeholder management process will need to focus on minimising the potential damage that opposing stakeholders can bring to the project. By acknowledging the concerns of stakeholders early on in the project life cycle a more active and communicative stakeholder management process can be achieved. In this way, trade-offs between benefits and negative impacts can be found and used to create an opportunity that reduces or eliminates potentially damaging conflicts with stakeholders. Even so, the aim should be to complete the project according to the requirements of the project owner. A clear challenge is then to find the trade-offs that satisfy as many stakeholder concerns as possible. A thorough stakeholder analysis should provide a basis for forthcoming project decisions. One definite source of controversy and conflict is that decisions on a course of action for the project were made without analysing the consequences the decision would have for external stakeholders. This tends to result in project manager not being prepared for the conflicts that could arise, and thus having no plan of how to resolve or to handle them.

15.4   Maintaining and increasing acceptance during project implementation

How stakeholders perceive risks and benefits largely depend on the information given to them by those responsible for the construction project. Thus, it is a relevant and rational response for stakeholders to question this information. Even if it cannot be proven to have been done deliberately, tendencies to overestimate the beneficial effects and underestimate the negative impacts can be seen in the projects studied. If the information given by those responsible for the project cannot be trusted by affected stakeholders there is the serious prospect of failure in the stakeholder management process. If it is assumed that the information given is correct from an objective viewpoint, the level of trust that stakeholders have of those responsible for the project will determine how they perceive the accuracy of the information. The problem arises that once the information given is perceived to be inaccurate, every effort to improve that information faces the risk of being met by suspicion from those stakeholders for whom the information is intended. In this case, a construction project manager will end up in a defensive position where any effort to promote his or hers choices are met with doubt and mistrust. A communicative process becomes impossible and the construction project manager is faced with defending a rearguard position with the strong likelihood of a poor image for the project and bad publicity.

In the projects studied, trust was important for the outcome of the stakeholder management process. In those projects where the communicative process was non-existent, there was strong distrust between those responsible for the project and affected stakeholders. Company officials are learning that without an open and ongoing communication process, they and the community in which they work often approach each other from a standpoint of mistrust (Young, 1990). Mistrust on the part of stakeholders may lead to their questioning the information given and lead to a perception of negative impacts that are exaggerated and different from the reality. The pros and cons of competing scenarios must be expressed and evaluated as early as possible in communication with different stakeholders in order to select the correct path for the project (Olander and Landin, 2005). Maximising fairness and equity, for all stakeholders, in decisions surrounding the construction project will circumvent many obstacles created by eroded trust in those responsible for the project (Grimes, 2000b). Opposition based on perceived negative impacts can be reduced through a communicative process that begins early on in the project life cycle, giving stakeholders increased knowledge and, possibly, some involvement in the decisions that affect their lives and business (Connor, 1998).

> *Failing to recognise or cooperate with adverse stakeholders may well hinder a successful project outcome. Indeed, strong and vociferous adverse stakeholders can force their particular interest on the project manager at some time, perhaps at a time least convenient to the project. Project stakeholder management is thus a necessity, allowing the project manager to set the timetable so that he can maintain better control* (Cleland, 1986).

Some lessons for construction project managers could be learned from the projects studied here. In this regard, it is important to investigate all possible alternatives and solutions to realise the objectives of the project, not only from the quantitative aspects of technology and economy, but also from the more qualitative aspects of potential influence from stakeholders. It will also be necessary to define clearly all positive and negative arguments about the selected alternative in relation to the other alternatives investigated, in order to be regarded as trustworthy by those stakeholders who are negatively affected by the project. The stakeholders' base of influence is not static and so stakeholder analysis must be conducted and updated during the entire life cycle of the project, with the purpose of gaining knowledge about the potential influence various stakeholders have at different stages. Prior to any major decision to proceed into a new phase of the project, there should be an analysis of how the decision affects the different stakeholders in order to be proactive in the stakeholder management process. Stakeholder management leading to stakeholder cooperation enhances project objective achievement, while stakeholder neglect hinders it (Cleland, 1986). Failure to consider the wider collection of stakeholders can result in extraordinary risks being ignored (Walker, 2000).

15.5 Conclusions

Most stakeholders have no formal power to affect the decision-making process for a construction project. Nonetheless, they often have an informal power that, when exercised, can press more powerful stakeholders into changing their position towards a project. A poorly executed communication process can urge weak oppositional stakeholders to create a stronger powerbase through the formation of alliances with potentially more powerful stakeholders – those they see as allies. A more proactive stakeholder strategy may result in a higher degree of acceptance, and even support, towards the project and its implementation thereby decreasing both the probability of a stakeholder group adversely affecting project decisions and the impact if they do. In contrast, lack of communication may encourage some members of a stakeholder group to organise in active opposition, increasing both the probability of their being engaged and their having an impact on the project from doing so.

It is important for construction project managers to analyse and acknowledge the concerns of different stakeholders and evaluate various scenarios for project implementation with respect to this knowledge, thereby raising the prospect of better communication throughout the stakeholder management process. The goals and the framework for the project must, however, be clearly defined. A communicative approach can have the weakness of being too optimistic in overcoming significant diversity of stakeholder values and interests. Thus, a stakeholder management process, and the communication involved in it, must be conducted with respect to the goals of the project and the possible trade-offs with various stakeholder concerns in order to obtain agreements that can be realised. In addition, construction project managers must accept that a project cannot be implemented without addressing the concerns of those stakeholders who are involved in it. Without this understanding, disputes will be stigmatised and confrontational, and the chance of achieving a communicative process of conflict resolution with stakeholders will be negligible.

For sufficient performance of the stakeholder management process, there needs to be an understanding of the complexity of stakeholder influences. The impact of stakeholders changes throughout the life of the project and depends largely on the perceptions they have of the project. The controversies seen in the projects studied are due mainly to miscommunication of impacts and mismanagement of stakeholder concerns. This indicates that construction project managers should acknowledge the stakeholder management process as an important task for which adequate resources have to be committed. The objective of the stakeholder management process should be to communicate the correct aspects of the project, be they good or bad, and to implement the project in such a way that the effects of negative impacts are minimised and, if possible, maximise the benefits for all stakeholders, or at least as many of them as possible.

References

Baier, M. (2003) *Norm och rättsregel – en undersökning av tunnelbygget genom Hallandsåsen [Norm and legislation – an investigation of the construction of the tunnel through Hallandsås]*, Lund Studies in Sociology of Law 17, Lund: Department of Sociology, Lund University.

Barthorpe, S. (2002) Enhancing project performance by implementing a societal stakeholder culture. *Proceedings of the 10th International Symposium on the Organization and Management of Construction*, University of Cincinnati, CIB, pp. 948–959.

Barthorpe, S. (2003) The considerate constructors scheme – a way to good relations? *Proceedings of the 3rd Nordic Conference – Construction Economics and Organization*, Lund: Division of Construction Management, Lund University, pp. 11–19.

Beekman, G.B. (2002) Social change and water resource planning and development. *Water Resources Development*, 18(1):183–195.

Boholm, Å. (ed) (2000) *National Objectives Local Objections – Railroad Modernization in Sweden*, CEFOS, Gothenburg: Gothenburg University.

Boholm, Å., Löfstedt, R. and Strandberg, U. (1998) *Tunnelbygget genom Hallandsås – Lokalsamhällets dilemma [Construction of the tunnel through Hallandsås – The dilemma of the local community]*, CEFOS, Gothenburg University, Gothenburg.

Bourne, L. and Walker, D.H.T. (2005) Visualising and mapping stakeholder influence. *Management Decision*, **43**(5):649–660.

Cleland, D.I. (1986) Project Stakeholder Management. *Project Management Journal*, **17**(4):36–45.

Cleland, D.I. (1999) *Project Management – Strategic Design and Implementation* (3rd edn)., New York: McGraw-Hill.

Connor, D.M. (1988) Breaking through the NIMBY syndrome. *Civil Engineering*, **58**(12):69–71.

Crawford, L., Pollack, J. and England, D. (2006) Uncovering the trends in project management: journal emphases over the last 10 years. *International Journal of Project Management*, **24**(2):175–184.

Danielsson, A. and Holmberg, I. (eds) (2002) *Ledarskapets olika skepnader – exemplet Hallandsås [The different appearances of leadership – the case of Hallandsås]*, Studentlitteratur, Lund.

Davy, B. (1997) *Essential Injustice – When Legal Institutions Cannot Resolve Environmental and Land Use Disputes*. Vienna: Springer-Verlag.

Dear, M. (1992) Understanding and overcoming the NIMBY syndrome. *Journal of American Planning Association*, **58**(3):288–300.

Dorshimer, K.R. (1996) Siting major projects & the NIMBY phenomenon: the Decker Energy Project In Charlotte, Michigan. *Economic Development Review*, **14**(1):60–62.

Engwall, M. (1995) *Jakten på det effektiva projektet [The hunt for the efficient project]*, Nerenius & Santérus Förlag AB, Stockholm.

Flyvbjerg, B. Skamris Holm, M.K. and Buhl, S.L. (2004) What causes cost overrun in transport infrastructure projects? *Transport Review*, **24**(1):3–18.

Freeman, R.E. (1984) *Strategic Management – A Stakeholder Approach*. Marshfield: Pitman Publishing Inc.

Gibson, K. (2000) The moral basis of stakeholder theory. *Journal of Business Ethics*, **26**(3):245–257.

Grimes, M. (2000a) Explaining public assessment of the new rail, Chapter 12. In: Boholm, Å. (ed) *National Objectives Local Objections – Railroad Modernization in Sweden*. CEFOS, Gothenburg: Gothenburg University.

Grimes, M. (2000b) Trends in research on siting controversy, Chapter 2. In Boholm, Å. (ed) *National Objectives Local Objections – Railroad Modernization in Sweden*. CEFOS, Gothenburg: Gothenburg University.

Hydén, H. and Baier M. (1998) När kunskapen blir onödig – om normative assymetri i fallet Hallandsåsen [When knowledge becomes unnecessary – the case of Hallandsås], In Statens offentliga utredningar [National public investigations] SOU 1998:137, *Miljö i grund och botten [Environment in ground and bottom]*, Swedish Government, Stockholm.

Hynds, P. and Martin, W. (1995) Atrisco Well #5: a case study of failure in professional communication. *IEEE Transaction on Professional Communication*, **38**(3):139–145.

Jawahar, I.M. and McLaughlin, G.L. (2001) Toward a descriptive stakeholder theory: an organizational life cycle approach. *The Academy of Management Review*, **26**(3):397–414.

Jergas, G.F. Williamson, E., Skulmoski, G.J. and Thomas, J.L. (2000) Stakeholder management on construction projects. *AACE International Transactions*, P12.1–P12.6.

Karlsen, J.T. (2002) Project stakeholder management. *Engineering Management Journal*, **14**(4):19–24.

Leiringer, R. (2003) *Technological innovations in the context of public private partnership projects*, Doctoral Thesis, Department of Industrial Economics and Management, Royal Institute of Technology, Stockholm.

Lemley, J.K. (1996) Image versus reality – Channel Tunnel image management. *Proceedings of the Institution of Civil Engineers*, Civil Engineering, **114**:12–17.

McAvoy, G.E. (1999) *Controlling Technocracy – Citizen rationality and the NIMBY Syndrome*. Washington DC: Georgetown University Press.

Mendelow, A. (1981) Environmental scanning: the impact of stakeholder concept. *Proceedings of the 2nd International Conference on Information Systems*, Cambridge, Mass.

Mitchell, R.K., Bradley, R.A. and Wood, D.J. (1997) Toward a theory of stakeholder identification and salience: defining the principle of who and what really counts. *The Academy of Management Review*, **22**(4):853–885.

Newcombe, R. (2003) From client to project stakeholders: a stakeholder mapping approach. *Construction Management and Economics*, **21**(8):841–848.

Olander, S. (2004) *Den omtänksamme byggaren – En studie av ett brittiskt handlingsprogram [The considerate constructor – A study of a British scheme]*, Lund: Division of Construction Management, Lund University.

Olander, S. (2007) Stakeholder impact analysis in construction project management. *Construction Management and Economics*, **25**(3):277–287.

Olander, S. and Landin, A. (2005) Evaluation of stakeholder influence in the implementation of construction projects. *International Journal of Project Management*, **23**(4):321–328.

Post, J.E. Preston, L.E. and Sachs, S. (2002) *Redefining the Corporation – Stakeholder Management and Organizational Wealth*. Stanford CA: Stanford University Press.

Rogers, G.O. (1998) Siting potentially hazardous facilities: what factors impact perceived and acceptable risk? *Landscape and Urban Planning*, **39**(4):265–281.

Sokolowska, J. and Pohorille, A. (2000) Models of risk and choice: challenge and danger. *Acta Psychologica*, **104**(3):339–369.

Susskind, L. and Cruikshank, J. (1987) *Breaking the Impasse – Consensual Approaches to Resolving Public Disputes*. New York: Basic Books Inc.

Susskind, L. and Field, P. (1996) *Dealing with an Angry Public – the Mutual Gains Approach to Resolving Disputes*. New York: The Free Press.

Walker, D.H.T. (2000) Client/customer or stakeholder focus? ISO 14000 EMS as a construction industry case study. *TQM Magazine*, **12**(1):18–25.

Watson, T., Osborne-Brown, S. and Longhurst, M. (2002) Issues negotiation – investing in stakeholders. *Corporate Communications: An International Journal*, **7**(1):54–61.

Winch, G. and Bonke, S. (2002) Project stakeholder mapping: analysing the interests of project stakeholders, Chapter 23. In: Slevin, D.P., Cleland, D.I. and Pinto, J.K. (eds) *The Frontiers of Project Management Research*. Newtown Square PA: Project Management Institute Inc.

Yin, R.K. (2003) *Case Study Research – Design and Methods* (3rd edn). Applied Social Research Method series, **5**, London: Sage Publications.

Young, S. (1990) Combating NIMBY with risk communication. *Public Relations Quarterly*, **35**(2):22–26

[2] Stakeholder Salience theory
See 2018 Joos, Aufseß & Pidun (Springer)
¢

16 Benefiting from Stakeholder Management – Electronic Archiving

Michaela M. Schaffhauser-Linzatti

16.1 Problem outline

The annual European Union benchmark survey, published on September 20, 2007, in Lisbon, confirmed that once again, Austria leads the online public service league with a nearly perfect score, standing out 'both on sophistication and full on-line availability, with scores of 99% and 100%, respectively' (EU, 2007). Austria also leads in providing a legally recognized secure electronic identity.

Austrian eGovernment targets include a full implementation of one stop shopping (BKA, 2007). A one stop shop is a single contact point for all administrative and organizational procedures between citizens, businesses, and public administration units as well as among public administration units. The communication via only one single interface accelerates administrative processes, saves financial resources, and minimizes possible transcription and communication errors, hence supporting customers' satisfaction and improving the business location.

Among others, the actual realization of the one stop shop principle depends on the authentication of electronic documents. Austria is the first and up to now single country which has solved the legal obstacle of realizing the equal footing of original paper documents and electronic documents. Electronic archiving of documents in fact comprises:

- archiving,
- forwarding, and
- providing access to

electronic documents which are legally recognized as originals. Implemented primarily in the field of eJustice, electronic archives support storage, electronic document transfer within courts (or any other public administration unit), and direct communication among lawyers, solicitors, or courts' employees and units. This reduces the courts' workload, costs, and time, and allows for an unbureaucratic data access independent of time and place (BKA, 2006).

In Austria, eGovernment and eJustice, in particular, contribute to an improved stakeholder management practice. Here, stakeholder management is understood as identifying stakeholders and – in a next step – balancing their interests.

The archiving of documents by the Austrian Ministry of Justice is used as a basis to demonstrate some benefits such as 'e'-stakeholder management. The following subsections comprise judicial, technical, and organizational aspects of Austria's successful electronic archiving solution by especially focusing on the multiple stakeholder aspects within the general framework. After examining the influencing legal regulations on the European Union and Austria in particular, a qualitative content analysis is applied to derive general lessons learnt from the best practice example of 'Archivium Dokumentenarchiv GmbH'.

16.2   Context

The European Union promotes eGovernment by various plans and programs. Amongst others, the eEurope 2002 Action Plan prepared for the Lisbon European Council enforces an acceleration of setting up an appropriate legal environment; the support of new infrastructure and services across Europe; efforts by public administrations at all levels to exploit new technologies; the promotion of the use of electronic signatures within the public sector; and an enhancement of consumer confidence (CEU, 2000). The succeeding eEurope 2005 program further demands interoperable eBusiness solutions for, among others, transactions, security, signatures, and standards (EC, 2002). The Interoperable Delivery of European eGovernment Services to Public Administrations, Businesses and Citizens (IDABC) program supports, among others, efficiency and collaboration improvements between European public administration units and reorganization efforts within public administration units (EC, 2004).

Austria reacted to the plans of the European Union by founding the Information and Communications Technique (ICT) Board within the Federal Chancellery in June, 2001. The ICT Board evaluates strategic proposals, regulates overall aspects in the field of information and communication technologies, and coordinates projects at federal, communal, and city level, while the federal ministries concerned are responsible for the actual realization of these projects (BKA, 2007).

According to Austria's eGovernment, electronic media shall provide all information to her citizens, businesses, and internal public administration units, and shall also support official electronic dealings via Web Accessibility Initiative (WAI) guidelines (CEU, 2002). Overall guidelines as well as general regulations on electronic signature and data protection are included in the Austrian e-Government Act (2004). For specific applications, particular laws and regulations have been released.

The Austrian Ministry of Justice offers the main field of eGovernment application and is indeed a key player in implementing Austria' ambitious eGovernment targets. This Ministry is headed by the Federal Minister of Justice who is also a member of the Federal Government. Being the third pillar of the constitutional state besides legislation and administration, justice is separated from administration at all levels. Hence, the Austrian Ministry of justice is mainly responsible for (BMJ, 2008):

- The ordinary courts, which are legally established governmental institutions and their decisions are made by independent, non-dismissible, irremovable, impartial judges.
- The offices of public prosecution, which are special authorities separated from the courts' to protect the public interest in criminal cases.
- The detention centers which are the competent institutions for the execution of prison sentences.
- The probation offices, which assist persons placed on probation.

In the field of eJustice as a part of eGovernment, the Austrian Ministry of Justice concentrates on the automation of some court procedures (BMJ, 2005), electronic legal communication, and electronic archiving of original documents (BMJ, 2007). The main areas of application are the companies register and the land register. Legal bases are the Gerichtsorganisationsgesetz (GOG, 2006) for regulating court organization; the Berufsrechtsänderungegesetz für

Notare, Rechtsanwälte und Ziviltechniker (BRÄG, 2006); the Signaturgesetz (SigG, 1999); the Signaturverordnung (SigV, 2000); Urkundenarchivverordnung (UAV, 2007) and Elektronischer Rechtsverkehr (ERV, 2006); and their corresponding directives. Further corresponding regulations are the Federal Act on Provisions Facilitating Electronic Communications with Public Bodies(E-Government-Gesetz-E-GovG, 2004), the Rechtsanwaltsordnung (RAO, 2007), and the Urkundenarchiv-Richtlinie (2007).

Most of all, the GOG revolutionizes electronic archiving and can be regarded as the precursor of this practice worldwide. Sections 89 et seqq. of the GOG allows for electronic petitions in general and demand predefined procedures to be performed electronically in particular; referring especially to lawyers and notaries. Further details are to be published by the Federal Ministry by specific orders. For instance, Section 91b of the GOG calls for the Federal Minister of Justice to archives documents that are subject to authentification or even notarization. Access to these archives then enables an electronic inspection as well as the creation of negotiable electronic documents. Significantly, section 91b subsection 7 states that the documents stored electronically in the archives should be regarded as original documents unless proven otherwise. While the archives themselves are to be kept by the Federal Minister, section 91c makes it legitimate for public bodies to establish electronic archives in their sphere of action; to be used for communicating with the courts under further judicial orders or authorization. Section 91d subsection 3 permits external service providers to implement and administer the databases of the archives, as long as they guarantee for the lawful and secure use of data.

The revolutionary aspect of the GOG is the first-time introduction of the so-called 'original fiction' wherein the documents stored in an electronic archive have the same legal status as the original paper documents, because the electronically stored documents are regarded as originals of the documents by law. The authenticity of these documents is proven by lawyers and notaries who import the documents into the archive by their secure digital signatures.

16.3   eJustice applications of archives

The long-term, electronic document archives of the Austrian Ministry of Justice can be seen as world champions in eJustice and eGovernment. They aim at replacing the multitude of physical archives being maintained by individual courts (Zisak, 2007). For an historical overview of eJustice in Austria and the companies and land register in particular see BMJ (2005).

The basic functions pertaining to these electronic archives are:

- administrating offices and authorizations
- scanning and importing of document images
- indexing documents
- maximum secure signing and archiving of documents
- searching documents for inspection
- investigating and downloading of documents by authorized persons
- administrating the index of authorized persons
- re-registering.

So far, the companies register and the land register have been available to the public since 2005 and 2006 respectively via the internet; and allow for the exchange of original documents due to the 'original fiction' (BMJ, 2006). The necessary secure digital signature is provided by 'a.trust Gesellschaft für Sicherheitssysteme im elektronischen Datenverkehr GmbH' which is the only accredited Austrian certification authority and belongs to a cooperation of Austrian banks, professional representations, and industrial firms (a.trust, 2007). The registers are intrinsically tied to the electronic document archives 'Archivium Dokumentenarchiv GmbH' of the Austrian Bar Association (OERAK, 2007) and 'cyberDOC GmbH & Co KG' of the Austrian Chamber of Civil Law Notaries (OENK, 2007). These two professional representations provide the entrance ports to the archives for all of about 5500 lawyers and about 500 notaries in Austria who have to bring in all documents referring to the companies register or land register in an electronic form.

The 'Archivium Dokumentenarchiv GmbH', simply called 'Archivium', was founded in 2006 as a limited company and actually represents a Public Private Partnership project. The two partners are the 'Siemens Austria AG' (2007) and the Austrian Bar Association via its 100% subsidiary 'RADOK GmbH' (RADOK, 2007). They each share 50% of the company and each nominate one executive director (Archivium, 2007). While the Austrian Bar Association acts as public corporation and is therefore allowed to provide an archive service according to section 89 of the GOG, the private company 'Siemens Austria AG' is responsible for the technical and organizational solutions. The objective of the company is the implementation and operation of a maximum secure, central archive. All Austrian lawyers have to use it after July 2007 for all matters regarding the companies register and land register as well as processes in which such documents are needed. The lawyers deposit their documents into 'Archivium', where their digital signatures are stored on the, identification cards. According to section 6 subsection 2 of the Urkundenarchiv-Richtlinie, corresponding to section 37 subsection 1 no 7 of the RAO and section 91c subsection 4 of the GOG, the documents have to be stored for 7 and 30 years, respectively; and these periods may be extended once.

'Archivium' offers the lawyers the possibility to:

- communicate with courts by sending maximum secured attachments
- recall documents from the lawyers' archive
- contact other lawyers directly
- recall special documents when acting as a liquidator.

There are no fees for recalling documents; but importing and saving documents costs 7€ excl. 35 cents fixed charge per document, when stored for 7 years; 15€ excl. 75 cents fixed charge per document, when stored for 30 years.

'CyberDOC GmbH & Co KG', further called 'cyberDOC', is a Public Private Partnership joint venture between the Austrian Chamber of Civil Law Notaries and the Siemens Austria AG (cyberDOC, 2007). It serves the Austrian notaries in a similar, however in a slightly different way than 'Archivium' the lawyers.

The success of the Austrian's eJustice efforts have been awarded several prizes, among others: by the European Union with the e-Government-Label in 2001 (BMJ, 2005), by the Austrian Trade Chambers with the Grand Price of the 'Amtsmanager Wettbewerb' in 2006 for the electronic document archive (BMJ, 2007), by the European Union with the 'best practice award' and with the 'European e-Government Award' in 2007, both for the electronic archive in cooperation with 'cyberDOC' and 'Archivium' (IDABC, 2007; Silhavy, 2007). Further, 'Archivium' won the first prize of the 'ebiz egovernment' award of the Federal Chancellery for Vienna and the third prize for Austria (Archivium, 2007).

16.4   The stakeholders' view

The electronic archive did not only cause technological and organizational changes for the stakeholders concerned, but also enlarged the number of stakeholders involved in document

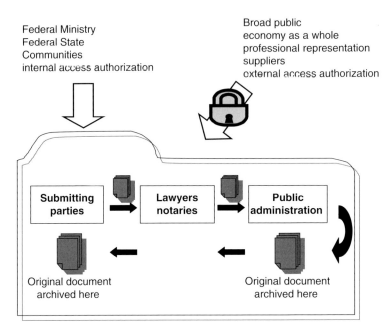

Federal Ministry
Federal State
Communities
internal access authorization

Broad public
economy as a whole
professional representation
suppliers
external access authorization

Submitting
parties

Lawyers
notaries

Public
administration

Original document
archived here

Original document
archived here

Figure 16.1   Traditional process of submitting and archiving original documents.

submission and archiving and improved – or even introduced – their interconnectedness, hence propelling stakeholder management. Figure 16.1 summarizes the former traditional archival process based on paper documents. Following the conclusion of a contract by a lawyer or a notary, the paper documents were submitted to the corresponding public administration; but at least one original paper document stayed with the submitting party as well as the public administration.

It is easy to show that the submission of an archival document manually did not require the comprehensive technical support of the three stakeholders: submitting party, lawyer or notary, and public administration. This manual process consumed resources inefficiently, mostly by manual inputs and rewritings. Beside the high administrative costs, the long waiting times involved were not conductive to the submitting parties, especially for the economy. Moreover, the traditional archiving of the original documents was expensive in the long run, did not support urgent enquiries, and had safety drawbacks.

Figure 16.2 summarizes the new electronic process of document submission and archiving in a generalized way. The submitting party still hands over paper documents to a lawyer or a notary who then transforms these into an electronic document and then forwards it to the public administration via the federal data center which again corresponds with the electronic archive. Authorized internal persons from the public administration as well as external persons, i.e., the submitting parties, lawyers and notaries, have access to specific documents archived. The electronic documents are archived by the electronic archive providers and the federal data center.

At first, beside judicial questions, the legislators who introduced electronic archiving mainly focussed on efficiency gains for the public administration. Secondly, their attention was on the first-hand appliers, i.e., the lawyers and the notaries as well as those they represent. Public and private stakeholders intrinsically profit from the electronic archive. In case of electronic archiving in the Austrian way, the parting line between public and private partners is sometimes

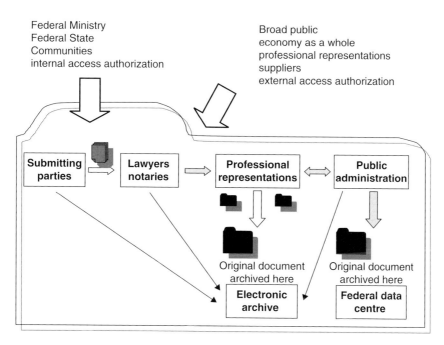

Figure 16.2   Electronic process of submitting and archiving original documents.

blurred, as many organizations are privately owned, but act on behalf of the public administration. In general, the public sector stakeholders involved in electronic archiving comprise the:

- Federal Ministries, federal states, communities
- Austrian Federal Computing Center
- professional representations
- persons with internal access authorization
- broad public

Private sector stakeholders include:

- electronic archive providers
- submitting parties
- lawyers and notaries
- certificate authorities and other suppliers
- persons with external access authorization
- economy as a whole

16.4.1   Public sector stakeholders

a. Federal Ministries, federal states, communities

Through electronic archiving, the public administration of all public bodies most of all gains time that has been inefficiently used by the manual handling of acts and procedures. Possible concerns that staff might be reduced in the long run seem unfounded as, at least in the Austrian case, clerks in the field of finance and justice can now concentrate on more enhanced operations, being unburdened by unnecessary routine work. Further, the reduction of mistakes (which unavoidably happenned with manual handling) leads to a higher customer satisfaction. Also statistically necessary analyses within data protection regulations are eased.

b. Austrian Federal Computing Center

The Austrian Federal Computing Center (Bundesrechenzentrum, 2007) is a typical Austrian public stakeholder. As the leading IT-service provider of the Austrian public administration it offers service-oriented interaction between the public administration and its customers and operates the commercial register and the land register via Portal Service Austria. Hence, the Austrian Federal Computing Center is only indirectly involved in electronic archiving; however, it gains in importance and volume of orders. In case public administration outsources such computing centers to a private operator, the financial benefits will increase as well.

c. Professional representations

From a legal point of view, professional representations as quasi-official public partners are only administrators of the archiving systems. However, as central organizations they might increase their influence towards their members as well as towards public administration. In Austria, the Austrian Bar Association and the Austrian Chamber of Civil Law Notaries are the professional representations involved so far.

d. Persons with internal access authorization

Public clerks concerned with electronically archived documents might get internal access authorization to use the electronic documents for further treatment. Their most important advantages are independence of place and time of document search, time savings, and reduction of routine work.

e. Broad public

Here, the term broad public subsumes all citizens and businesses not yet personally concerned with electronic archiving. They profit by an efficient administration and cost reductions, i.e., more efficient use of taxes; support of the overall economy; and, most important of all, trust in the public administration and eGovernance. The frequent argument of data abuse can be easily vitiated by the fact that criminal intentions can never be totally excluded; however, the technical solutions applied in Austria guarantee significantly more data safety than the traditional paper documents. From today's point of view, the electronic archives cannot be hacked. Also, illegal drag net investigations of public administration are heavily restricted by legal regulations.

16.4.2 *Private sector stakeholders*

f. Electronic archive providers

The electronic archive providers act on behalf of the public administration. Being profit-oriented, their private owners benefit from the official fees. Additionally, the providers offer instruction classes to initially teach lawyers and notaries the correct technical and organizational application of the systems. At least, successful electronic archives are showcases with an excellent reputation with all associated providers and owners.

g. Submitting parties

Documents of private citizens as well as businesses are submitted and archived electronically. In some cases, they both are charged a negligibly higher fee than for a paper submission, however, the advantages of time savings and less administrative efforts outweigh this cost.

h. Lawyers and notaries

Both stakeholder groups, at least partially competitors in Austria, face a new organization of archiving within their offices, changing from paper archiving to electronic solutions. Their only requirements are an adequate IT office equipment and, at the very beginning, instruction classes.

i. Certificate authorities and other suppliers

Like the electronic archive providers, the certificate authorities are private organizations working on behalf of the public administration. They profit by an increased demand for certification. Among the other suppliers, software providers for offices benefit from an increased demand for new hardware and software for use by lawyers and notaries.

j. Persons with external access authorization

Like persons with internal access authorization, documents can now be more easily treated independent of place and time by persons with external access authorization which, again, helps to save time and organizational efforts. Such persons are mainly the clients who have mandated their lawyers and notaries to transact their on their behalf and have electronically submitted and archived documents obligatorily.

k. Economy

Not to forget the economy as a whole. Due to accelerated processes and the one stop shop principle, Austria is an attractive business location for domestic and foreign enterprises. The electronic archive helps to reduce the so far previous overly complex administrative procedures and thereby saves time and money.

16.5 Lessons and benefits

The Austrian archiving solution for eJustice is an ideal case study to demonstrate successful stakeholder management by innovative eGovernment solutions. Some of the lessons learnt out of this experience are:

- All stakeholders directly involved in submitting and archiving documents should be involved in the judicial, technical, and organizational aspects right from the beginning of a project.
  - Judicial solutions comprise or at least regard all stakeholder interests and not only focus on the (ostensibly) main stakeholders.
  - Technical solutions should ensure the interoperability of all existing systems.
  - Organizational solutions should be more efficient and obviously more simpler than the former ones.
- Long-term solutions take should priority over financial or/and political short-term profits and this should also involve the choice of partners for the electronic archive organizations.
- The instruction classes offered by the electronic archive providers are a major part of the systems' success as the applicants are motivated to use electronic archiving even for acts which need not be submitted mandatorily and this reduces possible prejudices and negative approaches towards electronic solutions.

Managing stakes can yield tremendous advantages to governmental institutions, organizations, and individuals. The Austrian example discussed above illustrates that stakeholder management by eGovernment and eJustice, in particular, leads to win-win situations without negative side effects. The electronic archiving system developed in Austria brings the following main benefits to all stakeholders:

- cost savings
- time savings
- efficiency gains
- reduction of process tasks
- reduction of organizational efforts
- consequent reduction of tax use
- independence of place and time while depositing or accessing documents
- increased data security
- increased transparency, leading to a reduction of possible mistakes and misuse

16.6   Conclusion

The benefits the electronic archive provides to the large group of stakeholders are obvious; they mainly refer to time reductions pertaining submissions as well as cost savings. As the ground braking innovation of electronic archiving in Austria has proven to be a success story, it is for sure that other countries will follow. This example may help to support successors in their strive to establish new solutions from a stakeholders' point of view. One of the key factors of success has been the foundation of archivium and cyberDOC as Public Private Partnerships which installed a high degree of trust as well as efficiency at the same time.

References

a.trust Gesellschaft für Sicherheitssysteme im elektronischen Datenverkehr GmbH (a.trust, 2007) www.a-trust.at (Accessed 29/10/2007).

Archivium Dokumentenarchiv GmbH (Archivium, 2007) www.archivium.at (Accessed 29/10/2007).

Austrian Bar Association (OERAK, 2007) http://www.rechtsanwaelte.at/ (Accessed 29/10/2007).

Austrian Chamber of Civil Law Notaries (OENK, 2007) http://www.notar.at/ (Accessed 29/10/2007).

Austrian Ministry of Justice (BMJ, 2008) Justice administration introduces itself, http://www.bmj.gv.at/EN/justiz/content.php?nav=15 (Accessed 9/4/2008).

Austrian Ministry of Justice (BMJ, 2005) Use of IT within Austrian Justice, Justice-online, Information, as at August 2005, Vienna.

Austrian Ministry of Justice (BMJ, 2006) Elektronisches Urkundenarchiv, http://www.justiz.gv.at/_cms_upload/_docs/e_urkundenarchiv.pdf (Accessed 29/10/2007).

Austrian Ministry of Justice (BMJ, 2007) News, http://www.justiz.gv.at/service/content.php?nav=66&id=383 (Accessed 29/10/2007).

Berufsrechtsaenderungsgesetz für Notare, Rechtsanwälte und Ziviltechniker (BRÄG, 2006) BGBl. I Nr. 164/2005.

Bundesrechenzentrum (Austrian Federal Computing Centre, 2007) Welcome to the BRZ – Austrian Federal Computing Centre, Your eGovernment partner, http://www.brz.gv.at/Portal.Node/brz/public (Accessed 11/11/2007).

Corrigendum to Commission Decision 2004/387/EC of 28 April 2004 (EU, 2004) Decision 2004/387/EC of the European Parliament and of the Council of 21 April 2004 on the interoperable delivery of pan-European eGovernment services to public administrations, businesses and citizens (IDABC), Official Journal of the European Union L 144 of 30 April 2004, http://eur-lex.europa.eu/LexUriServ/site/en/oj/2004/l_181/l_18120040518en00250035.pdf (Accessed 25/10/2007).

Council of the European Union (CEU, 2002) Council Resolution of 25 March 2002 on the eEurope Action Plan 2002: accessibility of public websites and their content (2002/C 86/02), in: Official Journal of the European Communities, April 10, 2002.

Council of the European Union, Commission of the European Communities (CEU, 2000) An Information Society For All, Action Plan, prepared by the Council and the European Commission for the Feira European Council 19–20 June 2000, Brussels, 14.6.2000, http://ec.europa.eu/information_society/eeurope/2002/action_plan/pdf/actionplan_en.pdf (Accessed 25/10/2007).

CyberDOC GmbH & Co KG (2007) http://www.notar.at/de/portal/dernotar/service/begriffslexikon/c/cyberdoc/ (Accessed 15/11/2007).

E-Government-Gesetz – E-Gov (2004) *Federal Act on Provisions Facilitating Electronic Communications with Public Bodies – The Austrian E-Government Act, 2004: Bundesgesetz über Regelungen zur Erleichterung des elektronischen Verkehrs mit öffentlichen Stellen, Art. 1* of the Act published in the Austrian Federal Law Gazette, part I, Nr. 10/2004, entered into force on 1 March 2004.

Elektronischer Rechtsverkehr (ERV, 2006) BGBl. II Nr. 481/2005.

European Commission (EC, 2002) eEurope 2005, http://europa.eu/scadplus/leg/de/lvb/l24226.htm (Accessed 25/10/2007).

EC (2004) The Programme. Available online at HYPERLINK 'https://exch.wlv.ac.uk/exchweb/bin/redir.asp?URL=http://ec.europa.eu/idabc/en/chapter/3' \t "_blank" http://ec.europa.eu/idabc/en/chapter/3 (Accessed 19/11/2007).

European Union (EC, 2007) EU Benchmark Survey confirms Member States making significant progress in eGovernment, http://ec.europa.eu/information_society/newsroom/cf/itemlongdetail.cfm?item_id=3634 (Accessed 24/10/2007).

Federal Chancellery of the Republic of Austria (BKA, 2006) Administration on the Net, An ABC Guide to E-Government in Austria. (Available online) http://www.buergerkarte.at/en/mumok/Administration_on_the_Net.pdf (Accessed 15/11/2007).

Federal Chancellery of the Republic of Austria (Bundeskanzleramt, BKA, 2007) E-Government, www.cio.gv.at/egovernment/ (Accessed 24/10/2007).

Gerichtsorganisationsgesetz (GOG, 2006) RGBl. Nr. 217/1896, at last changed by BGBl. I Nr. 92/2006.

Heidrun Silhavy (2007) Österreich ist erneut E-Government Europameister, http://www.digitales.oesterreich.gv.at/ (Accessed 29/10/2007).

IDABC (2007) 4th Ministerial eGovernment Conference – Lisbon, http://europa.eu.int/idabc/ (Accessed 29/10/2007).

RADOK Gesellschaft für Organisation, Dokumentation und Kommunikation Gesellschaft m.b.H. (Online) www.radok.at (Accessed 29/10/2007).

RAO (2007) Rechtsanwaltsordnung (Available online at) www.rechtsanwaelte.at/downloads/rao01012007.pdf (Accessed 29/10/2007).

Siemens Austria AG (2007) www.siemens.com (Accessed 29/10/2007).

Signaturgesetz (SigG,1999) BGBl. I Nr. 190/1999.

Signaturverordnung (SigV, 2000) BGBl. II Nr. 30/2000, at last changed by BGBl. II Nr. 527/2004.

Urkundenarchiv-Richtlinie (2007) http://www.rechtsanwaelte.at/downloads/urkundenvarchiv_rl01072007.pdf (Accessed 29/10/2007). {AQ4}

Urkundenarchivverordnung (UAV, 2007) BGBl. II Nr. 481/2006.

Zisak, Anja (2007) *Electronic Document Archives of courts in Austria*, http://www.epractice.eu/cases/1002 (Accessed 29/10/2007).

17 Managing Stakeholders Conflicts

Helder M. Moura and José Cardoso Teixeira

17.1 Conflicts: a behavioural process

It is not unusual that a chapter on conflict management, which is about, generally speaking, helping people or organizations that are in conflict with each other to deal with their differences, opens up with this statement (Moore, 1986 cited in Gordon, 1966):

> *All societies, communities, organizations, and interpersonal relationships experience conflict at one time or another in the process of day-to-day interaction. Conflict is not necessarily bad, abnormal, or dysfunctional; it is a fact of life.*

Typically, a conflict situation results from resource shortage and antagonistic feeling. Conflicts may arise between individuals, between groups of individuals and between organizations. Conflict situations between people are subjective, meaning that although objective reasons may exist, conflict only breaks out if those reasons are perceived. Conflicts between people may trigger out organizational conflicts for the simple reason that organizations are (still) governed by people. However, organizational conflicts may have other root reasons, for example resource interdependency. A common definition of conflict is *a process that begins whenever an individual or a group feels negatively affected by another individual or group*. In other words, *people are in conflict anytime one's actions obstruct or by any means, make other's performance, less efficient*.

Individual conflicts exist in all human relations and those within the construction activity are no exception. Due to the great diversity of people involved in construction projects and to the enormous variety of situations emerging from the construction process, individual conflicts in this activity deserve particular attention. Groton (1997) found that conflicts between people in construction arise as a result of *poor interpersonal skills, inefficient communication, lack of responsiveness and unethical or opportunist behaviour*. Conflicts of this nature may remain within the individual sphere of people involved or build up to the organizations they work for if not adequately handled. This may easily develop into organizational conflicts, affecting several organizations participating in a construction project.

On the other hand, because organizations act through individuals, then conflict events emanate from key actors within the organizations, due to their different perceptions on a particular aspect about which they are unable to agree. This applies both to group conflicts within an organization (for example, inter-department conflicts within a contractor's organization) and to conflicts affecting several organizations working together in a construction project (for example, between the owner and the contractor). Beyond people issues mentioned above, Groton (1997) suggested two main identifiable group causes for organization conflicts: project uncertainty and process problems. Examples of the former are pre-existing conditions and outside forces and of the latter are incomplete project scope definition and poor performance.

In view of the above, the following definition for a conflict applies: *an interaction of independent people who acknowledge different objectives, wishes and values in the other part, capable of interfering with their own.* In this statement, there are three ingredients that seem to be present in every conflict: *interdependence, perception* (at least by one part) and *antagonism.*

In fact, conflict has always been present in organizations taking part in construction ventures since ancient times. One of the first reported conflict management practice between construction stakeholders took place in Ancient Greece (some authors identify the first civil engineer in European history as the Greek Eupalinos, responsible for the construction of the Samos island tunnel in the year 550 BC), where, after public interviews for selecting the contractor of a new statue or building, citizens were informed about actual progress and the cost of the construction project. This and other type of measures related to conflict avoidance tactics were also used in Ancient Rome, where construction contracts ought to define technical specifications, materials, guarantees or payment schedules, in order to adequately distribute risk between the owner and the contractor. Moreover, some conflicts have so deep historical, cultural and political roots that they are seeded by unstructured discernment and cannot be managed by the traditional way (for instance, the resolution of stakeholders' conflicts resulting from construction errors were already a concern in the ancient Hammurabi Code of Laws dated back to about the 18th century BC).

The first perception is that a conflict in construction has negative consequences but it may not be the case. Literature generally distinguishes between *functional* and *dysfunctional* conflicts. A *functional conflict* leads to the improvement of the production process or to a better outcome than would otherwise be expected. On the other hand, a *dysfunctional conflict* prevents progress, has negative effects in production and conducts to poor outcomes. The former is positive or productive while the latter is destructive and generally leads to disputes. Additionally, a functional conflict may lead to or degenerate into a dysfunctional conflict if inadequately managed.

Generally speaking, construction stakeholders aim at preventing disputes because of their possible harsh consequences. Some people will fiercely try to avoid conflicts because of the fear that conflict escalation will lead to unpredictable effects or retaliation. But there are also people who may benefit from disputes in terms of financial advantages, identity, status or power. For those people, a dispute may be viewed as an opportunity to engage diverse opinions and ideas from people holding different perspectives on the conflict issue.

The best way for solving conflicts (either individual or institutional) is transforming them into problems – or preventing them to evolve from these. The fundamental difference between these two concepts is that unlike in conflicts, there is no negative attitude or even hostility between parties involved in problems. In a conflict, there are antagonistic parties whereas in a problem there is a set of people working together to reach a solution. For problem solving, it is paramount that each individual feels he or she is part of the solution, not part of the problem.

The first step for conflict solving is conflict analysis, encompassing conflict types and conflict causes. Although these have broadly been addressed above, root causes for conflicts must be further investigated. Additionally, as with any other social course of action, there is a conflict process bringing in results and effects. The following sections present the most common types of conflicts between construction stakeholders and explain how they can be effectively managed.

17.2   Types of conflicts in the construction activity

Conflicts vary in terms of their legal, political and institutional framework, economic con-strains and pressures, people's culture, social structure, stakeholder interests, technical knowledge, environment and history just to name a few influencing factors.

Conflicts may involve stakeholders external or internal to the project or a combination of those. Conflicts between external stakeholders may be the most difficult to resolve because of their diversity and because of the lack of established procedures for tack-ling most of them. For example, in developed societies, public opinion tends to be more opponent than supporter of a construction project encompassing some environmen-tal impact, although it may respond to a specified public need; on the contrary, in less developed or poorer countries, public may be more keen to accept the project if it aims at solving important infrastructure needs (transportation, sewage, pipelines, water treat-ment, etc.).

Conflicts may involve two parties or several parties for the same reason or for a diver-sity of reasons. For example, a construction project may trigger out conflicts between the contractor and the client for lack of quality and excessive cost of the output, between the public administration and both the promoter and the contractor for noise impact and between the contractor and the site neighbours for lack of parking places on the sur-rounding area. The first step for adequately managing those conflicts is looking at them one by one. Still, a specific conflict may involve several parties from each side. Figure 17.1 shows the conflict logics between a set of m parties (individuals, institutions, etc.) from one side and a set of n opponents from the other side split into m × n conflicts between each pair of parties.

In the following subsections, it is assumed that conflicts involving several parties may always be broken up in a set of conflicts between two parties taking place at the same time.

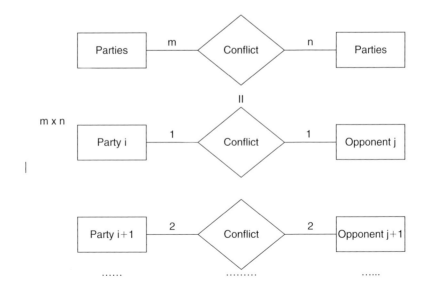

Figure 17.1   Splitting conflicts among several parties into a set of conflicts between two parties.

Conflicts may also be categorized in respect of the typology in:

- **Open conflicts**, when it is everyone's knowledge.
- **Hidden conflicts**, when it is known of some people only.
- **Latent conflicts**, when the conflict comes to the surface if, and only if, something changes the status quo.

With respect to the **clearness of possible solutions**, conflicts can be defined as (Rijsberman, 1999):

- **Well defined,** if they have sharp boundaries and the constraints are well defined; typically, clear solutions to the problem exist.
- **Fuzzy or ill defined,** if they have unclear objectives, important variables are not quantifiable and the values held by the parties may not be clearly defined. As a result, it may be very difficult to identify feasible solutions.

Actually, conflicts may have facets of both well- and ill-defined problems. In broad terms, well-defined conflicts can be viewed as a subset of ill-defined conflicts. The following subsections point out some solving strategies applicable to different conflict types.

17.2.1 Conflict level

The huge diversity of construction conflicts has propelled classification efforts from several authors. One of the simplest classification system found in the literature is by **level** of occurrence (Gordon, 1996) distinguishing *intra-personal* conflicts (within the individual), *inter-personal* (between individuals), *intra-grouping* (conflict deflates in a restricted group), *intra-organizational* (within an organization), *inter-group* (between different groups) and *inter-organizational* (between organizations).

Intra-personal conflicts obviously exist within construction project teams and construction stakeholders' organizations, as it happens in any organization, because there are people involved. These problems aim to be solved through psychology and are not dealt with in this book.

This chapter focuses on inter-group and inter-organizational conflicts because construction stakeholders (media, authorities, contractors) normally act either as organizations or as groups of individuals with the same interests (users, land owners, nearby residents). Moreover, inter-personal conflicts may lead to conflicts involving several groups and organizations, therefore deserving attention.

Finally, this chapter will not deal with conflicts taking place within construction stakeholder groups and organizations unless they are construction specific. If this is not the case, current approaches for conflict settlement apply and are not treated here. However, intra-group and intra-organizational conflicts may lead to conflicts involving several groups and organizations therefore deserving attention as well.

17.2.2 Conflict causes

Inter-personal conflict causes are diverse. Generally speaking, subjective causes for conflicts exist when someone recognizes interference on his or her individual sphere. The individual sphere is a set of tangible and intangible values which we claim the right to detain. Examples of the former are capital assets, property issues like cars or houses and examples of the latter are safety, comfort, pride, time or recognition from the others.

Published literature on the topic seems to agree on four main causes for inter-group and inter-organizational conflicts:

- **Cognitive conflicts**, resulting from different evaluation of empirical data or facts, given that parties interpret, combine or deduce different conclusions from the same basic elements. In most cases, it may be realized that data available is insufficient or facts not clear enough to properly assess a situation. This can be resolved by the technical team through additional studies to clarify facts or to obtain additional and more reliable data. And this may contribute for changing the conflict situation into a structured problem.
- **Conflicts of objectives or interests**, applicable to sharing benefits, rejecting negative consequences of something, financing external costs or allocating disposable resources. These conflicts ought to be solved through sound conflict management techniques.
- **Normative conflicts**, resulting from divergences about values, behaviours and norms that should prevail in socially adequate conducts. Root causes for these conflicts are ethical and moral principles that are not negotiable. The best approach to this situation is by reformulating the conflicting issue in order to transform it in a conflict of interests.
- **Conflicts of relationships** are caused by personality or behaviour of stakeholders' representatives. These can normally be solved either by negotiation or, when the conflict has raised over, through mediation by an independent third party.

A more restrictive taxonomy of conflict causes includes conflicts of objectives or interests into cognitive conflicts and merges normative conflicts and conflicts of relationships into a single cause class:

- **Cognitive conflicts**, normally task-oriented and focused on differences of judgement about data or facts and on the way to achieve objectives.
- Affective or **socio-emotive conflicts** relating to personality differences, irritability, frictions and animosity, and tend to be more dysfunctional and less constructive.

17.2.3   Conflict object

Another way of differentiating conflicts is by **object**, or basic core (prime matter) of the dispute. Accordingly, the following apply to inter-group and inter-organizational conflicts:

- Conflicts over **objectives, needs or interests**, when one party perceives that its needs, concerns or objectives are incompatible with the other party's.
- Conflicts over **processes** arise when people, groups, organizations or institutions use different approaches for solving the same problems.
- **Structural** conflicts occur due to the way society is structured in terms of social, legal, economical and cultural arrangements, and the relative position and power of each stakeholder within that social order.

17.2.4   Conflict life cycle

In a construction project, conflicts may occur in one stage and evolve to the next. In fact, a conflict is a dynamic phenomenon with a specific life cycle, like any construction product. According to some authors (Groton, 1997), there are generally four stages in every conflict: first, conflict progresses from initiation to escalation, then to controlled maintenance, abatement and finally to termination/resolution (Figure 17.2).

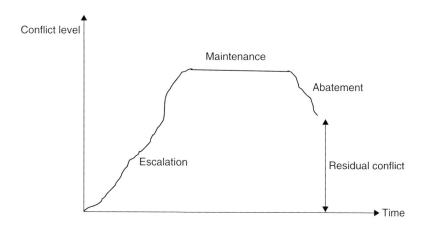

Figure 17.2   Conflict life cycle.
(*Source*: Adapted from Groton, 1997)

For each one of these stages, there is an appropriate management approach:

- Potential or dormant conflicts: develop power matrix and incorporate needs/interests.
- Erupting conflict, with positions being developed: range the options, depending on the conflict nature and the relationships among parties.
- Evolving conflict:
  - Towards litigation: use third party assistance (mediation or arbitration).
  - Towards resolution/abatement: no assistance or facilitation may be needed.
- Resolved conflicts: depends on the situation.

Because construction is an interactive process involving several people, with different needs and perceptions during a long time, it is plausible that all types of conflicts may occur. This justifies the need for an adequate management approach for dealing with conflicts in construction. Table 17.1 summarizes the conflict classification described above.

17.3   Construction stakeholders and potential conflicts

Project stakeholders are persons, groups or organizations having any interest in the project and who may influence the project planning, design, implementation and future use. Although all projects have its set of different stakeholders, some common classification of stakeholders may be established:

- Key stakeholders like the project owner, suppliers, performing organizations, the project management team and others depending on the project.
- Internal stakeholders, directly participating in the project.
- External stakeholders comprehending people affected by the project, interested parties and statutory authorities.

According to Olander (2003), the stakeholders depicted in Figure 17.3 may be found in a construction project.

Identifying potential conflicts between project stakeholders is an important step towards conflict anticipation and conflict management. Therefore, once project stakeholders are

Table 17.1 Classification of conflicts

Type				Level		Causes				
Well defined **Fuzzy**										
Typology	Open	Hidden	Latent	Inter	Personal	Cognitive	Objectives or interests	Normative	Relationships	**Causes**
					Group					
					Organization					
				Intra	Personal					
					Group					
					Organization					
						Cognitive		Socio-emotive		
						Object				
						Objectives, needs or interests				
						Processes				
						Structural				

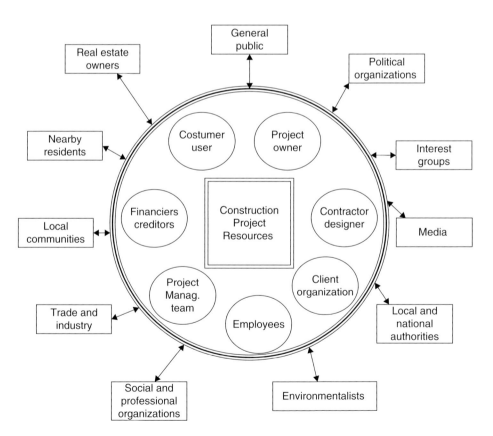

Figure 17.3   Construction project stakeholders.
(*Source*: Adapted from Olander, 2003)

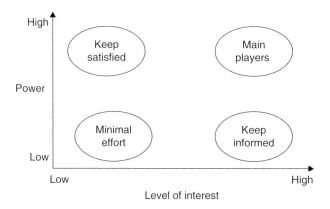

Figure 17.4   Power/interest matrix.
(*Source*: Adapted from Johnson and Scholes, 1999)

recognized, the next stage is to assess their needs and expectations in relation to the project and the conflicts each stakeholder may possibly become involved in. This depends on:

- the stakeholder (power/interest);
- the type of project (public/private, building/road/industrial);
- the stage of the construction process (design/implementation/use).

17.3.1   Stakeholder power

Solving conflicting interests between stakeholders must take into account their relative influence on the implementation of the project. Mapping the power and the interest of stakeholders is an interesting technique for this purpose. Figure 17.4 shows the power/interest matrix (Johnson and Scholes, 1999), which aims to answer the following questions:

- How interested is each stakeholder on the project decisions?
- Are they willing to interfere?
- Do they have enough power to act?

17.3.2   Key stakeholders

Key stakeholders are the main players in the project, holding high power to influence the project and high level of interest in it. Main players are more likely than any other stakeholders to create difficulties in solving conflicts if their needs are not attended. **Keep-informed** stakeholders have great interest in the project (land owners, nearby residents, public in general, groups of interests, environmental bodies) and may be severe opponents to it but have limited power to influence project decisions. **Keep-satisfied** stakeholders, on the contrary, hold high power to influence the project decisions (investors, authorities, legislative bodies, investors, media), but are often passive, meaning that conflicts with them may be avoided if they feel happy with the implementation of the project. The last set of stakeholders is the **minimal effort** group, who have low interest in the project, therefore raising few conflicts, and are not able to have a great impact on the decisions (trade and industry, for instance). It is worth noting however that the examples given above between brackets correspond to typical stakeholder positions in mainstream construction projects but they may assume other positions sometimes. For example, an environmental body may easily become a key player in a sensitive project.

17.3.3 Conflicts between external stakeholders

Real estate owners

These are people affected by land acquisition for implementing the project. The intensity of conflicts involving real estate owners depends on whether the project is private or public, and if it is implemented in an urban area or in a rural area.

In a private project, real estate owners may be normally included in the 'keep satisfied' group, because the development of the project depends on the successful negotiations with land owners.

In the case of a public project, the project promoter may get the land either by common agreement with the land owners or through an expropriation process. In either case, both parties aim to reach fair and timely compensation for the land but the perception of each party on the meaning of this may prompt a conflicting situation. The former approach is more straightforward than the second, therefore some reduction on conflict intensity may be achieved if this option is invoked by one party while negotiating a common agreement with the other party. This strategy particularly applies to buildings or parcels of land that do not directly affect the home or the livelihood of people, in which case other aspects than money and time may prevail. Sometimes, only by finding an alternative home with equivalent location or alternative farming with similar conditions (such as access and water) contributes for decreasing conflict intensity with families affected. But obviously, this will also depend on the solvency of people concerned.

Public in general

Public in general may be affected by the project and because of that may be considered an external stakeholder. Any project has positive and negative externalities mostly coming about during the construction and the utilization stages.

Some of these externalities are environmental impacts and may be related to the construction stage, to the utilization stage or to both of them. In some countries, it is mandatory to develop an EIA (Environment Impact Assessment) for large projects like roads, bridges and shopping malls. The EIA often addresses the utilization stage of those projects and sometimes the construction phase as well if the site is expected to create significant impacts (in a dam project, for example). One of the steps of the EIA is the public consultation process, where an accurate identification of stakeholders and the presentation of their concerns and expectations should be considered.

Although in a smaller scale, the construction stage of any project may cause relevant environmental impacts, for example dust, mud, ruined accesses, long traffic deviations, noise, risk of injury while crossing the site, damages to private property, etc.

Externalities should always be considered before construction commences and preferably at the pre-design stage where all the negative and positive impacts of possible solutions should be addressed and compared, including the zero option (doing nothing). This is essential for avoiding conflicts. Additionally, there are cases where public participation adds significant value to the project. For example, in a public transportation project, different solutions may be proposed to the people affected by the project. Because of their knowledge about the project location, their suggestions may be advantageously considered and possibly adopted at the design phase if they prove technically feasible, therefore avoiding future conflicts.

Public in general includes real estate owners, nearby residents, local communities and so on, each of those may be viewed as an independent subcategory of stakeholders. Nearby residents are people living or working close to the project and expect receiving information about construction evolution, respect for their daily life and that everything on site is being done, to solve and minimize the problems. The difference for local communities is that these are organized, include small businesses and can be categorized (for instance, a specific neighbourhood). Normally, local communities are organized and have their own representatives to whom the construction project should be first explained and communication directed. If the local community is affected by the project, there will be the risk of potential conflicts, if that is not regarded in the project decision process.

Local trade and industry

Local trade and industry are usually considered independent stakeholders from the general public because they may be affected by different project externalities. Trade and industry are businesses of several kinds (small manufacturing activities, restaurants, shops, etc.) that may be positively or negatively affected by the project. For example, they may possibly benefit from a new infrastructure project (improved communications) or loose clients to a new shopping mall (reduced attractiveness).

Building up the project may have similar impacts. For example, local restaurants may possibly gain new clients (working on or visiting the site) or loose used to be clients (because of traffic deviation). In order to properly manage these potential conflicts, negotiated solutions with trade and industry representatives must be found, for instance through the betterment of local facilities or by compensating expected profit decrease.

Environmentalists

Like other Non Governmental Organizations, environmentalists have the power to influence project decisions, as their aim is to alert the public opinion to the negative project consequences to the environment. Conflicts can arise if the project management team neglects or detracts their views with poor structured arguments. For a project where the EIA is not compulsory but for which environmental impacts are claimed, it is adequate to conduct a consultation process to the active environmentalists and to negotiate alternative project solutions although not compromising the main objectives of the project.

Local and national authorities

These are very important stakeholders because they have the power to influence project decisions by issuing final approvals on the project. These stakeholders are ruled by civil servants and politicians (mayor, minister, secretary of state, directors, etc.), therefore project conformance with rules and regulations partially depends on their interpretation on those rules and regulations and on the directives they must comply with in order to sustain strategic political decisions.

Conflict avoidance with these stakeholders is decisive for the project success, and may be achieved by maintaining informal contacts with them in all stages of project development. This is particularly important during the design phase and the pre-construction phase of the project in order to anticipate their decisions.

Media

According to some authors, media may not actually be considered a stakeholder as they have no stake in the project. However, media can have a decisive power and capability of influencing other stakeholders in the project decision process. Furthermore, it is common that some stakeholders use the media for influencing other stakeholders' decisions on the project (for example, politicians or national authorities relevant to the project approval or rejection). Taking into account the power of media, conflicts should be avoided through the implementation of adequate communication.

Political organizations and interest groups

Especially for large public projects, it is crucial that the main political parties converge on the project aims, the main technical solutions and the sources of funding. In recent years, interest groups have taken the lead in some projects but political organizations still hold significant power to influence decisions on projects holding regional and national relevance.

Interest groups are also called lobby groups and may act both locally and nationally as proponents or opponents to a project. Interest groups can be formed in many different ways and have different power to influence project decisions. Normally, they act in the pre-construction phase of the project, with the aim of conducting the process to fit their interests of location, dimension, accessibility or user facilities.

Conflicts arise when the decisions are opposite to the interests of the above stakeholders. They may then attempt to use their power and political influence, to discredit the decision and eventually change it according to their interests.

As for large private projects, the support from political parties and interest groups may avoid strategic opposition to the necessary approvals during the design and pre-construction phase. On the contrary, their opposition may lead to major difficulties for conducting the project, cause delays and possibly lead to the project abortion.

One particular way of reducing conflicts with these stakeholders is to carefully sustain the decisions with sound technical background, to present them truly and clearly and to personally interact with the relevant actors in order to better explain them.

Social and professional organizations

Trade unions are examples of social organizations that may have some influence on the project. They may act as supporters during the feasibility phase of the project, help during the design phase and influence political decisions during the pre-construction phase (trade associations, for example). But they may also act as project opponents during the construction stage if site impacts are significant or site conditions are not acceptable for workers (trade unions, for example).

Managing conflicts with these stakeholders during the design and pre-construction stages is similar to the described above for other similar organizations. Conflicts arising due to poor work conditions on site, low wages and excessive extra-working hours can be avoided or minimized if the project management team keeps regular meetings with worker representatives in order to understand their concerns and explain the decisions that affect workers' salary and safety conditions.

17.4   Conflicts between internal stakeholders

Unlike external stakeholders, internal stakeholders to construction projects are usually tied by mutual contract arrangements, whereby rights and duties of the parties are set, as well as the risks each party ought to bear and if these can be insured. Additionally, contracts usually establish the resolution procedures of conflicts possibly arising from their relationships.

Project owner

The owner is the most relevant project stakeholder and is mostly affected by the project success. Accordingly, the owner is expected to develop all the necessary efforts to avoid project conflicts or, at least, to minimize them to a controllable level, by using adequate conflict management techniques. The owner may also play the role of sponsor, promoter and client for the other project stakeholders.

Conflicts may start at the owner organization. Especially in public entities, some internal opposition to a particular project may arise, due to either resource dispute or conflicting approaches on investment priorities between different sectors. These conflicts should be adequately managed by using sound decision criteria, adequate diffusion of needs and expected benefits for end-users and previous alternative solutions under scrutiny.

Costumers and end-users

The ultimate reason to launch a construction project obviously depends on the needs of these stakeholders (either assigned or not, either directly or indirectly), thus evidencing their importance. For private investments (houses, offices, stores), needs are usually evaluated through market research techniques. The costumer is the end-user of the facility and who directly pays for it (either by purchase or rental in its multiple forms). For public projects, however, the end-user may not directly pay for the facility (as in public concessions) but indirectly through taxes. Accordingly, end-users' needs should be properly identified during the conception and the design phase, in order to avoid conflicts during construction, due to mistaking their expectations.

Financiers and creditors

Financing institutions need to ensure the return of investment and adequate profitability if the funds are private, and the achievement of the project goals of scope, time, cost and quality, if the funds are public. If project costs escalate, incomes may reduce or the project profitability may be at risk, therefore financiers may stop capital allocation or creditors may claim for the payment of debts, therefore endangering the project conclusion. This evidences their importance as project stakeholders. Adequately managing internal conflicts implicates accurate and permanent monitoring of the project's cash flow, as well as the use of risk analysis techniques in order to ensure alternative solutions if, for instance, the planned revenues are not achievable.

Designers, suppliers, contractors and subcontractors

This group of stakeholders contributes with products and services to the implementation of the project. Before construction begins, the designer is the most important stakeholder, while afterwards the contractor becomes the most relevant. Depending on the type of delivery, procurement process, distribution of risk chosen by the project owner and the type of contract arrangement (traditional, cost plus, construction management, etc.), conflicts may be more or less manageable and more or less contained.

Conflicts between the contractor and the client frequently arise from different site conditions, change orders, delays, suspension of works, defective contract documents, among others. Normally, conflict resolution procedures are disposed in contractual documents, and, depending on different legal systems, they ought to include direct negotiation, mediation, adjudication boards, dispute review boards, etc. The main measures to avoid conflicts must be implemented in the design and the pre-construction stages, as they are focused on improving the quality of contract documents and include, for instance, geotechnical baseline reports, constructability reviews and partnering approaches.

Employees

Employees of any stakeholder organization can obstacle the project success, if they are not sufficiently motivated by the project, or if they have any kind of conflict within their employer (salary, promotions, work conditions, etc). This type of conflicts often arise, through strikes, organized meetings or written claims to the board of directors, and should be tackled by the Human Resources Department of the organization concerned.

Project management team

The project management team is the instrument used by the project owner to achieve the project goals and objectives. In order to meet all the specifications and requirements established for the project, the team should hold sufficient empowerment and embrace all necessary competences. Conflicts frequently emerge from different views and perspectives on the assignment of responsibilities and emergent relationships. These should be properly established prior to the beginning of the project, through responsibility matrixes and adequate communication channels.

17.5   Relationships between stakeholders

Depending on the way they react to conflict situations the relationships among stakeholders identified above can be classified as **unitary**, **pluralist**, or **coercive** (Rijsberman, 1999):

- A **unitary relationship** refers to conflicts where a (probably small) number of stakeholders have similar values, and the parties to the conflict are likely to agree on objectives, but may still have conflicts of interest.
- In a **pluralist situation**, stakeholders do not agree or do not share each other's value systems, neither one of the stakeholders dominates, even though parties ought to reach compromises on objectives and values.
- A **coercive relationship** among stakeholders describes a situation in which parties do not share a common value system, but one of the stakeholders is powerful enough to make its own value system dominant (and coerce the other stakeholders to accept it).

The relationships established between stakeholders and the balance of power among parties is an important issue in every conflict. Actually, those relationships can range from a basic agreement on objectives, but conflicting interests, to situations where one of the stakeholders is powerful enough to coerce the other to acquiesce to them. For that reason, and, as will be seen further in this chapter, alternative resolution methods like mediation or arbitration are more effective when biased or unbalanced power is present because the most powerful party tries to force its reason.

17.6   Conflict management

International literature essentially deals with conflicts in construction under the following approaches:

- by analyzing conflicts between the client and the contractor (particularly under traditional contract arrangements), usually by adequately managing construction claims;
- by identifying possible conflicts with external stakeholders (this being an undesirable phenomenon to be reduced and ideally eliminated from the construction process) through understanding and mitigating their underlying causes.

The efficacy of the preventive approach predominating in the construction industry has been challenged by those who consider conflict inevitability and view the problem in the way it is managed rather than in its existence. Therefore, the trend nowadays is developing efficient conflict management approaches, rather than minimizing possible conflict sources. Furthermore, conflicts may lead to the best possible solution to the problems under discussion and become an opportunity for organizational learning, creativity and fulfilment of organizational and individual potential (Hughes, 1994).

Actually, the goal of conflict management is not to avoid conflicts, but to develop the skills and methods to help conflicting people, groups or organizations in conflict, to express their differences and solve their problems in a collaborative and constructive way. Therefore, it is essential that conflicting stakeholders are fully involved in the resolution process as a way of developing effective methods for dealing with their differences. Moreover, disputes may result in litigation which instead of being fair and equitable, may not be the best solution for all conflicting parties. Avoiding this outbreak is one of the most decisive reasons for implementing conflict management techniques.

The classical approach for successful conflict management maximizes the integrative function of the two parties in conflict (the win–win approach) although it ignores the consequences of the solution for third parties affected by the dispute. However, the correct framework for conflict management should also take into account the maximization of outcomes for all the other stakeholders, through a utility function like (Wall and Callister, 1995):

$$\text{Max } Y = a + b_1 x_1 + b_2 x_2 = b_3 x_3 + \cdots + b_n x_n + b_{n+1} x_1 x_2 + b_{n+2} x_1 x_3 + \cdots, \qquad \text{eq. (17.1)}$$

where
x_1 – utility to disputant 1
x_2 – utility to disputant 2
$x_3, \ldots x_n$ – utility to third parties affected by the dispute between 1 and 2.

Additionally, adequate conflict management techniques should not just maximize all parties' outcomes, but also increment relationships between them, as a way for reducing future disputes (this assignment is represented by the term $b_{n+1} x_1 x_2$) and to increase the joint utility represented by the above function.

The assessment of conflict between stakeholders in construction projects depends on four essential factors, the first three of them being endogenous to the conflict and the last exogenous (describing the surrounding context):

- the type/power, characterization and relationships of the stakeholders (internal, external, authority/public/contractor);
- the stage of the construction project cycle (pre-contractual, execution, exploration);
- the type, nature and stage of the conflict (behaviour, data, needs, values, latent, potential, processes, etc.);
- the legal and institutional context of the project (public, private, environmental, transportation, developmental, etc.).

The success of conflict management depends on the adequate interaction of the above factors through six basic steps (see Figure 17.5):

1. identification of the threat, type, stage and dimensions of conflict;
2. identification of underlying facts, perceptions, social needs and cause effect relationships;

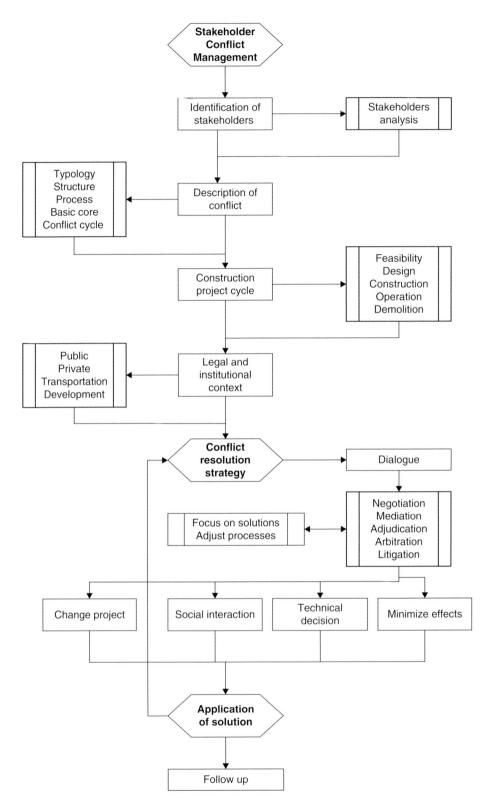

Figure 17.5   Conflict management phases.

Table 17.2   Stakeholder analysis matrix

Stakeholder	Power analysis			Importance/interest		Conflict/impact		
	Source	Level	Cooperation	Importance	Interests	Affect	Way	Details
Nearby resident	Votes/ meetings	Medium	Good	Medium	High	Directly	Positively	Reduce distance
Commercial owner	Pressing authorities	Low	Weak	Low	Low	Indirectly	Negatively	Reducing incomes
. . . ..	. . .	. . .	. . . .	. . .	. . . .	. . .	. . .	. . .

3. identification of all stakeholders, their interests, powers, fears and needs;
4. involvement of all recognized parties in the management process;
5. identification of suitable conflict management resolution strategy and choice of guidelines;
6. application of chosen resolution strategy and guidelines;
7. measuring the success of conflict management.

One of the most important phases of this process is the stakeholder analysis. This is used to identify, assess the importance and anticipate the influence (either positive or negative), that each stakeholder will have on the project. The results of this analysis are used to develop strategies for supporting effective conflict management procedures, by minimizing possible conflicts and reducing obstacles to its successful implementation.

The stakeholder analysis may be done by using the example matrix of Table 17.2, where the cells are filled with quantitative indicators.

17.7   Resolution of disputes between internal stakeholders

Parties to a dispute must first decide whether to seek resolution to a conflict through a non-consensual process, like litigation or arbitration, or through more collaborative means like direct negotiation or conflict prevention techniques.

Once the decision has been taken, the parties must choose which approach to employ, since there is no methodology that will be effective in all cases, and indeed more than one may be used. The circumstances, conflict assessment and therefore the obstacles to agreement vary from one case to another. Disputes may involve many parties or only a few; the problem may be more or less urgent; the emotional investment and the power of the stakeholders may vary; the public interest may or may not be at stake and the factors involved may be well understood or more uncertain. Gaining expertise in conflict management includes learning about the specific advantages and disadvantages of the various strategies, and assessing which one is best for addressing a particular conflict situation.

Conflict resolution techniques may be viewed as a set of approaches ranging from those in which all initiatives and authority remain within the parties in conflict to those in which there is a strong intervention from a third party. On the other hand, although considerable differences may exist from one approach to the other, they may also overlap. Moving from the least to the most directive, most approaches will involve some element of relationships (direct negotiation), procedural assistance, substantive assistance, advisory non-binding assistance (mediation) and binding assistance as well (arbitration).

The use of conflict prevention does not imply that conflicts between parties will not prevail. Similarly, the use of third party aid like arbitration does not imply that it will be less effective than the personal efforts of arbitrator to get the parties to cooperate as much as possible. When neutral parties play, the results will depend on the qualities and experience of the person or persons chosen. This implies that the selection of a conflict resolution technique is not independent from the external party involved.

17.7.1 Abandonment

Dispute resolution generally assumes that some pathway will be followed towards a settlement or acceptable decision. However, during the process, one party may possibly decide to discontinue with the dispute. Although they may not be considered dispute resolution techniques, avoidance or abandonment by one party are possible dispute actions and are quite frequent in construction. Reasons for this are diverse, including low expectations on positive results, lack of funds to pursue, commercial reasons, lack of assertiveness or passivity.

Avoidance, on the other hand, is normally structured as a procedure. Generally speaking, the best way to avoid conflicts and disputes with stakeholders is to promote dialogue. This assumption is reflected on the processes for dispute resolution between internal stakeholders as can be seen below.

Although abandonment always derives from the voluntary or involuntary decision of one side, avoidance is often induced by the nature of construction stockholder's procurement relationships. More specifically, contract forms have been developed in several European countries as a means to avoiding disputes between parties involved in construction contracts (and as a pathfinder for dispute resolution, as well). In some cases, the role of avoiding conflicts is assigned to one stakeholder. An example of this is the certification process carried out by the contract administrator or the project manager in most contract arrangements: although being a client agent, he or she is expected to fairly assess the amounts payable to the contractor by balancing its possible optimistic demands with the potentially restrictive views of the owner.

17.7.2 Negotiation

Negotiation is possibly the most common and inexpensive form of dispute resolution in construction, whereby the control of the dispute process remains within the parties involved. In order to achieve a good negotiated settlement for a conflict, four characteristics should be met: fairness, efficiency, wisdom and stability.

With the implementation of these type of measures, conflict resolution probably drives from win–lose situations to win–gain solutions, where all participants try to find new ways to reach their goals, and, at the same time, meet the goals of the opponents. In this process, parties may act by their own as in direct negotiation or may introduce an advisor or a facilitator.

The graph of Figure 17.6 shows the five common conflict handling styles that may be found during stakeholders' negotiation, in relation to individual or mutual satisfaction (Thomas, 1992; Loosemore *et al.*, 2000). Essentially, two main approaches may be identified in Figure 17.6: competitive and cooperative.

Competitive negotiation applies when the party using it is insensible (or, at least, partially insensitive) to the needs and wishes of the other party. In this case, the insensitive party will do anything to get concessions, irrespective of the costs implied to the other party.

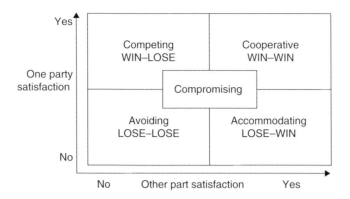

Figure 17.6   Conflict management styles.
(*Source*: Adapted from Thomas, 1992)

In order to achieve this, a variety of tactics may be employed that may be divided into the following three categories (Meltsner and Schrag, 1973):

- **Positional tactics**, aim at providing some form of advantage over the other party or at placing the other party at a psychological disadvantage. Examples of this are insisting that meetings take place at a place you may feel more comfortable than your opponent or trying to outnumber the number of participants of the other side for some functional reason.
- **Initial tactics** are used in order to try to achieve a favourable bargaining position from the start. Examples are to place your major demand first in the agenda or to start the negotiation with a higher demand than you really expect to obtain.
- A range of **general tactics** may also be used, the most common of which are: flattering, persuasion, promises or threats and irreversible decisions.

Cooperative negotiation is sometimes called win/win negotiation and as the name suggests pre-empts a very different approach from the competitive negotiation. According to Fisher and Ury (1991), there are five basic elements to this approach:

- *Separate people from the problem:* The opponent should be regarded not as someone you do not like and wishing to cause damage to you personally but as someone with whom you will have the chance of solving a problem through a mutual advantageous solution. Accordingly, negotiators should focus on the problem rather than in each others.
- *Focus on interests not on positions*: In an organization, interests are what really matters for problem solving not the victory of your position on the problem. Accordingly, negotiators should focus on the reasons for their demands.
- *Generate options for mutual gains*: Generally, it is better for you to find ways of increasing mutual benefits than to discuss with your opponent how to share it. Competitive negotiators will seek to obtain as much as possible during the negotiation process.
- *Insist on using objective criteria*: If criteria used during a negotiation process are validated by all parties involved, then chances will increase of getting a good agreement. Win/win negotiators will adopt mutual recognized criteria to measure the outcome of their bargain.
- *Consider the best alternative to a negotiated agreement*: Negotiators should evaluate the consequences of not reaching an agreement through the negotiation process they are carrying.

Clearly, the cooperative approach assumes conflict resolution through problem solving but some authors claim that this is not always possible. Arguably, ultimate hard negotiation tends to occur in every conflict-resolution process; this placing a competitive negotiator in a better position to win over a cooperative negotiator and not the contrary. However, Fisher and Ury (1991) contradict this view and state that it is possible to bring a competitive negotiator to a problem-solving negotiation by using the following five step approach:

1. Avoid attacking the other side and assume the position of an objective onlooker.
2. Listen and acknowledge the other side's point of view and agree whenever possible.
3. Direct the other's attention to the problem of meeting each side's interests.
4. Build on the other side's ideas in order to make it easy and convenient for him or her to agree.
5. Make it hard for the other to say no.

The negotiation process normally starts with an initial meeting where participants establish the game rules, get acquainted to each other's initial position and attitude, highlight main areas of agreement and disagreement and settle what type of records or data will be accepted.

According to the same authors, the negotiation process should be conducted as follows:

Establish commonly accepted facts.

Separate people from problems.

Base your position in principles, even if your opponent tries to make it personal.

Stress equality principles.

Ask questions instead of making statements.

Explore the principles of the other side.

Listen, rephrase and clarify your points.

Do not decide at once but give yourself time to think about the problem and to prepare a reply.

Expose your reasons before proposing.

Show your proposition as a fair solution.

Present your views on the consequences of reaching an agreement or otherwise.

Give the other side a chance of influencing the outcome of the negotiation process.

End up negotiation conciliatorily even if you do not completely feel it.

17.7.3   *Mediation and conciliation*

Mediation and conciliation is an ancient dispute resolution process that originated from China some 3000 years but its widespread use in construction is much more recent dating from the middle years of the last century. Mediation may be viewed as a negotiation process between disputing parties carried out with the help of a neutral and independent third party. It is essentially an informal process by which parties seek assistance from an independent consultant for solving their dispute. Therefore, the fundamental role of the mediator is to facilitate the decision making of the parties involved in the dispute. This is achieved by impartially advising and consulting them, helping the parties to understand their own and their opponent's position better, exploring alternative solutions and so forth. Accordingly, the mediator may act not just as an advisor but as a manager of the dispute process. Conciliation has been used quite interchangeably with mediation but tends to mean a more proactive attitude in some instances. In practice, the process of mediation or conciliation may be more facilitative or more evaluative depending on whether the consultant merely tries to aid communications between parties or if he or she comments on the subject matter and makes recommendations towards the outcome.

Three stages of the mediation process are usually identified in the literature, namely: pre-mediation, mediation and post-mediation.

Pre-mediation corresponds to the preparation of the mediation process and includes the initial inquiry, procuring and contracting the mediator, preliminary preparations and first communications among parties (Brown and Marriot, 1993). The initial inquiry aims at agreeing on the process itself which may involve explanation and persuasion of the opponent parties. The procurement of the mediator is an important step towards mediation success because all parties must be convicted of the confidential, impartial, unprejudiced and legal nature of his or her performance. On the other hand, the mediation contract should establish the costs and the schedule of the process. Immediately after appointment, the mediator will become a party in the dispute process and as such should launch communications among parties and clarify their positions on the dispute. This may involve obtaining written summaries and **supporting** documents from each side for consideration.

Proper **mediation** involves, in the first place, meeting the parties and listening to their positions. Private meetings between the mediator and each party may be necessary, in order to build mutual confidence relationships, clarify some issues, identify needs and expectations and potential settlement options of each side, and so forth. These meetings are essentially consultative, not binding and are sometimes called *caucus*. In this process, the mediator may need to conduct several meetings with the parties, ask for advice from an expert or consultant, overcome impasse situations if they occur, etc. In the scope of this, he or she may perform more facilitative or more proactive of the ongoing negotiation. Finally, the mediator will record the agreements reached.

If agreements are met through the mediation process, then **post-mediation** is about their practical implementation. Otherwise, post-mediation will be the preparation for other forms of settling the dispute. However, it must be noted that successful mediation does not necessarily mean full agreement between sides nor the mediator should raise so high the expectations at the start. After mediation, the parties may have gained better insights into their dispute or may have limited the disputing issues thereby narrowing the field between them, and leading to a positive result.

In view of the above, the skills of the mediator and how efficiently he or she performs his or her functions play a decisive role in the mediation process. The role of the mediator has been described above and essentially involves managing the mediation process, collecting and analyzing information, facilitating communications, exploring possible solutions and promoting agreements between the parties. The skills required for a mediator to effectively achieve this are somewhat more difficult to state. Some of these skills derive from the nature of the functions and relate to the ability of being neutral and acting confidentially. Other skills are interpersonal mainly the ability to communicate and to achieve settlements. As for the latter, the mediator should base his or her decision/action on thorough investigation of each side's position and on inventing plausible solutions, while employing strategies like empathy, persuasion and distraction.

Some European countries have developed mediation and conciliation systems ranging from the facilitative end to the evaluative end of the spectrum, but in most cases the process is mainly informal. Informal mediation is essentially a flexible process conducted by a neutral agent engaged at a certain point of a dispute. This may take place by mutual agreement or may follow from the initiative of one side, after concluding that the negotiation process being carried needs help from a third party. A senior person recognized by the industry is often selected so that the chances of being accepted by all sides may increase. Initially, the mediator serves as a facilitator, informally discussing the dispute with the parties. In some cases, the facilitative role continues until both parties settlean agreement but in other cases, it ends up with some form of non-binding recommendation that the parties may possibly accept. Opposite to the informal mediation, following anofficial mediation system implies in some countries contracting mediators from a pre- assessed list.

17.7.4   *Expert assessment*

Expert assessment or determination is a process by which parties in a dispute commonly agree on asking a third party to decide a particular issue. Unlike mediation which is a non-binding process, at least until some agreement may be reached, expert determination implies mutual acceptance of the expert decision. The use of this form of dispute resolution is very common in construction. Examples are real estate valuation, technical valuation in a number of circumstances and rent review. Actually, expert assessment often occurs in other forms of dispute resolution like in arbitration and in court litigation.

   Unlike the mediator who need not be an expert (which in fact, may bias the mediator's view), the expert is by definition a specialist on the issue to deal with. Selecting and contracting the expert therefore assumes a paramount importance. Firstly, the subject of the expertise should be clearly and precisely expressed. Secondly, the costs and the schedule of the work to be performed should be established. Thirdly, the expert decision should be accepted to be the final and binding by all parties involved, unless one party further decides to challenge it through arbitration or litigation.

17.7.5   *Adjudication*

Adjudication may be defined as a process where a neutral third party gives a decision on some issue which is binding on the parties in dispute, unless or until revised in arbitration or litigation (Gould *et al.*, 1999). Under traditional arbitration, the disputing parties must agree on the adjudicator who will thereafter act empowered by that agreement but unlike in mediation, the arbitrator's decision does not require the cooperation of the parties. The difference for expert determination is that the adjudicator may investigate the circumstances of the dispute and more freely interrelate with the parties than the expert tends to do.

   Statutory adjudication was introduced in British construction following the recommendations in the Latham report (1994) and has no parallel in other European countries. Statutory adjudication is covered by the Housing Grants, Construction and Regeneration Act (Part II). Under this Act, a party to a construction contract is unilaterally given the right to refer to adjudication in order to solve a dispute arising from a contract. Therefore, unlike in traditional adjudication, no previous agreement between parties is required to start up a statutory adjudication procedure.

   Section 108 of the Act sets out the minimum requirements for an adjudication procedure which may be summarized as follows:

- *Notice*: A party to a construction contract must have the right to give a notice at any time of its intention to refer a particular dispute to the adjudicator.

- *Appointment*: A method of securing the appointment of an adjudicator and furnishing him or her with details of the dispute within 7 days of the notice is mandatory.
- *Timescales*: The adjudicator is then required to reach a decision within 28 days of this referral.
- *Impartial action*: The adjudicator is required to act impartially.
- *Inquisitorial action*: The adjudicator is required to take the initiative in ascertaining facts and the law.
- *Binding nature*: The adjudicator's decision is binding until the dispute is finally determined by legal procedures, by arbitration or by agreement.
- *Immunity*: The adjudicator cannot be held liable for anything done or omitted in the discharge of his or her function unless acting in bad faith.

In addition to this basic procedural framework, the Act further requires that all construction contracts comply with the provisions of the Scheme for Construction Contracts.

Following the publication of the Act, the standard forms of contract mainly used in Great Britain incorporated the adjudication process in different ways. The Joint Contracts Tribunal (JCT) standard form, mostly used for building, was amended with a set of adjudication procedures whereas the Institute of Civil Engineering (ICE) issued a stand-alone adjudication procedure to be used together with its standard form, largely disseminated in civil engineering contracts. This complies with Latham's (1994) recommendation that standard forms should embrace a system for adjudication.

There are no restrictions on who may possibly be appointed as an adjudicator either acting informally or under a statutory procedure. Many British construction bodies have established lists or panels of their own approved adjudicators but there are few restrictions on who can join the lists. An Academy of Construction Adjudicators has also been created.

17.7.6 Arbitration

In the last few years, construction industry has encouraged the use of international commerce practices by **arbitration** as an attempt to avoid costly and timely courtroom litigation procedures for solving disputes. This is a formal dispute resolution procedure subject to statutory controls, whereby disputes are solved by a private arbitrator selected by common agreement, or by a private tribunal, normally made up of three arbitrators, one appointed by each party and the third by common agreement. Arbitrators must have appropriate qualifications, minimum work experience and act impartially. During the arbitration process, arbitrators can call witnesses, require expert opinions and call the parties to testify, as well as other formal courtroom procedures.

Arbitration is voluntary, but once accepted by the disputing parties, the final decision is binding, unless a break of procedures, fraud or conflict of interest can be proved, in which case the decision may be revised by a court of law. Accordingly, the final decision may be enforced by the courts if necessary. Nonetheless, there are arbitration systems where the decision can be submitted to a court of appeal.

The advantages of arbitration when compared to litigation have been well acknowledged by the industry and include flexibility, economy, expedition, privacy, freedom of choice of arbitrators and finality (Gould *et al.*, 1999). Besides being less costly and time consuming than the court of law, the arbitration tribunal is composed of recognized experts in the construction field, who are familiar with the industry practices, which may not be the case with common judges. Therefore, disputing parties feel that arbitrators can best understand their problems and are best prepared to reach an equitable solution for the dispute. However, some disadvantages have been recognized when more than two contenders are involved.

Four stages of the arbitration process are usually identified in the literature:

- The arbitration agreement, whereby parties agree they will submit to arbitration present and future disputes.
- Selection of the arbitrator or arbitrators, by common agreement of the conflicting parties.
- The arbitration procedure, starting with the initiative of one of the parties after recognizing that a dispute has arisen.
- Award and enforcement which is final and binding on the parties, unless on the cases mentioned above.

There are ofcourse limits to the type of disputes that can be handled through arbitration and in some cases there are established limits to the maximum amount of money that can possibly be involved. Moreover, the arbitration process presents a set of variants:

- Documents only, by which the arbitration tribunal decides upon the documents submitted by the parties without hearings.
- Amiable or *ex aequo et bono*, if the arbitrator aims at arriving at a an equitable solution for the conflict.
- Last offer arbitration which is based on the offer of each party in monetary terms for a specific conflict.

Arbitration itself may have different names depending on the specific construction legal systems, like adjudication boards, dispute review boards, mini-trials, private judging, etc.

17.7.7 Litigation

If no agreement is achieved between conflicting parties to use any of the previous systems for resolving their disputes, then they can apply to the courts of law. **Litigation** is the current name for disputes dealt with in the courts of law.

The procedure followed by the courts starts with the claimant issuing a case and the claim particulars. Then the defendant is given the opportunity to admit the claim, defend from the claim particulars or merely acknowledge receipt of the claim form. The defendant may also decide to make a counterclaim. The legal system of each country follows a specific litigation track but it is not unusual that different tracks are adopted according to the nature of the claim and to the financial amount claimed.

The next step is the trail whereby the court of law will decide the merit of the case, in terms of legal evidence, sustained facts and quantum evaluation. However, very few proceedings result in trial and subsequent judgement in the European tradition, mainly because parties often reach an agreement just before the trial. In other occasions, the claimant may obtain judgement without trial because of any process flaw of the other side.

The complexity of some construction disputes often requires courts to contract experts for case assessment. Usually, a commission of three experts is appointed, one by each side and the third designated by the courts. The report of this commission is then added to the process but in most countries it is not binding. Once starting their functions, the experts must act neutral and professionally but their role has been criticized for supporting their client's position rather than providing an independent opinion. For this reason, many reports are not unanimous and do not give confidence to the court's decision.

17.7.8   *Building bridges*

Despite the great variety of dispute resolution processes presented above, there is a trend in the literature for conflict prevention essentially focused on the anticipation of disputes from the construction stage to the design stage of project development. This has advantages of gaining time and benefiting from the dispute outcomes to enhance the project.

This can be done by improving the quality of contract documents as most construction disputes actually result from problems originating in the design or in the pre-construction phase of projects. Providing for independent design–quality review, mandatory insurance against errors and omissions and demanding for total quality management of architectural and engineering firms are some examples of good practice of dispute avoidance.

One common source of claims and disputes, representing a significant risk of cost and time overruns is the different site conditions encountered by contractors. In most cases this results from insufficient geotechnical subsurface studies, or from misunderstanding of site surveying reports. The concept of Geotechnical Baseline Report has been introduced to overcome this problem, whereby the designer states the ground conditions admitted for the design, on the basis of site investigation (if any) performed. Furthermore, the report anticipates the subsurface behaviour in respect to the most likely construction method to be used by contractor, as well the necessary information about slope stability, dewatering methods and strength of bedrock levels.

Another type of measure which can be implemented on the design and pre-construction phase of a project is the constructability reviews and the value engineering incentive clauses. Constructability can be defined as the optimum use of construction knowledge and experience in planning, design, procurement and field operations to achieve overall project objectives. Normally, constructability reviews are done in the last stage of design process, by the construction staff experienced in past projects and familiar with claims and disputes presented in those projects. Value engineering is the assurance that the constructed facility is adequate for its function at the lowest reasonable life-cycle cost, where the value index refers to the ratio of the worth of materials or methods required to provide the function, against their cost. This approach can be used either in the design phase, where the implementation can provide higher savings, or in the construction phase through an incentive clause in the contract, where the savings from value engineering studies are apportioned between client and contractor. Normally, this contract clause motivates the presentation of better cost-effective solutions, but still meeting the objectives of the project, once it guarantees rewards for discovering value engineering improvements.

Another bridge for dispute resolution comes from the concept of amicable settlement first introduced in a clause of the 1987 Fourth Edition to the FIDIC Red Book. Basically, the clause is a pre-requisite to arbitration and states that no notified arbitration may commence *unless an attempt has first made by the parties to settle the dispute amicably*. Other codes of practice throughout Europe, although not specifically referring to the amicable settlement of disputes, may recommend that other forms of dispute resolution may be attempted by conflicting parties prior to embarking on an arbitration procedure, for example negotiation and mediation. In the scope of these, the role of consultants and mediators is above all to facilitate the settlement, therefore bridging the gap between contenders.

Moreover, contract forms may encompass dispute avoidance techniques by endorsing the power to resolve differences between parties to a specific stakeholder, as in the certification example mentioned earlier on this section. Another example is partnering contracting which can be defined as the establishment of a working team among the parties, for mutually beneficial resolution of the ongoing difficulties and problems that typically arise on a construction project. The objective of this technique is to set up a climate of cooperation, communication, fair play and mutual confidence between client, contractor, designer and other stakeholders, which can start even before the bidding stage. This process consists of voluntary workshops, seminars and meetings that help the parties establishing working relationships in a non-adversarial atmosphere, where arising problems can be discussed and resolved, therefore avoiding the build up to formal claims.

Under the same logic, in an effort to prevent the escalation of a dispute between the owner and the contractor, an independent third party may be agreed for settling disagreements soon after they occur. This neutral exists either formally or informally in a set of European countries. In Britain, for example, this was named the Dispute Resolution Adviser after the work of Chartered Institute of Arbitrators.

Similar to this concept are the Dispute Resolution Board developed in the United States and the Dispute Review Boards in other countries. Essentially, they are boards of three people who evaluate the disputes as they arise during the project and make recommendations to the parties. The boards are constituted and agreed for each construction contract (each party selects one board member and agree on the third) at the commencement date. Like arbitrators, the members of the boards should be construction industry experts recognized by both parties, acting impartially and objectively, whose main assignment is to get information about project progress and observe construction problems as they occur, being able to encourage parties to deal with them promptly and realistically in cooperation with each other. Another example is the Dispute Adjudication Boards provided in the FIDIC General Conditions.

Multi-stage approaches for dispute resolution are also common. Accordingly, parties try the mediation approach in the first place and step to other more formal approaches if it fails, like arbitration. Currently, the mediator is later appointed as arbitrator, therefore benefiting from his or her previous knowledge on the dispute. A compulsory multi-stage approach may also be stated in the contract whereby, according to the financial amount claimed, complaints may be tackled through arbitration or must otherwise follow a litigation procedure.

Mini-trials and executive trials are a new trend. All the dispute resolution process takes place in a short period of time (say 1 or 2 days) and is conducted by an independent third party with recognized law expertise acting as a facilitator. The parties are represented by executives of the disputing parties holding decision power that have the opportunity of settling out their differences in a private look like courtroom.

17.8   Resolution of disputes between external stakeholders

Applicable procedures for dispute resolution involving external stakeholders are much less structured than above and the risk of dispute escalation to litigation is much higher. Generally speaking, dialoguing is the easiest path to conflict avoidance.

17.8.1   Dialoguing

Stakeholder dialogue is based on the principle that people affected by decisions ought to have an effective participation in the decision-making process. This should be done at an early stage when all the options for the construction project are still possible, and the opinions of participants can influence the outcome. This means the earlier stages of the design phase.

The effectiveness of dialoguing increases as stakeholders are prepared to accept changing their views if adequately convicted. Moreover, dialoguing encourages people to step down from positional argument with win/lose outcomes and focus on cooperative and creative problem solving by working together in a consensual process, using the body of knowledge and ideas that each stakeholder brings up. Best practice includes the promotion of workshops, workgroups, seminars, study circles, open houses and so forth. Olander (2003) identifies the aims of these initiatives as follows:

- to provide stakeholders with all the relevant information about the construction facility or process;
- to justify why they are being involved;
- to explain why and how their input is useful and where it fits in the overall decision-making process;
- to clarify the extent to which it may influence the final decision of the project solution.

The outputs of the dialoguing process are often well informed and technically acceptable solutions acknowledged as the best possible options by the majority of the stakeholders. More importantly, the process tends to generate active support for the project and improved relationships within the community.

However, relations with external stakeholders must be carefully established. Depending on the significance of the construction project, the origin of the funds (public or private), the nature of the facility being developed and the type of stakeholders, there are different ways relations may be built with the aim to avoid possible conflicts. Table 17.3 depicts different ways of relating, and the expected level of influence in the project decision-making process.

Despite conflict avoidance efforts, conflict situations may possibly break out involving external stakeholders, due to:

- strong stakeholder interests (inflexible, unwilling to modify demands);
- lack of information or poor communication strategy;
- inter-personal conflicts;
- lack of consultation process.

The next sections deal with negotiation and an example of strategy for overcoming a specific conflicting situation.

17.8.2   *Negotiation*

Most of what has been said about negotiation between internal stakeholders is applicable with some adaptation to conflicts involving external stakeholders as well. Additionally, in most countries, this is the only possible approach to conflict settlement before litigation when external stakeholders are concerned.

The same two main approaches as before may be identified in negotiation: competitive and cooperative. The following lists a set of good practices and tactics that may help to maintain conflict situations under control, either for general interactions (Wall and Callister, 1995) or for construction industry specific interactions:

- *Foresee issues*: undertake conflict analysis and mapping
- Evaluate the extend of stakeholders' participation in the final solution (there are different cultural backgrounds)

Table 17.3   Ways of relating to stakeholders

	Type of relations	Description	Stakeholders influence in decisions	Actions	When to use
Least	Giving information	Stakeholders are informed about the project	None	– Press release, TV – Newsletters	Information not controversial/trust
	Gathering information	Stakeholders provide information to help decisions, but don't participate	Very little	– Questionnaires – Interviews – Surveys	Reliance on the use of information
Level of involvement	Consultation	Stakeholders are consulted but don't participate in decisions	Limited	– Written comments – Interactive meetings	Stakeholders trust in decision-making process
	Participation	Decision-making process is shared with some specific stakeholders	Can influence specific subject or issue	– Workshops – Topic groups – Round table meetings	Willingness and ability to accept influence of outcome
	Bounder dialogue	Decision is taken together after dialoguing within some pre-set conditions	Stakeholders fully involved with some pre-set constrains	– The above processes in a pre-planned and coherent way, eventually facilitated by mediators	– All solutions are possible, within pre-fixed parameters
Most	Open dialogue	Decisions is taken together	Stakeholders fully involved in decisions		– Wider and complex problems, with open outcomes

Source: Adapted from Eurosite, 2003.

- Keep players involved and processes transparent and clear
- Maintain the stakeholders enthusiastic and capability in participation
- Ensure that institutions have legitimacy, are trustable and inspire confidence
- Assure that stakeholders are aware of conflict boundaries (rights, roles, responsibilities, legitimacy)
- Adopt appropriate leadership styles
- Structure organization to avoid conflict
- Address conflict causes, diagnoses and implement corrections
- Promote meetings (workshops, seminars)
- Negotiate, mediate, and arbitrate
- Enforce truth
- Expand group boundaries
- Guide communications between disputants
- Set up formalized appeal systems
- Act as decision makers
- Offer incentives
- Enforce cooperative problem-solving attitude between disputants
- Recognize women as stakeholders and peacemakers
- Address implications for youth and children

- Conduct relevant stakeholder analysis and conflict perceptions
- Build and maintain effective partnerships
- Recognize the primacy of local people
- Widen and deepen dialogue
- Recognize the potential and the limits of external influence
- Be transparent and communicate intentions
- Act in timely flexible ways and think long term
- Respect cultural diversity
- Recognize and act only so far as legitimacy allows and remains impartial
- Be accountable
- Enable institutional learning
- Use creative, incentive-driven approaches for construction engagement
- Act on lessons about the need for coordinated, coherent action and policy

On the other hand, according to Susskind and Field (1996), collaborative negotiation should follow the following six main principles for consensus building with external stakeholders:

1. acknowledge the concerns of the other side;
2. encourage joint fact finding;
3. offer contingent commitments to minimize impacts if they occur, promising to compensate knowable but unintended impacts;
4. accept responsibility, admit mistakes and share power;
5. act in a trustworthy fashion all the times;
6. focus on building long-term relationships.

17.8.3   *Not in my backyard*

One of the most common conflicts within external stakeholders is related to the location of new public facilities that local communities and residents do not want close to their homes. This is called the NIMBY (not in my backyard) syndrome. The symptoms of the syndrome are group actions, sometimes vigorous, by local communities to stop the implementation of the controversial project affecting their livings or the environment (industrial facilities, dams, waste treatment and so on). The question is how to solve this conflict, when even though the proposed construction meets all economic, legal and environmental requirements, is still not accepted by the public. Table 17.4 summarizes a set of proposed guidelines to help solve this conflict (Kunreuther and Susskind, 1991 cited in Olander, 2003).

17.9   Conclusions

Conflict is a complex behavioural process existing in all relations between individuals, groups of individuals and organizations. Conflict management is an important management function in the construction activity and should deserve considerable attention from all construction stakeholders. This is because construction projects involve a great diversity of people and organizations, experiencing a variety of situations throughout the project life cycle being tackled in the scope of different and possibly antagonistic interests of stakeholders.

Conflicts do not always have negative consequences but may lead to improvement in construction performance. Moreover, some stakeholders may benefit from conflicts and view them as opportunities for innovation.

Table 17.4   Guidelines to help solve the NIMBY syndrome

Action	Explanation
Institute a broad-based participatory process	Representatives of all affected groups should be invited to participate and assist each stage of the decision process. All those affected should have a chance to review the criteria for site selection. Groups with different points-of-view should have a chance to criticize the recommendation of facility proponents and the analyses upon which their proposals are based. A joint fact-finding process should be used so that all stakeholders can play a role in specifying the information about risks, costs and benefits that they need in order to make informed decisions.
Achieve agreement that the status quo is unacceptable	The sitting process must begin with an agreement that the facility is needed. The relevant stakeholders need to understand the consequences of doing nothing.
Seek consensus	A serious attempt should be made to involve all the relevant stakeholders to address their values: concerns, potential needs and wants. Differences can be addressed by searching for new ways of framing questions or different ways of packaging trade-offs.
Work to develop trust	Lack of trust is perhaps the most important barrier to reaching consensus. Those attempting to site a facility must recognize potential sources of mistrust, including lack of support for the project, previous negative experiences and suspicions towards the government and other institutions. One way to establish trust is to admit past mistakes and avoid exaggerated claims and promises that cannot be fulfilled.
Choose the solution that best addresses the problem	Problems must be addressed with a design and solution of the facility that stakeholders can agree is appropriate. A comprehensive list of alternative approaches and their long- and short-term implication, including the option of taking no action, should be made public in non-technical language. The choice of alternatives and technology should be based on input from the residents of the community who may well know more about the problem 'on the ground' than many experts.
Guarantee that stringent safety standards will be met	No community should be asked to compromise its basic health or safety so that a facility can be built. Preventive measures for reducing the hazard should be encouraged and the proposed facility must meet all health, safety and environmental standards, Interested parties should also have an opportunity to specify any additional standards that could be met through mitigation, such as changes in the design of the facility, substitute technologies, operational modifications and training of operators. Monitoring and control procedures involving the host community are important in minimizing risks and maintaining standards.
Fully address all negative aspects of the facility	When impacts cannot be prevented or mitigated to the satisfaction of the affected parties, various forms of compensation, specified by the stakeholders involved, can be negotiated. These agreements may include property value guarantees, creation of equivalent habitats when loss is unavoidable, and the offer of service when impact occurs.
Make the host community better off	The applicant should put a package of benefits together so that the host community feels that it is better off with the facility than without it.
Use contingent agreements	Some concerns about the management of facilities can be resolved by specifying contingent agreements that spell out what will be done in case of accidents, interruption of services, changes in standards or the emergence of new scientific information about risks and impacts, and provide means of guaranteeing that contingent promises will be met at no cost to those likely to be adversely affected.

(Continued)

Table 17.4   (Continued)

Action	Explanation
Seek acceptable sites through a volunteer process	Encourage communities to volunteer sites indicating that it is not an irreversible commitment and that there are potential benefit packages that come with the facility.
Consider a competitive sitting process	Assuming that multiple, acceptable volunteer sites are found, the sponsors of the facility should consider a competitive process of site selection.
Work for geographic fairness	It is inappropriate to locate too many noxious facilities in a single locale even if a community is willing to accept them.
Set realistic timetables	It is appropriate and helpful to set and enforce realistic deadlines. However, a good process allows all parties adequate time to consider the full range of options and weigh technical evidence as it is gathered. Opponents have administrative and legal means of slowing, even halting, siting processes that they feel have excluded them. It may be necessary to 'go slowly in order to go fast'.
Keep multiple options open at all times	It is never a good idea to have only one possible site even at the final stage of the process. Negotiations regarding possible incentive packages are more likely to produce reasonable results if a facility sponsor does not feel 'held hostage' by the only possible site.

Accurate conflict classification is the first step towards a timely and proper resolution of disputes. Accordingly, conflicts in construction have been categorized in various ways, namely by type, typology, level, object, stage in the project life cycle and underlying cause. The classification of construction stakeholders in the conflict context is also important for conflict analysis. These are named internal if they have direct participation in the project and external otherwise. Moreover, according to their interest, willingness and power to act, stakeholder groups may also be categorized in key players, keep-informed, keep-satisfied and minimal effort groups. Following the classification of construction stakeholders, several conflicts arising between them have been indicated, and corrective actions for conflict avoidance have been approached.

Recent trends in construction conflict analysis points up to conflict managing approaches rather than conflict avoidance techniques because it has been widely recognized that conflicts may lead to better and innovative solutions. In view of this, factors influencing the success of conflict management have been identified, and the phases towards adequate conflict resolution have been surveyed.

Additionally, different conflict resolution techniques have been analyzed, applicable either to internal or external stakeholders. The former include the opposite approaches of non-consensual processes like arbitration and litigation, whereas the latter include the more collaborative ones, like mediation and direct negotiation between parties. It has also been concluded that dialoguing and negotiating during the decision-making process is the best way to avoid future uncontrolled conflicts.

References

Brown, H. and Marriot, A. (1993) *ADR Principles and Practice*. London: Sweet and Maxwell.
Eurosite (2003) *73rd Eurosite Workshop, Natura 2000 – Conflict management and Resolution*. Park Interrégional du Marais Poitevin, France, 2–5 April de 2003. Available at http://www.eurosite-nature.org/IMG/pdf/73_conflict_pt01.pdf (Accessed 17 November 2007).

Fisher, R. and Ury, W. (1991) *Getting to 'Yes': Negotiating an Agreement without Giving In* (2nd edn). London: Century Business.

Gordon, J.R. (1996) *Organizational Behaviour: A Diagnostic Approach* (5th edn). Englewood Cliffs, NJ: Prentice-Hall.

Gould, N., Capper, P., Dixon, G. and Cohen, M. (1999) *Dispute Resolution in the Construction Industry*. London: Thomas Telford, Limited.

Groton, J.P. (1997) Alternative dispute resolution in the construction industry. *Dispute Resolution Journal*, **52**(3):49–57.

Hughes, W. (1994) Improving the relationship between construction law and construction management. In: Fenn, P. (ed) *Proceedings of TG15 Conference on Construction Conflict: Management and Resolution*. Lexington, USA: CIB Publication No.171, 16–19 October 1994.

Johnson, G. and Scholes, K. (1999) *Exploring Corporate Strategy: Text and Cases* (5th edn). London: Prentice Hall Europe.

Kunreuther H. and Susskind, L. (1991) The facility sitting credo – Guidelines for an effective facility sitting process. *EIA Review*, University of Pennsylvania: Philadelphia.

Latham, M. (1994) *Constructing the Team*. London: HMSO.

Loosemore, M., Nguyen, B. and Denis, N. (2000) An investigation into the merits of encouraging conflict in the construction industry. *Construction Management and Economics*, **18**(4):447–456.

Meltsner, M. and Schrag, P. (1973) Negotiating tactics for legal services lawyers. *Clearing House Review*, **7**:259–263.

Moore, C. (1986) *The Mediation Process, Practical Strategies for Resolving Conflict*. San Francisco, CA: Jossey-Bass.

Olander, S. (2003) *External Stakeholder Management in the Construction Process*, Licentiate dissertation, Department of Building and Architecture, Lund Institute of Technology, Lund, Sweden.

Rijsberman, F. (ed.) (1999) *Conflict management and consensus building for integrated coastal management in Latin America and the Caribbean*. Resource Analysis Report for Inter-American Development Bank. Delft, Nederland. Available at http://idbdocs.iadb.org/wsdocs/getdocument.aspx?docnum=359595 (Accessed 9 December 2007).

Susskind, L. and Field, P. (1996) *Dealing with an Angry Public: The Mutual Gains Approach to Resolving Disputes*. New York: The Free Press.

Thomas, K. (1992) Conflict and negotiation processes in organizations. In: Dunnette, M. and Hough, L. (eds) *Handbook of Industrial and Organizational Psychology*. Palo Alto, CA: Davies-Black® Publishing, pp. 651–717.

Wall Jr., J. and Callister, R.R. (1995) Conflicts and its management. *Journal of Management*, **21**(3):515–558.

18 Environmental Stakeholder Management

Felix Nikoi Hammond and Colin A. Booth

18.1 Introduction

This chapter is unique. Unlike the others, its principal subject of analyses is the environment because the ongoing eco-crises – worsening atmospheric, land, biodiversity and water pollution – have given rise to serious concerns about the long-term viability of the natural environment. The infinitely challenging task of this chapter is to address the questions: Are managers of construction and property enterprises likely to voluntarily, without government interventions, manage the environment as a primary stakeholder? And crucially, how may they do so? The central argument here is that, there are great chances that business managers will overlook or give the environment less than adequate attention if not compelled by superior reasons other than moral suasion and encouragements to be discretionarily socially responsible. It is now widely accepted that the environment is too vital a subject to be left out of business considerations, dialogues and treatises. The reality, however, is that, because the environment is non-human, it is not represented in person to influence the course of business decisions and actions that affects it or even those that it affects. Besides, taking account of the environment in business decisions entails costs. In spite of that, the contributions that good environmental management makes to the profit maximisation objectives of business enterprises are not always self-evident, they are also difficult, in some cases impossible, to fathom in monetary terms. Consequently, when businesses, in the bid to maximise profits or secure their survival, look for opportunities to cut costs, the costs associated with their environmental management practices tend to be one of the sure candidates for liquidation. This point is amply exemplified by the fact that the oft-cited definitions of stakeholder constituencies converge largely on humans – individuals and groups of individuals; in particular, those whose interests are served or damaged by activities of business undertakings. Whether or not business managers, such as managers of construction and property enterprises would be voluntarily committed to the environment, depends on whether they are obliged or have the incentive to do so.

To better understand the issues in this chapter, it is useful to start with an overview of the general concept of stakeholder management. Cognisance is, however, paid to the fact that the preceding and subsequent chapters of this book have examined the stakeholder concept into sufficient detail; thus, of necessity, only a highly condensed version of the increasingly complex stakeholder management concept is discussed here.

18.2 The stakeholder management concept

Modern stakeholder doctrine is relatively recent. It represents a significant shift in business governance paradigm (Clarke and Clegg, 1998; Jones and Wicks, 1999) from mainstream management principles (Jonker and Foster, n.d. 2002). The origins of modern stakeholder management doctrine are traceable to the 1984 classic by R.E. Freeman (Freeman, 1984), even though its historical roots are traceable to early organisation theorist such as Barnard (1938), March and Simon (1958) (see further: Andriof and Waddock, 2002:19–42). In the last decade or so, the doctrine have been well received among both academics and practitioners (Kolk and Pinkse, 2006). Much of the contributions made to the doctrine took place in the 1990s and have come from experts like Starik (1994), Clarkson (1995), Donaldson and Preston (1995), amongst others (see further: Clarke and Clegg, 1998). So strong is the acceptance of stakeholder management in modern business management ethos that Evans and Freeman (1988) could contend that businesses should be managed for the benefit of its stakeholders.

This position is challenged by advocates of free enterprise like Milton Freidman who, in what became a contentious article in 1970, vehemently contests and disapproves any view that purports that businesses are managed for reasons other than profits. Freidman's argument would be returned to later. To be able to manage a business, such as a construction and property enterprise for the sake of its stakeholders, knowledge and understanding of the nature of stakeholders and the way and manner that they may be managed must be gained. Before turning to the rather vex issue of whether the environment falls within the stakeholder constituency of construction and property business enterprise, approaches to stakeholder management is first outlined.

18.3   Approaches to stakeholder management

A variety of stakeholder management approaches are identifiable. These presume that stakeholders of organisation are known with some certainty. These approaches have basically taken two routes. One route concentrates on how organisations manage their stakeholders in practice. The second route prescribes best practices in stakeholder management principles. Experts like Carroll (1979) and Clarkson (1995) identified four main ways by which organisations manage their stakeholders namely, 'reactive, defensive, accommodative or proactive'. Logsdon (1994:127–128) avers that, where the core values of the stakeholder is opposed to, or incompatible with that of the organisation, then it would be legitimate for the organisation to employ a reactive or defensive strategy. Conversely, where the core values are compatible with that of the organisation, an accommodative or proactive strategy may be adopted.

Savage *et al.* (1991:5) offer a slightly modified version of these ideas. As Table 18.1 illustrates, they prescribe 'collaborative strategy' in situations in which the stakeholders' potential to corporate with the organisation as well as their potential to threaten the organisation are both high; 'defensive strategy' when the potential to threaten is high, but the potential to cooperate is low; 'involvement strategy' where the potential for threat is low but that for cooperation is high; 'monitoring strategy', when both potential for threat and cooperation are low. It is unclear how the core values of a stakeholder could be incompatible with that of the organisation when indeed core values of organisations are supposed to be a fair mixture

Table 18.1   Stakeholder management strategies

	High cooperation	Low cooperation
High threat	Collaborative	Defensive
Low threat	Involvement	Monitoring

Source: After Savage *et al.* (1991:5).

the respective values of its stakeholders. Besides, it is unclear what particular type of reactive, defensive, accommodative or proactive strategies will be applicable in any given situation.

The seven stakeholder management principles developed and popularised by the Centre for Business Ethics and Board Effectiveness (CC (BE)) between 1993 and 1998 through a series of colloquy is widely accepted by experts in this field as best practice[1]. These principles are reproduced in Table 18.2.

It is not possible to expand on all these points in evaluating how the environment may be managed as a stakeholder in construction and property enterprises. In this chapter, the Clarkson Centre's principle is relied on to analyse the management of the environment as a stakeholder in construction and property enterprises. Consequently, only the fourth principle is applied to the environment as a stakeholder.

To be useful, organisations must have a clear sense of who their stakeholders are, and to whom these principles must be applied. That is, they must be in no doubt as to nature of those in whose interest and for whose benefits the business must be managed. Given the wide acceptance of the stakeholder doctrine in the management discipline, one could be forgiven for presuming that some agreed and referable definitive list of, or formula for identifying, stakeholders exist. Surprisingly, no such list or formula has been agreed. Mainstream management consider employees, suppliers, shareholders and customers as the main entities in whose interests and for whose benefits businesses are to be managed (Jonker and Foster, n.d). Freeman (1984:53), the originator of the stakeholder idea, prescribed a formula by which the stakeholders of a business enterprise may be identified: that is, 'any groups or individual who can affect or be affected by the achievement of an organisation's purpose. . .'

Table 18.2   The Clarkson principles of stakeholder management

Principles	Description
Principle 1	Managers should acknowledge and actively monitor the concerns of all legitimate stakeholders, and should take their interests appropriately into account in decision-making and operations.
Principle 2	Managers should listen to and openly communicate with stakeholders about their respective concerns and contributions, and about the risks that they assume because of their involvement with the corporation.
Principle 3	Managers should adopt processes and modes of behaviour that are sensitive to the concerns and capabilities of each stakeholder constituency.
Principle 4	Managers should recognise the interdependence of efforts and rewards among stakeholders, and should attempt to achieve a fair distribution of the benefits and burdens of corporate activity among them, taking into account their respective risks and vulnerabilities.
Principle 5	Managers should work cooperatively with other entities, both public and private, to ensure that risks and harms arising from corporate activities are minimised and, where they cannot be avoided, appropriately compensated.
Principle 6	Managers should avoid altogether activities that might jeopardise inalienable human rights (e.g. the right to life) or give rise to risks which, if clearly understood, would be patently unacceptable to relevant stakeholders.
Principle 7	Managers should acknowledge the potential conflicts between (a) their own role as corporate stakeholders, and (b) their legal and moral responsibilities for the interests of stakeholders, and should address such conflicts through open communication, appropriate reporting and incentive systems and, where necessary, third-party review.

Source: Clarkson Centre for Business Ethics & Board Effectiveness.

This definition remains the oft cited in the stakeholder literature (Mitchell *et al.*, 1997). However, some adjustments to this definition have been introduced. Starik (1994:90) summed up the various definitions of stakeholder in the following words: '. . .there may be numerous levels of specificity as to what the term "stakeholder" mean, depending on what the user is referring to. The range appears to be bounded in this case, on one end, by those entities which can and are making their actual stakes known (sometime called "voice") and, on the other end, by those which are or might be influenced by, or are or potentially are influencers of, some organisation'. The main addition of this definition is to include in the stakeholder constituencies, those that may *prospectively* affect or be affected by the activities of the organisation. Shankman (1999) widened this definition further to include, so to speak, the entire society. Yet Cochran (1994) chose to confine the definition of stakeholder narrowly to those with direct economic connections to the organisation. Clarkson (1998) further categorises stakeholders into primary and secondary. According to Clarkson (1998), primary stakeholders consist of those with 'formal, official or contractual' connections with the organisation . . . without whose continuing participation the corporation cannot survive as a going concern'. Secondary stakeholders, conversely, according to Clarkson (1995) are those that affect or are affected by the organisation, but not in any formal contractual relationship and, hence, not key to its survival.

It is observable from this definition that stakeholders, at least the primary ones, are basically human beings and may include, in the case of a construction and property enterprise, financial claim-holders (shares, creditors and bond-holders), suppliers, employees, customers, property occupiers, the community, government departments and so forth. It follows that, since the environment is non-human, it does not constitute a stakeholder by this definition. Perhaps, if Freeman's definition is widened to include, 'and things' after the word 'individuals', the environment could have been construed as a stakeholder. As it stands, the best that can be done is to take the human stakeholders as proxies for the environment. But since, by these definitions, the litmus test for qualification is '. . . can affect or be affected . . .' it seems most strange that, the environment that is most affected and that affects businesses like construction and property enterprises should be left out of consideration. To this realisation, Starik (1994) launched a call for the elevation of the environment to the status of a stakeholder. Consequent upon this call, Starik (1994:92) widened Freeman's definition to encase '. . . any naturally occurring entity who is affected or affects organisational performance'. Now, it is time to take a look at the stakes that the environment has in construction and property enterprises.

18.4   The natural environment as a stakeholder

The point has been made earlier to the effect that, to constitute a stakeholder, the entity in question must affect or be affected by the activities of the enterprise in question. It does not take much to realise that the activities of construction and property enterprises affects and are affected by the natural environment. From the economic perspective, these effects and affections are relevant only to the extent that they affect the wealth levels of those affected. The discussions that follow concentrate on the channels through which the activities of enterprises within the construction and property industry so affects (not how they are affected by) the natural environment. Any deliberation of the natural environment must begin with the nature and importance of the natural environment.

The natural environment (usually referred to, simply as, the environment) underpins and insures the viability of human, animal and plant life. It consist of the air [atmosphere], water and land in or on which people, animals and plants live[2]. The environment provides both the means and basis for sustenance, provision of necessaries, enjoyment of life and survival. The materials and non-human inputs used in construction and property enterprises are extracted from the environment. Accordingly, it is the environment that gives the sector its social and economic merits. The pursuit of profits without adequate attention to the environment from which the means for such pursuits are obtained is clearly a self-defeating undertaking. Besides, quality of life of human agents of construction and property enterprises as well as their customers and clients are inseparably linked to the quality of the environment. Particularly, 'environmental degradation has been demonstrably linked to human health problems, including some types of cancers, vector-borne diseases, emerging animal to human disease transfer, nutritional deficits and respiratory illnesses' (UNEP, 2007).

The bottom line is that, life ceases when the environment is damaged. Bearing this point in mind coupled with the fact that damages to the environment, 'are to a considerable extent irreversible and may be catastrophic . . . if unchecked' (IMF, 2008b), it appears most strange that anyone would entertain the thoughts of excluding the environment from its primary stakeholders. For when the environment is destroyed the enterprise ceases to exist.

Different activities and phases of construction and property undertakings impact the environment differently. For the sake of convenience, and to facilitate understanding of the ways the construction and property enterprises affects the environmental, the environment is subdivided into four main subcomponents – the atmosphere, land, water and biodiversity[3]. This subdivision is based on a classification used in the 'global environment outlook' of the United Nations Environment Program (UNEP, 2007). Additionally, construction and property continuum of activities have been subdivided into four subcategories. The possible impacts of activities of construction and property undertakings on each are then analysed.

18.5 Atmosphere

Construction and property enterprises generate considerable effects on the atmosphere. The major atmospheric concerns today are climate change[4] and air pollution. Consequently, the effects of construction and property activities on the environment are looked at from these two perspectives. Scientific evidence consisting of computer models suggests that anthropogenic green house gas (GHG) emissions (principally carbon dioxide) are the main drivers of climate change, in particular global warming (UNCHE, 1972; CBO, 2005; IPCC, 2007a; UNEP, 2007). As such, the effects are looked at from the sector's (i) anthropogenic emission of GHGs[5] and (ii) air pollution. The sources of construction and property driven GHGs and pollution are detailed in Table 18.3.

Evidence that the climate system is warming is now unequivocal (IPCC, 2007c:30; UNEP, 2007). The Earth's average temperature is estimated to have increased by ~0.74°C over the past century (UNEP, 2007) and is expected to increase to between 1.8°C and 4°C this century (Meehl *et al.*, 2007). For the next two decades, a warming of ~0.2°C per decade is projected (IPCC, 2007a).

Of course, not all carbon dioxide in the atmosphere is anthropogenically generated; some are generated by natural processes such as volcanoes, earthquakes (UNEP, 2007) and emissions from the ocean, but it is those generated by human causes that is fuelling the recent significant rise[6]. For detailed discussions on climate change, see IPCC (2007b). Also, see IMF (2008a) and Stern (2007) for the economic costs and fiscal implications of climate change.

The main source of anthropogenic carbon dioxide emissions is fossil fuel combustion. Energy (particularly from electricity, oil, coal and natural gas) is essential to the construction

Table 18.3 Construction and property effects on the atmosphere

Affected component of the environmental	Pre-construction	The construction process	Post-construction	Operational
Atmosphere	**Anthropogenic emission of GHGs (mainly carbon dioxide) through:** fossil fuel based energy (mainly electricity) use during the extraction and production of construction materials and products. fossil fuel based energy (mainly hydrocarbon fuel – diesel fuel) use during the transportation of construction materials and products through the supply chain to construction sites **Emission of noise, gaseous, chemical and dust pollutants as well as hazardous substances during:** extraction and production of construction materials and products transportation of construction materials and products through the supply chain.	**Anthropogenic emission of Green House Gases (GHGs) through the use of[7]:** electricity to power cranes and other hoisting equipment, welding equipment, site lighting, small tools, etc. hydrocarbon liquid fuels (mainly diesel fuel) to power air compressors, dump trucks, etc. propane gas used for heating of site offices, workshops heating, remote lightning and buildings under construction. **Emission of noise, gaseous, chemical and dust pollutants as well as hazardous substances during:** demolition, digging, excavations and site preparation.	**Anthropogenic emission of Green House Gases (GHGs) through the use of:** electricity and hydrocarbon energy in construction waste disposal, management and recycling **Emission of noise, gaseous, chemical and dust pollutants as well as hazardous substances during:** Waste disposal, management and recycling.	**Anthropogenic emission of Green House Gases (GHGs) through the use of:** propane gas and electricity by occupiers of completed buildings and structures to heat internal spaces provide fire for cooking, power for lightening and equipment.

and property processes – from the production of construction materials, through the construction processes, waste management and recycling of construction waste, to the operation, use and occupation of the buildings, facilities or structure (Table 18.3). A significant proportion of these energy requirements are met through the combustion (burning) of fossil fuels because fossil fuels contain hydrocarbons when combusted to generate carbon dioxide (Osman and David, 2007). In the process, carbon dioxide is emitted into the atmosphere. In most countries, construction and property sector generate a considerable percentage of fossil fuel driven carbon dioxide emissions.

For example, in the United Kingdom, the Environmental Agency (EA) estimates that 'building, maintaining and occupying homes accounts for almost 50 per cent of UK's carbon dioxide emissions' (EA, 2008). The production of construction materials (such as steel and cement) accounts for a further 10% of carbon dioxide emissions in the United Kingdom (Seager, 2007). Table 18.3 summarises the means by which these occur.

In addition to GHGs, construction activities also emit gaseous, air particles, chemical and noise pollutants into the atmosphere. These are very dangerous to human health as some can enter the bloodstream, irritate the lungs and cause a wide range of diseases; they can also exist in the air for an indefinite time (Osman and David, 2007; UNEP, 2007). As Table 18.3 shows the emission of pollutants occur mainly from the pre-construction stage (production of materials) throughout the lifespan of the facility.

18.6   Land

Land is the most basic input in construction. Chemical substances are increasingly used during the construction process. It is not always possible to prevent these chemicals from contaminating the land and soils. For instance, chemicals used to preserve timber from insect infestations and destruction most often come into direct contact with the land and soil to pollute them. Then again, the embedment of steel and blocks into land in the course of the construction process are important sources of chemical contamination of land. So also, excavation of contaminated land for foundation and earthworks, demolition of contaminated buildings and recycling or burning of contaminated waste and decaying of materials installed in land during the construction phases represent important sources of ground pollution and contamination of ground gases (CIRIA, 1993). Contamination of land may also result from spills and improper disposals of chemicals used by occupiers of, particularly, industrial buildings and infrastructure. Contaminated lands are harmful to human health and a danger to the stability of the superstructure. Energy use in existing buildings represents significant emitter of carbon dioxide and hence a precipitator of climate changes.

18.7   Biodiversity

Biodiversity, according to UNEP (2007), 'is the variety of life on Earth. It includes diversity at the genetic level, such as that between individuals in a population or between plant varieties, the diversity of species, and the diversity of ecosystems and habitats'. Primarily, construction materials such as coarse aggregates (stones and gravels) and fine aggregates (sand and clinkers used in the production of cement) are extracted or mined from land. Extraction and mining of construction materials from land is a sure cause of loss of habitation and disturbance of biological cycles of a vast number of species or even destruction of many species.

Chemicals and explosives used in the mining and extraction processes also terminate or shorten the lifespan and, in some cases, annihilate certain species altogether. Moreover, the expansion of construction activities may lead to encroachments on forest, agricultural and biologically sensitive lands, which again may lead to loss of habitats and so forth. Furthermore, the GHGs emitted into the atmosphere may generate considerable increase in temperature beyond the tolerable limits of many biological lives leading to the destruction of such species.

18.8   Water

Water is vital to life. The ocean exerts a cooling effect on the earth's temperature. Oceans, lakes and other water bodies provide the habitat for aquatic life as well as the source of freshwater for humans and land-based animals. Construction materials and the construction process account for a significant percentage of global freshwater use. The use, occupation and operation of buildings and structures require water for cleaning, cooking, drinking and washing. Thus, degradation in water quality can have profound effects on all life forms as well as on the construction industry itself. Spillage of pollutants and hazardous substances emitted from construction and property enterprise affects freshwater availability and use and the conservation of aquatic life. This could also lead to growth in waterborne diseases. The excessive emission of carbon dioxide into the atmosphere to which the construction and property sector makes significant contributions affect 'ocean salinity and acidification, sea levels, precipitation patterns, extreme weather events and possibly the ocean's circulatory regime' (UNEP, 2007).

Having gained some basic insights of the ways that construction and property enterprises affect the environment, it is now opportune to look at how the environment may be managed as a stakeholder using the available stakeholder management tools.

18.9   Stakeholder management of the environment

Two points are worth making before delving into this section discussion. First, it is not just the construction and property enterprises that affect the natural environment, as the above discussions may erroneously impress on readers. The natural environment also affects construction and property enterprises. For instance, extreme weather conditions such as floods, hurricanes and tornadoes destroy buildings and structures. Moreover, ground and soil contamination deteriorate chemical strengths of elements of buildings and structures constructed on land. Secondly, the effects of construction and property enterprises on the environment are not always negative; there are conceivable positive effects as well. These positive effects may come in the form of environmentally friendlier land uses, landscaping, efficient drainage systems to convey excess floodwater and so forth. The point has already been made to the effect that, a limitation on space has necessitated the concentration on the negative effects only. But this restriction is justified also on the grounds that it is the negative effects that constitute the collective concern of the world today. That said, in evaluating the balance of economic burdens and benefits associated with construction and property enterprises these economic benefits are brought into the analysis.

This section of the chapter now turns to the stakeholder approach to the management of the environment. In particular, as made clear earlier, the chapter applies the fourth of the seven point stakeholder management principle outlined in Table 18.2 above to the management of the environment as a stakeholder.

18.10   Fair distribution of benefits and burdens

A clear appreciation of the nature and sources of costs and benefits of construction and property enterprises is a prerequisite to understanding how best to fairly distribute them. The first point to grasp in the quest to fairly distribute costs and benefits between the business and the environment is that construction and property enterprises, like any other productive economic venture, are input–output based. It is in this very input–output relationship that the costs and benefits of activities of construction and property enterprises reside. Deeper exploration into this inputs–output relationship holds the key to an understanding of the orders of the magnitude of costs and benefits associated with such enterprises. This will also point to the sources from which these costs and benefits are emitted.

Essentially, construction and property undertakings entail the adjustment, rearrangement, modification or conversion of inputs such as construction materials, human resources and land into a wide range of land-based products, such as landed properties and infrastructure. Costs are incurred in the process. As Table 18.4 shows, these costs fall into two main categories – private and social costs. Each category also has direct (transaction costs) and indirect (opportunity costs) dimensions.

The range of payments made in acquiring the inputs – land, labour, equipments and financial capital – and investing in the methods and techniques of converting these inputs into desired landed products constitute the direct private costs of construction and property enterprises (quad 1 in Table 18.4). These costs are, also, sometimes referred to as expenses, payments or private transaction costs. It is noteworthy that, direct private costs are the only expenses borne by, and paid for, directly from the coffers of the business. They are, thus, the only costs that construction and property business enterprises typically take account of in computing their profits and the only costs considered in construction and property management tenders: that is, the only costs relevant to the construction costs estimator and accountant in preparing profits and loss financial statements. Profits computed based exclusively on direct private costs are usually termed accounting profits and they are obtained by subtracting the direct private costs from the revenue or turnover associated with the undertaken. An appreciation of this point is vital to an understanding of how the environment may be managed as a stakeholder in construction and property enterprises. This is primary because, accounting profits are the universally accepted indicator of economic performance of businesses. Thus, all decisions of business managers are directed towards improving their accounting profits on an ongoing basis (Smith, 1776). This is realised by expanding their outputs as much as possible and keeping their direct private costs considerably below their revenue or turnover.

Since the only costs that matter in assessing the economic performance of business enterprises is their direct private costs, one of the favourite pastimes of business managers is to avoid paying for all non-direct private costs. This is because paying for them reduces their profits and, hence, their economic performance. Thus, if there is a non-direct private cost to distribute, business managers, if given the chance, will bear none (zero) of it. They will instead prefer and, in fact, ensure as much as possible that entire cost is borne by the other sharing party. Where the sharing party is non-human like the environment, then it is the environment that will bear it all or at least take the lion's share of it. It follows that, unless costs, such as environmental costs are brought within the direct private costs, they are unlikely to be taken into account and, hence, will be difficult, even impossible to achieve a fair distribution of costs between construction and property enterprises on one hand and the environment on the other. Put differently, if a construction and property enterprise is to participate in the sharing of environmental costs at all, those costs must be forced into their private costs. Latter sections will consider how this may be done.

Table 18.4 The costs concept

Costs borne by	Construction firms	The society (excluding the investor)	The society (including the investor)
Types of costs	Private costs	External costs (or externality)	Social costs
Direct costs (Transaction costs)	(*quad 1*) Financial payments made by the construction firm: **Fixed costs** Costs of fixed assets such as office buildings and equipments **Variable costs** Payments for materials and land Salaries and wages Overheads Professional fees Assets Taxation Statutory fees	(*quad 3*) All costs of environmental resources and government budgetary resources incurred as a result of the construction: Costs of air pollution Costs of water pollution GHG emission Depletion of environmental resources Destruction of species Time of employees of government waste management and sanitation agencies (measured by salaries and benefits) necessitated	(*quad 5*) Sum of direct private costs and indirect social costs
Indirect costs (opportunity costs)	(*quad 2*) After-tax income forgone by engaging in the particular project. That is the next best alternative use of time and money spent on the project: Earnings forgone Value of business opportunity forgone by engaging in the particular project	(*quad 4*) Tax revenue forgone by government as a result of the forgone alternative: Foregone tax revenue from the earnings forgone	(*quad 6*) Sum of indirect private costs and indirect social costs

For now, it is worth establishing the range of costs that must be brought within the private costs of construction and property enterprises.

As Table 18.4 depicts, accounting profit, given by turnover minus direct private costs, leaves a whole range of costs, very important costs indeed, out of account. One such costs, is the indirect private costs, usually referred to as private opportunity costs or shortly as, opportunity costs. By definition, the opportunity costs of an undertaken are the next best income earning opportunity that a business losses because it employed its resources in an alternative venture instead. To illustrate, suppose that the resources used to execute say *project-A* could instead have been used to execute *project-B* to earn a profit of say £100. In that case, the opportunity cost of project-A is £100 (i.e. the income earning opportunity sacrificed). This assumes that *project-B* is the next most profitable alternative project that could have been embarked on, had *project-A* not been chosen first. The opportunity costs are important costs that must be brought into the computation of profits. Accordingly, the economists compute profits as turnover minus both the direct and indirect (opportunity) costs: that is, economic profits equal accounting profits minus opportunity costs.

Even so, economic profits itself omit the whole of the external costs category, a costs category of utmost significance in environmental discussions. External costs must necessarily be brought into the calculus if the environment is to be properly managed as a stakeholder (quads 3 and 4 of Table 18.4). This is because it is the costs that must be fairly shared between the environment and the enterprise that caused them.

18.11   Nature of the burden

What are external costs? As Professor A.C. Pigou (1962:174) rightly states, 'in some occupation, a part of the product of a unit of resources consists of something which instead of coming in the first instance to the person who invests the unit, comes instead, in the first instance (i.e. prior to sale taken place), as a positive or negative item, to other people'. Simply put, the external costs of an activity are the costs associated with the harmful consequences suffered by others other than those whose decisions it were to undertake the activity. With respect to construction and property enterprises, the negative things which should have been sustained, that is brought within the direct private costs of the investors, but are instead sustained by other people (neighbours and society as a whole) include, as discussed above, anthropogenic emissions of GHGs and pollutant into the atmosphere, land and water together with the destruction of biodiversity as explained earlier. This may be direct or indirect. The external costs are incidental to any production activity such as land development. To be factored into the calculation of costs, it should be possible to translate these harmful consequences into their monetary equivalents. Sadly, this is not an easy endeavour. This is partly because of the great variety of externalities or harmful consequences that may accompany a single construction and property activity. Some of these have been indicated in the costs table above. The true monetary equivalent of an emission of a pollutant from a construction and property activity may include the direct costs associated with an ailment that it may cause including the medical bills borne both by the patient and the government as well as the loss of income and productive labour hours during the ailment. On a much larger scale, this could lead to the loss of fertility of agricultural lands, loss of habitats, high taxation to deal with extreme weather conditions, food crises and so forth. The same calculation would have to be made for GHG emission, land contamination, water pollution and indeed any identifiable negative harm to the natural environment. This is nearly impossible to compute and any attempt at estimation can only offer indications of the orders of the magnitude rather than accurate knowledge of

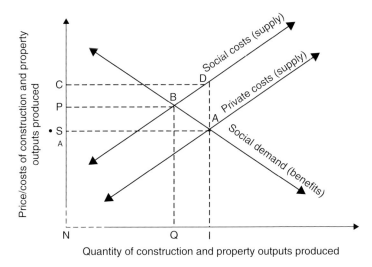

Figure 18.1   **Environmentally harmful construction and property externalities.**

the state of affair. However, it leaves no one in doubt that the costs can be enormous. For examples of such estimates see Stern (2007).

Figure 18.1 illustrates this point. To simplify the discussions, assume that the entire benefits from construction and property undertakings accrue exclusively to the responsible enterprises. That is, the social benefits of the undertaking are the same as the private benefits, which accrues entirely to the investor. In terms of economics, this assumption is stated as the equalisation of private and social benefits. It has to be borne in mind also that the total costs associated with any undertaking is represented by its social costs; that is, it is private costs (direct and indirect) plus the external costs (direct and indirect). Consequently, if the investor bears the total costs of the enterprise, there should not be any difference or divergence between the private costs and social costs associated with the undertaking.

On the other hand, if the investor bears just the private costs of the activity, then the private costs will diverge from the social costs associated, an indication that someone else is bearing the external costs. In Figure 18.1, take P to be the minimum unit price at which landed products of construction and property enterprises is likely to be sold if the investor bore the total costs (private costs + external costs), and not just the private costs into account. P also represents the reasonable costs that consumers should be prepared to incur on a unit of the products, which is the price that consumers would consider reasonable for the products. At this price and costs, the optimal quantity that investors can afford to produce is Q.

This price, costs and quantity represent the social optimum input–output arrangement. This is because, with this arrangement the maximum costs incurred by the investors, represented by the areas of the rectangle PBQN, are identical to the minimum possible benefits they receive. Conversely, if investors ignore the external costs of their activities, they are likely to perceive, quite erroneously, the price G as the minimum unit price for their outputs. Likewise, if consumers ignore the external costs they sustain from the activity, which is normally the case in practice, then they are likely also, like the investors, to perceive the price G as the reasonable unit price to pay for the output. However, consumers are only able to ignore the external costs in their imagination and not in actual fact since the external costs (e.g. pollution) are imposed on them and, hence, have either paid for it already or will inevitably pay for it – they have no choice in the matter.

They cannot choose to suffer from an associated pollution, global warming or extreme weather conditions. The position only enables construction and property enterprises avoid paying for their external costs. With less costs to pay, the construction and property enterprises are able to expand their outputs beyond the social optimum quantity of Q to I. If it is reasonably presumed that associated harmful consequences increase as outputs are expanded, the failure to take account of the external costs would lead necessarily to an expansion in the environmentally harmful consequences of the construction and property sector. At this point, construction and property enterprises achieve benefits measured by the area covered by the rectangle GAIN. The maximum costs likely to be incurred by the investor to achieve this are also GAIN. Meanwhile, taking the external costs into account, the actual costs that the investor should actually pay to be able to produce that level of output is represented by the area of the rectangle CDIN. This means further that, by producing at this quantity and price arrangement, construction and property enterprises are able to avoid the external costs represented by the area of the rectangle CDAG and force society to pay for them.

This is certainly an unfair distribution of the costs associated with construction and property activity. The fairer way of distributing these external costs is for the responsible entity to compensate the victims of the external costs. Under normal free enterprise operations, firms are likely to avoid compensating society for such damages done. This is because doing so will mean taking account of costs that are outside of their direct private costs. As Adam Smith (1776) explained 'every individual endeavours to employ his capital so that its produce may be of greatest value. He generally neither intends to promote the public interest, nor know how much he is promoting it. He intends only his own security, only his own gain . . .' More recently, IMF (2008a) has said, 'emitters of greenhouse gases (GHGs) fail to recognize the aggregate damage they cause, so emit more than is collectively desirable'. Thus, in a free market situation, external costs are unlikely to be fairly distributed between the environment and the responsible entity.

Since this unfairness occurs because the private costs diverge from the social costs, the solution lies in eliminating the wedge between the private costs and the social costs: that is forcing a merger between the private and social costs to bring the external costs within the private costs. The available economic instrument by which this may be achieved is explained below.

18.12   Nature of the gains

Construction and property enterprises generate benefits that accrue to others – neighbours and society (external benefits) instead of to them that they are unable to charge for under normal market conditions. Examples of such benefits have already been provided above. The total benefits they generate are called the social benefits. Like social costs, this is made up of private benefits and external benefits. The private benefits are those that accrue from the activity to the responsible enterprise and the consumers of outputs of the enterprise. The external benefits are those that accrue to others (neighbours and society) other than the responsible producers and consumers, as illustrated in Figure 18.2. Ideally, since it was the decision of the business to embark on the enterprise, it is reasonable to expect the total benefits to accrue to it and no one else. Unfortunately, we do not leave in an ideal world. To illustrate the point, we shall presume for now, unrealistically though, that construction and property businesses bear the entire costs of any construction and property undertaking.

If construction and property businesses take account of the entire costs associated with their activity, then they will be operating along the social supply line, that is their private costs will coincide with the social costs. This means that the quantity of outputs they produce

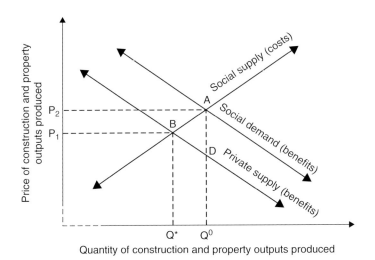

Figure 18.2   Environmentally beneficial construction and property externalities.

and the price at which these are sold will be determined by the point of intersection of the social demand; that is, the entire benefits from the undertaking on one hand and the social costs – that is, the entire costs of the undertaking on the other. At this point of intersection, the social optimum quantity, the socially desirable output level from construction and property activities will be Q^0 at a corresponding price of P_2. At this output level, relatively more of the positive side effects of construction and property activities would be produced. In practice, however, it is not always possible for construction and property firms to capture the full benefits of their activities. For instance, a private construction and property business is unable to cream-off or charge for the property value appreciation that its activities bring on neighbouring properties. Consequently, they concentrate on the benefits they can directly capture through efficient pricing of their products. They most often operate at the point of intersection between the private demand and social supply. Again, if it is reasonably presumed that incidental beneficial outputs decreases with contraction in outputs, then the inability of construction and property enterprises to capture the external benefits of their operations would ultimately lead to a reduction in the production of these good side effects, as they will be operating at reduced output levels of Q^* and a corresponding reduced price level of P_1. The only way that construction and property businesses could be incentives to expand the positive side effects of their activities is for them to capture the external benefits they generate as well.

18.13   Fair distribution of gains and burdens

The basis of a fair distribution of the gains and burdens of construction and property enterprises with the natural environment must be a comparison between the external benefits and costs associated with the undertaken. This is because, from the economic viewpoint, the stakeholder status of the environment is based primarily on the external benefits and costs; the environment is affected by the imposition of external costs on it; it affects construction and property enterprise by denying it the external benefits. If the external costs imposed on the environment are equal to the external benefits that the environment denies the business, then the benefits and costs are fairly distributed; otherwise it is an unfair distribution. The real problem is how the costs and benefits may be distributed fairly if external costs are more than external benefits or vice versa.

Before turning to how fairness in the distribution of the burden and benefits may be achieved, it is important to sum up the points made so far on the externalities of construction and property businesses. The presence of negative externalities implies that construction and property businesses underprice their products. This is, essentially, because the costs of these products to the business do not reflect their total costs – it excludes their associated external costs – the costs of the damage to the environment. When goods are underpriced, their consumption tends to increase which in turn leads to overproduction. Increase production, in turn, leads to increases in damages to the environment. However, more importantly, the increase consumption and underpricing of construction and property products discourages investments in technologies to mitigate the negative impacts of construction and property activities on the environment. The presence of positive externalities, on the converse, leads to a reduction in the profitability of construction and property businesses. This, in turn, discourages construction and property businesses from engaging in activities that generate external benefits. The balance between the uncompensated harm caused and the uncharged benefits received by society, as a result of construction and property activities, provides an indication of the extent to which costs and benefits of construction and property are fairly distributed.

18.14   Instruments for fair distribution of gains and burdens

It is worth bearing in mind that unless the costs and gains are brought within the private costs and benefits of the responsible entity, it would be unlikely that the gains and burden associated with construction and property activities would be fairly distributed. It is against this backdrop that the available economic instruments for the fair distribution of the gains and benefits of construction and property activities are evaluated.

18.15   Pigovian pricing – polluter pays principles

The first and perhaps traditional instrument of ensuring the fair distribution is through government taxation and subsidies. This is sometimes referred to as Pigovian pricing. Though the modern statement of the Pigovian pricing instrument is attributed to the British economists A.C. Pigou, its historical roots can be traced to the 1700 BCE code of Hammurabi, which specifies that:

(53) If any one be too lazy to keep his dam in proper condition, and does not so keep it; if then the dam break and all the fields be flooded, then shall he in whose dam the break occurred be sold for money, and the money shall replace the corn which he has caused to be ruined.

or

(55) If any one open his ditches to water his crop, but is careless, and the water flood the field of his neighbour, then he shall pay his neighbour corn for his loss.

The basic principle is that, without the intervention of government to force polluters and emitters of carbon dioxide, for instance, to bear the costs of their pollution as part of their direct private costs, no other feasible mechanism can ensure the fair distribution. Essentially, Pigou (1962) suggested that, it is the responsibility of the state to oversee the fair distribution of burdens and benefits associated with economic undertakings. Pigou offered 'extraordinary encouragement' and 'extraordinary restraints' upon investments as the main instrument of distribution. According to Pigou (1962) 'the most obvious forms, which these encouragement and restraints may assume are, of course, those of bounties and taxes'. In order words, a price has to be attached to the emissions in such a way that polluters are taxed an amount equal to the external costs of their pollution or reimbursed for the external benefits they generate. This, if implemented fully, will ensure that construction and property businesses bear, as far as money can pay, the full costs (private and external costs) of their operations whilst receiving the full benefits of their operations. A pollution tax, such as carbon tax, which is a fee levy for each ton of emission of GHGs, is an example of a negative externality tax. Such taxes constitute restraints and they make it costly to emit pollutants; thereby, acting to discourage actors from engaging in activities, or using means that emits pollutants. This will, in turn, encourage investments in technologies that lead to the accomplishment of the undertaking at much reduced rate of pollutant emissions. It also forces businesses to take explicit account of the external costs of their operations; thereby, forcing businesses to price their products appropriately and, hence, produce at the social optimum level. Pollution taxes will lead to fair distribution of costs only when the tax amount is equal to the external costs. When the tax amount falls short of the actual external costs associated with the undertaking, construction and property businesses will still not be bearing the full external costs of their operations and, hence, continue to underprice their products and continue to emit significant tons of pollutants.

The Pigovian pricing when properly implemented truly brings the external costs within the direct private costs. There are, however, several problems with the Pigovian pricing approach. Firstly, in those sectors where it is possible for the businessperson to pass on all its costs to the consumer by building it into the price of its products, the tax imposed will ultimately be borne by the same consumer who suffered the external costs. The tax could, thus, compound the costs faced by consumers at the benefits of the business. In such cases, the tax will not discourage damages to the environment. Then again, precise estimates of the actual external costs associated with the emission of a unit of each identifiable pollutant is required for pollution tax to work efficiently. As observable from the above discussion, the actual costs of the damage associated with particular pollutants are uncertain and, hence, difficult to quantify in monetary terms. Besides, construction and property activities emit several different pollutant types – some may be easily detected, many may still remain unknown for a considerable length of time. The danger with this is that, the tax imposed could also be in excess of the actual external costs associated with the activities. Where it is not possible to pass such taxes onto the consumer, this may lead to a reduction in outputs. For example, the construction sector relies heavily on fossil fuel to generate energy. Yet fossil fuel, as pointed out earlier, is the prime cause of carbon dioxide emission. Thus, a carbon tax that cannot be passed on to the consumer will necessarily lead to a reduction in the use of fossil fuel leading to a reduction in energy use and, hence, outputs until alternative sources of energy are found.

18.16   Cap-and-trade

The cap-and-trade instrument sets an upper limit or an allowance on the total emission of a given pollutant that an entity is allowed to emit in a given time period. Each allowance entitles companies to emit or use one ton of carbon dioxide (CBO, 2008:vii) or any other pollutant to which it applies. The emission allocations are couched in the form of transferable rights.

This means that, once an allowance is issued to a given entity that entity is free to sell it to another for profits. Likewise, other entities are free to buy more allowances from those that are willing to sell theirs. The cap-and-trade approach is used in the United States to reduce emissions that cause acid rain and is currently being used in the European Union to limit CO_2. This approach does not bring the costs of emission directly within the private costs of the responsible entity and, hence, unless the costs of emitting beyond the allowable limit are very high, most entities are unlikely to keep within their allowable limits. There are also monitoring costs that have to be borne by the regulator. This monetary cost adds to the external costs and could end up causing more financial harm to society.

18.17   Coasian contracts

In 1960, Ronald Coase questioned the efficiency of the Pigovian pricing as a way of balancing the costs and benefits of externalities. Coase argued that under given circumstances, parties can effect a fair distribution of the external costs and benefits of any given business activities provided the parties are willing and able to engage in voluntary negotiation and trade without government interventions. This will happen if the rights of the parties not to be harmed are well defined and costs of transactions are also relatively low. Where these conditions are met, parties affected by externalities will be able to negotiate with those causing the harm on the appropriate compensation package. Failing to reach agreement, those affected will be able to seek the authority of the courts to enforce their rights through actions in negligence and to claim damages. Anderson (2004:448) provides a useful analogy:

> *Suppose there is an apartment building with two apartments. In one apartment lives a person who enjoys music and values louder and louder music (more decibels) at a declining marginal rate . . . In other words, additional decibels provide more value to the music lover, but the marginal value of decibels declines until it reaches zero at the maximum number of decibels that can be produced by his equipment. In the other apartment lives a person who values quiet such that fewer decibels of noise are worth more with the marginal value of quiet declining until it reaches zero with no noise. There is an optimal level of noise at the point where the marginal value of a decibel to the music lover is just equal to the marginal value of a decibel of quiet to the quiet lover . . . Consider a case where there are no rules regarding noise in the apartment building and where the quiet lover moves in first. When the music lover moves in and turns his stereo up to full volume, the quiet lover will clearly have reduced value of quiet. He is likely to respond by knocking on the door of the music lover asserting a first possession right to be free of noise . . . Assuming that he can defend this right both morally and legally and sell it, the costs are fully accounted for when the music lover compensates the quiet lover for the costs he bears or ceases producing music . . . If the quiet lover cannot defend his right to quiet, there will be too much noise because the music lover is not bearing the cost of lost quiet. [If the music lover moves in first] . . . If the quiet lover could force the music lover to reduce the volume without compensation; that is, the music lover cannot defend his rights, there will be too much quiet because the quiet lover is not bearing the cost of reduced decibels.*

The practicality of Coasian contracts in the context of construction and property undertakings is doubtful. This may involve construction and property enterprises entering into contracts with countless hosts of affected individuals and institutions at considerable costs. Will they actually do that? Will they be that socially responsible?

In September 1970, Milton Friedman, an eminent economist, published an article in New York Times Magazine in which he argues that business enterprises have no other responsibility than the direct responsibility they have towards their employers. That is 'to conduct the business in accordance with their desires, which generally will be to make as much money as possible while conforming to the basic rules of the society, both those embodied in law and those embodied in ethical custom'. 'Insofar as his actions accord with his "social responsibility" reduces returns to stockholders, he is spending their money. Insofar as his actions raise the price to customers, he is spending the customers' money. Insofar as his actions lower the wages of some employees, he is spending their money'. Thus, unless it is imposed on them, private enterprises are unlikely to entertain such contracts and the externality they create may remain unresolved.

18.18   Sustainable strategies for the environment as a stakeholder

European and national government interventions to manage the environment as a stakeholder provide guidance and directives for construction and property businesses to address environmental considerations as part of their activities – to balance environment, economic and social perspectives for sustainable futures. This has occurred because of increased awareness of global environmental problems.

Sustainable development is the most notable environmental strategy for future construction and property enterprises. The first major worldwide conference on environmental issues was convened by the United Nations (1972) in Stockholm, Sweden. This highlighted the world as a place of limited natural resources with a finite capacity to support human life and, as a consequence, the UNEP was created and, moreover, many governments set up national ministries and agencies for the environment. Subsequently, the publication of the Brundtland report (1987), 'Our Common Future' by the World Commission on Environment and Development (WCED), drew international attention to the acute environmental issues facing the global community. This laid the groundwork for the United Nations 'Earth Summit' conference (1992) in Rio de Janerio, Brazil. Its greatest achievement was that the summit resulted in the production of Agenda 21, which is a global partnership that addresses the problems of today and concomitantly prepares the world for the challenges of the future.

Agenda 21 implementation reflects a global consensus and political commitment on development and environment cooperation. In essence, it addresses the development of societies and economies by focusing on conservation and preservation of our environments and natural resources. However, the success of Agenda 21 is the responsibility of Governments and Local Authorities. That said, sustainable development has become an established research discipline and, moreover, as a major consideration in the shaping of government and corporate policy. National strategies, plans, policies and processes are crucial in its achievement. In the United Kingdom, for instance, sustainability is increasingly being applied to all areas of construction. In fact, the government has now introduced a range of fiscal incentives aimed at encouraging the take-up of sustainable design and construction measures (e.g. landfill tax, aggregates levy, land use incentives, renewable grants scheme and the energy efficiency commitment scheme).

In recent years, to promote the adoption of more sustainable construction practices and for these to be supported through the planning system, a wealth of government reports has been published. These include: 'Building a better quality of life: a strategy for more sustainable development in the UK', which demonstrates the strategy for policy development in the United Kingdom, and the subsequent 'Building a better quality of life: a strategy for more sustainable construction', which established key themes for action by the construction industry. Other similar themed works include: 'The construction industry: progress towards more sustainable construction'; 'Better building summit – issues paper'; 'Sustainable construction brief'; 'Demonstrations of sustainability' and 'Achieving sustainability in construction procurement: sustainability action plan'. Details can be accessed from http://www.berr.gov.uk/.

The most noteworthy report is 'The code for sustainable homes', which was introduced (2007) to improve the overall sustainability of new homes. In an era, where the general public is more environmentally conscious, there is a growing appetite amongst consumers for more sustainable products and services. Therefore, the code offers a framework within which the home building industry can design and construct homes to higher standards and, at the same time, offer a tool for developers to differentiate themselves from their competitors. The code measures the sustainability of a home against nine design categories: (i) energy and carbon dioxide emissions; (ii) water; (iii) materials; (iv) surface water run-off; (v) waste; (vi) pollution; (vii) health and well being; (viii) management and (ix) ecology.

18.19   Conclusion

Nowadays, environmental protection and sustainable futures are no longer things that only dreadlock tree-huggers worry about. People all over the country are now considering their own energy–resource waste footprints on the environment. However, that said, the environment is an often-overlooked stakeholder of construction and property business enterprises. Managing the environment as a stakeholder entails among others the fair distribution of the benefits and costs associated with the undertaking. The environmental damages caused by the activities of construction and property firms represent the external costs component of their activities. Profit motives lead managers of these enterprises to avoid paying for these costs. They will only pay monetary equivalent for costs that fall within their direct private costs. Thus to foster fair distribution of the gains and burdens of construction and property enterprises, a way has to be found to bring the external costs within their private costs. Pollution taxation and subsidies are the most effective instruments for realising this end.

Endnotes

1. http://www.rotman.utoronto.ca/~stake/principles.htm.
2. See Cambridge Advance Learner's Dictionary (2003), Version 1.0, Cambridge University Press.
3. For detailed description of these components of the environment see (UNEP), U. N. E. P. (2007) *Global Environment Outlook: GEO4. Environment for Development*. Nairobi: United Nations Environment Programme.
4. Climate change is defined by the IPCC as 'a change in the state of the climate that can be identified (e.g. using statistical tests) by changes in the mean and/or the variability of its properties, and that persists for an extended period, typically decades or longer. It refers to any change in climate over time, whether due to natural variability or as a result of human activity. This usage differs from that in the United Nations Framework Convention on Climate Change (UNFCCC), where climate change refers to a change of climate that is attributed directly or indirectly to human activity that alters the composition of the global atmosphere and that is in addition to natural climate variability observed over comparable time periods'.
5. According to the Kyoto Protocol to the United Nations Framework convention on climate change (1998) defined green house gases to include carbon dioxide (CO_2), methane (CH_4), nitrous oxide (N_2O), hydrofluorocarbons (HFCs), perfluorocarbons (PFCs), sulphur hexafluoride (SF_6).

6. It is worth noting that the scientific evidence that the recent significant rise in temperature is caused by human activities is not yet incontrovertible. (CBO), C. B. O. (2005) *Uncertainty in Analysing Climate Change: Policy Implications*. Washington: The Congress of the United States. Other eminent scientists have questioned the science behind this which was most forthrightly conclusion drawn mainly by the intergovernmental Panel on Climate Change. However, given potential costs that society will bear for not taking any action if it turns out that the IPCC conclusion is valid, the need for caution and risk hedging demands that, at least for now, from policy making perspective, IPCC conclusion must form the basis of climate based actions, until they are conclusively overturned. It must not be forgotten however that, if the IPPC conclusions are found to be incorrect, then it will beg the question whether the considerable resources devoted to addressing climate change by both governmental bodies and individuals could not have been channelled into more productive ventures.

7. See CIRIA (1993) Environmental issues in construction: A review of issues and initiatives relevant to the building, construction and related industries, Vol. 2 – Technical Review

References

Anderson, T.L. (2004) Donning coase-coloured glasses: a property rights view of natural resource economics. *Australian Journal of Agricultural & Resource Economics*, **48**(3):445–462.

Barnard, C.I. (1938) *The Functions of the Executive*. New York: Basic Books.

Carroll, A. (1979) A three-dimensional conceptual model of corporate performance. *Academy of Management Review*, **4**(4), 497–505.

CBO (2005) *Uncertainty in Analysing Climate Change: Policy Implications*. Washington: The Congress of the United States.

CIRIA (1993) *Environmental Issues in Construction: A Review of Issues and Initiatives Relevant to the Building, Construction and Related Industries*. London: Construction Industry Research and Information Association.

Clarke, T. and Clegg, S. (1998) *Changing Paradigms: The Transformation of Management Knowledge for the 21st Century*. London: Harper Collins.

Clarkson, M.B.E. (1995) A stakeholder framework for analysing and evaluating corporate social performance. *Academy of Management Review*, **20**(1):92–117.

Clarkson, M.B.E. (1998) Corporate social performance in Canada, 1976–1986. In: Preston, L.E. (ed. *Research on Corporate Social Performance and Policy*. Greenwich: JAI Press.

Cochran, P. (1994) The Toronto conference: Reflections on stakeholder theory. *Business and Society*, **33**(1):95–98.

CBO (Congressional Budget Office) (2008) *Containing the Cost of a Cap-and-Trade Program for Carbon Dioxide Emissions – Testimony/Statement of Peter R. Orszag*. Washington: The Congress of the United States.

Donaldson, T. and Preston, L.E. (1995) The stakeholder theory of the corporation: Concepts, evidence and implications. *Academy of Management Review*, **20**(1):65–91.

EA (2008) *Sustainable Construction*. Environment Agency. [Online] Available at http://www.environment-agency.gov.uk/business/sectors/32705.aspx (Accessed on 3 October on 2008)

Evans, W.M. and Freeman, R.E. (1988) A stakeholder theory of the modern corporation: Kantian capitalism. In: Beaucham, T. and Bowie, N. (eds) *Ethical Theory and Business*. NJ: Prentice Hall.

Freeman, R.E. (1984) *Strategic Management: A Stakeholder Approach*. Boston: Pittman Books Limited.

IMF (2008a) *The Fiscal Implications of Climate Change*. Washington: An IMF report. [Online] Available at: http://www.imf.org/external/np/pp/eng/2008/022208.pdf (Accessed on 7 October 2008).

IMF (2008b) *World Economic Outlook: Housing and the Business Cycle.* Washington: IMF. [Online] Available at http://www.imf.org/external/pubs/ft/weo/2008/01/ (Accessed on 15 October 2008).

Jones, T.M. and Wicks, A.C. (1999) Convergent stakeholder theory. *Academy of Management Review,* **24**(2):206–221.

IPCC (2007a) Summary for policymakers. In: Solomon, S.D., Qin, M., Manning, Z., Chen, M., Marquis, K.B., Averyt, M., Tignor and Miller, H.L. (eds) *Climate Change 2007: The Physical Science Basis. Contribution of Working Group I to the Fourth Assessment Report of the Intergovernmental Panel on Climate Change.* Cambridge: Cambridge University Press.

IPCC (2007b) *Climate Change 2007: Mitigation. Contribution of Working Group III to the Fourth Assessment Report of the Intergovernmental Panel on Climate Change.* Cambridge: United Kingdom, Cambridge University Press.

IPCC (2007c) *Climate Change 2007: Synthesis Report. Contribution of Working Groups I, II and III to the Fourth Assessment Report of the Intergovernmental Panel on Climate Change.* Cambridge: Cambridge University Press.

Jonker, J. and Foster, D. (2002) Stakeholder excellence? Framing the evolution and complexity of a stakeholder perspective of the firm. *Corporate Social Responsibility and Environmental Management,* **9**(4):187–195.

Kolk, A. and Pinkse, J. (2006) Stakeholder mismanagement and corporate social responsibility crises. *European Management Journal,* **24**(1):59–72.

Logsdon, J.M. (1994) The Toronto Conference: reflections on stakeholder theory: Essay by Jeanne M. Logsdon. *Business and Society,* **33**(1):124–128.

March, J. and Simon, H. (1958) *Organisations.* New York, Wiley.

Meehl, G.A., Stocker, T.F., Collins, W.D. *et al.* (2007) Global climate projections. In: Solomon, S., Qin, D., Manning, M., Chen, Z., Marquis, M., Averyt, K.B., Tignor, M. and Miller, H.L. (eds) *Climate Change 2007: The Physical Science Basis. Contribution of Working Group I to the Fourth Assessment Report of the Intergovernmental Panel on Climate Change.* Cambridge: Cambridge University Press.

Mitchell, R.K., Agle, B.R. and Wood, D.J. (1997) Toward a theory of stakeholder identification and salience: Defining the principle of who and what really counts. *Academy of Management Review,* **22**(4):853–886.

Osman, C. and David, S. (2007) *Fossil Fuels.* Ann Arbor: University of Michigan.

Pigou, A.C. (1962) *The Economic of Welfare* (4th edn.). London: Macmillan & Co. Ltd.

Savage, G.T., Nix, C., Whitehead, C. and Blair, J. (1991) Strategies for assessing and managing organizational stakeholders. *Academy of Management Executive,* **5**(2):61–75.

Seager, A. (2007) Construction sector rises to challenge of building eco-friendly homes of the future. The Guardian, 28 February 2007 [Online].

Shankman, N. (1999) Reframing the debate between agency and stakeholder theories of the firm. *Journal of Business Ethics,* **19**(4): 319–334.

Smith, A. (1776) *An Inquiry into the Nature and Causes of The Wealth of Nations.* London: Methuen & Co., Ltd.

Starik, M. (1994) The Toronto conference: Reflections on stakeholder theory. *Business and Society,* **33**(1):82–131.

Stern, N. (2007) *The Economics of Climate Change: The Stern Review.* London: HM Treasury.

UNEP (2007) *Global Environment Outlook: GEO4. Environment for Development.* Nairobi: United Nations Environment Programme.

UNCHE (1972) *Report of the United Nations Conference on the Human Environment. –* Held in Stockholm, 5–16 June 1972. New York: United Nations.

19 Using Change Management to Support Stakeholder Management

Nidhi Shah and Philip T. Harris

19.1 Introduction

The stakes that an organisation contends with are often multivariate and incongruent. More so, responding to stakeholders can sometimes warrant change within an organisation, e.g. change in procedures, change in products, and so on. The need to change can be driven by external factors such as new legislation or increased competition and by internal factors such as the implementation of new technologies (Price and Chahal, 2006). The need for an organisation to reposition itself due to competition and other considerations can also induce change. Stakeholders are one of the sources that can trigger change. Thus, a stakeholder-conscious organisation should be ready to change to some degree at some point in time.

This chapter discusses the concept of change and how to implement it effectively as an essential feature of organisational practice. It proffers an action plan for change management in organisations and covers the assumptions and inconsistencies of the concept as well as highlighting its inherent paradoxes. Before these 10 steps are discussed, change and its management are explained.

19.1.1 The concept of change

Organisational change is not an end in itself. It is only a means of adjusting to new conditions and sustaining or increasing competitiveness, performance and effectiveness (Kubr, 2002:90). This explanation represents most of the definitions of 'change'. Change comes in different magnitudes and types. Change can range from the simple to the drastic, from the operational to the strategic type, etc. An example of change can be a relocation of a function, department or business. Change can also come in the form of a different service, product, management style, etc.

Change seems to stem largely from 'rapid globalisation', 'liberalisation', 'growth of technology' and 'competition'. While the rapid rate of technological development has led to an increase in the pace of change, globalisation and deregulation have also given birth to increased competition. To survive these turmoils, organisations need to continuously strive for developing and maintaining a creative and innovative edge over its competitors. With the ever-increasing pace of social, technological and political changes, there is the need to maintain a relative stability and continuity in organisations, i.e. stability in terms of workforce and shared values and change in terms of structure, workflow, productivity, etc. One of the vital tasks of management is to be able to strike this balance between desired change and stability in an organisation to ensure organisational success (Markham, 1991; Morgan, 1997; Sadler, 2001; Kubr, 2002).

People may sometimes not be susceptible to change, at least not quickly. So organisations should prepare for change by adopting a management programme that will be effective. The discussion of this chapter provides a platform for implementing a successful change programme.

19.2   Change management

Organisational change management is concerned with facilitating the process of change through modification of strategies, structures and processes (Jimmieson *et al.*, 2008).

> *Change Management is the process of developing a planned approach to change in an organisation. Change Management can either be reactive, in which case management is responding to changes in the macro environment (that is, the source of the change is external), or proactive, in which case management is initiating the change in order to achieve a desired goal (that is, the source of the change is internal) . . . Change management can be approached from a number of angles and applied to numerous organizational processes*

> (Paton and McCalman, 2000:22)

Change management, today, is as an essential task of the top management in any organisation. It has become relevant of late (since early 20th century) that this issue has been identified and dissected by managers, consultants and business academics and is an imperative for organisational success (Clemmer, 2006). If managing change is an essential task, then why was it not so three decades ago?

The approach to change depends upon the need for change in organisations. Different needs warrant different approaches. Typically the objective in change management is to minimise disruptions to the normal flow of activity (Paton and McCalman, 2000). Literature exhorts those who have been successful at changing their organisations.

The change-process seems like a magical elixir for solving organisational bottlenecks. This elixir is sold to organisations, enticing its management to buy it, especially as organisational practices have become very fluid in the last two decades. Consequently, in the frenetic search for organisational success and survival in 'today's rapidly changing environment' managers have succumbed to its *Midas touch*.

If organisations do have an easy recipe for instant success by adopting transformation processes then why do studies on the success of such initiatives show that almost two out of every three tend to fail? (Stewart, 1993). If there is a high failure rate in bringing about change, then change should not be taken lightly at all. We have charted a process-map[1] for change management (Figure 19.1). This process-map provides a framework for implementing change in an organisation. The attributes shown in this process-map are discussed in the book with the aim of enabling more effective change management in organisations.

19.2.1   A process-map for understanding change and its management

The process of change management in an organisation is triggered by changes in the external environment. Often, the 'external environment' will affect the operations of organisations either directly or indirectly.

Once the need for change is identified by the top management of an organisation, it appoints a transition team to draft a plan of action including the identification of the exact 'change points' in the organisation. This team usually comprises members from top

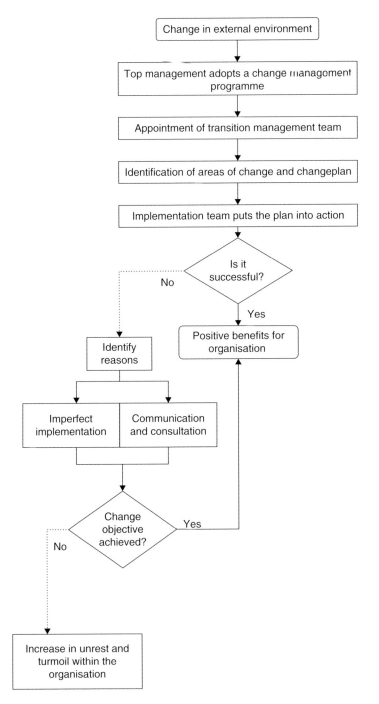

Figure 19.1   Understanding change management.

management and/or management consultants. Once the plan is ready, an implementation team comprising the front-level managers is appointed to execute the plan. On the basis of the response to the execution plan, the success or failure of the change programme is determined.

Success in change management is measured in terms of pre-determined criteria which are chalked out by the planning team. If a transformation plan fails, imperfect implementation and resistance to change are usually identified as the reasons for it. If an organisation feels its poor leadership was responsible for the failure, then more effective leaders are appointed to execute the plan yet again. On the other hand, if an organisation thinks that by encouraging participation by its employees and/or advice from a management consultant will solve the problem, then this step is adopted.

Post-implementation of solutions just discussed, the firm determines whether its change objectives have been achieved or not. If they are achieved then the organisation is able to reap the full benefits of a successful change programme. However, if the attempt to effect change is unsuccessful, then the turmoil in the organisation will increase. The quest of an organisation is to ensure that an attempt to change is successful.

19.3 A management plan for introducing change in organisations

An organisation should approach change in a careful and planned manner, given that success is not always guaranteed. The concept should be thought-through from beginning to end before it is embarked upon. In this chapter we discuss 10 aspects which can be considered while implementing change in an organisation. These 10 aspects are as follows:

1. Recognising change in the external environment
2. Change management in organisations
3. Adopting a change management strategy
4. Implementing a change strategy
5. Coping with success and failure
6. Dealing with imperfections
7. Dealing with resistance to change
8. Leadership
9. Communication and culture
10. Using consultation

What follows is a discussion of each of these aspects. Rhetorical questions are used occasionally to clarify meaning.

19.3.1 Recognising change in external environment

Repeated discussion on an ever-changing environment has given the concept of change a taken-for-granted outlook and this is why it is regarded today as one akin to a universal truth. Grey (2003:1) goes a step further in his criticism of this worldwide ranting of need for change in managing organisations by describing it as 'a totem before which we must prostrate ourselves and in the face of which we are powerless'. In support of Greys' (2003) description of change as a convenient explicit or implicit justification for specific change programmes in organisations, the whole notion of change has been given a threatening outlook, leading to the need for change management.

Grey (2003) notes that the importance of a changing environment has gained momentum in the last three decades, i.e. the time-span people refer to as 'today's times'. This notion carries with it the assumption that there was a relatively stable environment in the past (pre-1960s). So can it be implied that organisations were not threatened by change before the last thirty years? Evidence from history shows otherwise.

Mankind has been exposed to changes beginning from pre-historic Stone Age through the Age of Human Civilisation. The world has witnessed the rise and fall of the Roman Empire, the Renaissance, the Reformation, the Enlightenment, the Industrial Revolution, the World Wars, shifts from religious to secular conceptions of the world, the post war periods (1945 until the oil crisis in 1974), the invention of computers and its growth, the space race, the Cold War, de-colonisation, the Arms Race, the Korean and Vietnam wars, etc. These developments hardly constitute a stable environment (*Human Civilisation: Modern culture*, www.ecotao.com/holism). Trade has been carried along all these changing times. If organisations have been able to cope in the past despite the changes in the external environment, then why do we need specialists to teach organisations how to manage change, and the need for change management?

Globalisation has been cited as the most common trigger for change in organisations today. However, evidence shows otherwise. A study by Hirst and Thompson (1996) on the increased economic globalisation by comparing ratio of trade to Gross Domestic Product (GDP) claimed that there has been a drop in the GDP of most industrialised countries in the 20th century. Hirst and Thompson (1996) asserted that there was more global integration during the Gold Standard period than in the 1980s and also that migration of people has reduced in the 20th century as compared to the 19th century. Their overall conclusion from their study was that the integration of the international economy has hardly improved since 1914.

Grey (2003) contented that the ambiguity of change dimensions and the identification of the exact group of people affected by such changes underlie the notion of change. The issue is how change is viewed, i.e. whether change is seen as a threat or as an opportunity for individual or organisational survival and success. The answer depends upon the readiness for change by each individual or organisation (Clemmer, 2006).

Changes in the external environment usually trigger the need for change in organisations. What follows is a discussion of this pre-cursor of change management programmes.

19.3.2 Change management in organisations

Clemmer (2006) describes change management as an oxymoron making as much sense as 'regularly irregular' or a 'mandatory option'. On the whole, change management is portrayed as a 'one size fits all' solution to management, based on the assumption that there is an orderly thinking and implementation process, i.e. a 'planned approach' to plot a course of action to counter changes in the external or internal environment of the organisation. But this contradicts the very nature of change because it is something which cannot be predicted accurately, so having a planned and systematic approach to combat an ambiguously foreseen change is doomed to meet adverse reactions. Moreover, the nature of change confronted by an organisation will differ with time and situations thereby requiring a new plan to meet the change. This may lead to numerous transformation programmes in organisations. Ironically, continuous/regular transformation efforts might not add to the capabilities of organisations to adapt to change. It is an objective approach to a subjective concept, which is the inherent paradox of the change management concept.

In the course of drawing up relatively elaborate change programmes, managers often undermine the role of employees in the process.

Despite the numerous reasons for change (external, internal or proactive), the common theme is that it requires changes in the organisational design, i.e. alterations in the culture and structure of organisations, and recent trends of flattening organisation structures, incorporating flexible procedures and responsive change mechanisms, characterise practices of 'post bureaucratisation' (Heckscher, 1994). However, it is the people who form these structures, and run them as well, so in effect they expect a change in peoples' behaviours and attitudes. Managers specify the kind of response they think would be suitable for the effective implementation of the transformation process.

19.3.3   Adopting a change management strategy

Adoption of a strategy, appointment of a transformation team, plotting change objectives and deciding the plan of action is done under the aegis of the top management. Once the plan is charted out, the task of implementing it is handed over to another team comprising of middle and lower-level managers. The underpinning notion is one of a hierarchal approach to change management. The organisations' need for change is basically the top-managements' need for it. As Clemmer (2006:12) explains: 'they (rigid leaders) aren't comfortable with letting their improvement plan and path to higher performance unfold and evolve toward their vision, values, purposes, goals and priorities … they think they can start with answers'.

Moreover, senior executives and/or external consultants do not usually oversee the implementation of the strategies and sometimes even detailed discussion on the rationale behind the change programme. This is evident from the definition of a consultants' role by Markham,

> as we say in our business, we can teach a client to fish and we can be a fishing guide who goes along on the fishing expedition, but we clearly are not the ones to do the fishing . . . that must be left up to the managers in the company that desires change (1991:32).

If consultants are not aiding the implementation of the plan they drew up, then there are bound to be areas which will be misinterpreted, adding to the confusion created by the announcement of change programmes in the organisation. The following paragraphs discuss the underlying assumptions of implementation programmes which in turn lead to either the success or failure of the programme.

19.3.4   Managing change/implementation of change strategy

According to Paton and McCalman (2000) management can follow either a participative or imposed approach to incorporate a transformation plan. Even though an organisation claims to follow a participative approach, the need for change can be taken-for-granted and imposed upon the employees by the top management. The employees' participation can be restricted to negotiating with the planning team regarding the utilisation of resources and methods to be used. Top management must not be seen to be paying lip service to this concept.

Can change can be managed at all? The Kurt Lewins Model suggests this is possible with the premise that there is a balance of forces for maintaining equilibrium in an organisation (Paton and McCalman, 2000). It is the management's responsibility to increase the forces for change or decrease the forces against change. Hence, change management consists of three phases: (1) unfreezing to adapt to change, (2) making the change and (3) refreezing to resume routine tasks. The underlying assumption is that an organisation can be moulded into different forms like a block of ice. The organisation is compared to a machine in which there are forces which stabilise or destabilise it.

The phenomena of freezing and unfreezing particles of water can take place under required temperatures and can be replicated countless number of times; however, this concept cannot be applied to human beings and organisations because neither reactions nor a particular situation can be replicated in different organisations. It is not easy for employees to abandon (unfreeze) their current working styles and attitudes towards their work, adopt a new approach (change) and then treat the new process(es) as if they were the ones they have been following all along (refreeze) to carry out their daily tasks. The inherent flaw with this approach lies in the issue of generalisabilty and mechanistic approach towards the employees (Grey, 2003; Dawson and Jones, 2005).

However while talking about an organisation, we are referring to the people who work there and every individual is unique by his/her personality which makes them unpredictable and cannot be subjected to generalisations as identical particles of water. Moreover, one of the most difficult and time-consuming changes in an organisation is the one related to altering the existing culture. The validity of a top-down approach to cultural change has been questioned widely (Smircich, 1983). It seems that some humans have an intrinsic nature to resist change. Hence, there is a need to recognise this resistance and to draw up plans for overcoming it.

19.3.5   Success or failure of a change programme

An organisation is likely to succeed with its change programme when prior planning is involved (Price and Chahal, 2006). Success is achieved when something planned, desired or attempted for is obtained (American Heritage Dictionary of the English Language, 2000). In this light, the desired outcomes for organisational transformation programmes vary depending upon the type of change required. For some organisations it may include such variables as achieving financial targets, achieving key performance improvement targets, the perception of acceptance by the workforce, establishing grounds for sustaining the change in the organisation, establishing an effective two-way communication channel and ability to 'walk the talk'(Found *et al.*, 2005).

What might have proved successful for one firm might have intensified the work conditions in another instead of solving the issue at hand. This issue of generalisation crops up once again. The ambiguity regarding the concept of success and what it means to different organisations raises the question of the validity regarding change and its management and prescribing ways to 'successfully' deal with it. The definition of a successful change programme is subjective, depending upon the type of change desired. This creates an ambiguity around the terms of success and failure related to change initiatives.

Studies indicate that nearly two-thirds of all change programmes fail and more often than not, an imperfect implementation of the change plan due to the resistance to change is often cited as reason for failure (Stewart, 1993). The following aspects elaborate on this further.

19.3.6   Imperfect implementation

Imperfect implementation is probably the most commonly cited reason for the failure of an organisational change programme. However, managers do not imply an acceptance of the inherent paradoxical assumptions associated with the generalisation of change management. In fact what it does mean is that social and organisational relations can be controlled and it is just a matter of doing things correctly. Moreover, there is no basis to evaluate what would have happened had 'such-and-such' not been done. Imperfect implementation has been the staple defence for failure right through management history. This can be seen when Taylor ascribed the failure of his scientific approach to management to inability to fully implement the approach. Recent examples are those related to the failure of total quality management and business process re-engineering in organisations (Dawson and Jones, 2005).

Despite the high rate of failure, management literature describing how and why organisations should undertake change management abounds. Why is so much effort and energy spent on an endeavour that is still susceptible to failure? Having theorised for decades on change management, why is there no single approach for implementing a change programme perfectly? Understandably, situations and contexts cannot be replicated in organisations and the success of an initiative depends upon the time and place of its implementation. However, why do change management gurus continue to stress on the generalisabilty of change management approaches? These are issues which question the very feasibility of projecting change management as an essential task for 'managers of today'.

19.3.7 Resistance to change

Evidence from literature indicates that although change is imperative, when people are confronted with it, they usually resist (Pendlebury *et al.*, 1998; Paton and McCalman, 2000; Senior, 2002). In fact, studies show that nearly two in every three change initiatives in large-scale corporations tend to fail (Stewart, 1993). The reason being that employees resist change, which in turn impedes on its implementation.

The proponents of change management identify the fear of failure, lack of respect for and trust in the change-agent(s), tendency to avoid emotional turmoil arising out of change initiatives and the dislike of imposed change as the primary reasons for resistance to change (Paton and McCalman, 2000). It is a natural tendency for humans to resist change if they feel their comfort is being threatened. This is explained by evolutionary psychologists that humans have an aversion for loss when they feel their positions are comfortable. This can be attributed to the fact that since the dawn of civilisation, man has suffered hardships in his quest for survival, his primary motive. So whenever he felt that he was in a place where he has an assured form of sustenance, he avoided wandering off in search of new land and when his territory was invaded by others, he defended it violently. The resistance to change can be analysed from this point of view. This will help managers to understand how their employees and even they themselves think and feel about change. The trait of personal defence has been passed down through generations and can be seen in the organisational context.

Nichelson (1998) suggests that when faced with resistance, managers should take it as a normal occurrence instead of closing-in on the concerned employees. Space, safety and support can enhance the generation of creative thoughts and actions from employees, consequently acting as stimuli for change and an employee-driven need for change. Such an approach is often not considered by most managers who desire to have a transformation in their organisation. Moreover, managers have a tendency to generalise the resistance to change into a simple five-stage process comprising

1. *Denial* for change,
2. *Defence* from change, i.e. trying to avoid it,
3. Employees begin to *discard* their old ways,
4. Employees start *adapting* to the new system and finally
5. the employees *internalise* the new system, making it their routine.

Here we see a glimpse of the tendency of change-agents to generalise the transformation process so as to incorporate it into simple and straightforward models which will occur in all situations.

People exist within a set of meaning, i.e. structures which influence their thoughts and opinions about their experiences. These constructs are open-ended and the consequences of their alterations are unpredictable. Organisations to which they belong form one of these constructs (Grey, 2003). This observation tends to question the validity of predicting the behaviour of the workforce through a step-by-step process because the effect of alterations to their meaning-structures cannot be predicted accurately. For example, one day, if the manager of an organisation announces that there will be a 25% rise in pay for everyone, it is highly unlikely that the change will be met with a denial-defence-discard-adapt-internalise cycle; thereby contradicting the aforementioned generalisation. Herein we see the importance of the nature of change and its context, which many change-models fail to incorporate, consequently portraying change management as a simple step-by-step planned approach.

People are often claimed to be the most important asset of an organisation; however, few behave as if this were true. This irony is also highlighted by the relegation of reactions to change as something which needs to be quashed down at all costs. No wonder nearly 50% of transformation programmes are deemed as failures (Found *et al.*, 2005). Leadership, improvement in communication and use of management consultants are often adopted for managing the implementation of change and resistance to the process.

19.3.8 Leadership

Similar to the ubiquitous identification of resistance to change as the reason for unsuccessful change programmes, leadership is cited as the means through which resistance can be controlled and overcome. What needs clarification is what is meant by leadership? For some it might mean an autocratic approach for imposing change on the workforce, yet other organisations might want a charismatic and dynamic leader who can influence the employees to win their commitment towards a change programme (Robbins and Finley, 1998).

The underlying notion is that of using one's power and authority in order to regulate random and chaotic events in the organisation. According to Mintzberg (1998) the issue of power and politics can hardly be undermined during the process of strategy formulation (in this case, change strategy) especially in times of huge change initiatives, when significant shifts in power relationships occur, consequently leading to conflicts in organisations which can sometimes lead to the ultimate failure or abandonment of the change programme. For example, senior executives play strategic candidates games[2] to promote their desired changes and on the other hand, the workforce may resort to mass movements[3] ranging from rebellion to protest to express their discontent about the change programmes. Such political conflicts cause undercurrents of negative emotions which can be detrimental to organisations. Instead of recognising the powerful impact of resistance to change, managers continue to emphasise the need for influential leadership to effect change management thereby highlighting the top-down approach of management towards transformation initiatives.

It is opined that leaders are often as resistant to change and prone to backsliding as their employees. Michael Eisner, CEO of Walt Disney is a most recent example of the phenomenon. Eisner was forced to rope in a high-profile number two executive Michael Ovitz to share his responsibilities as the CEO. However, Eisner failed to grasp the crux of this change in his duties, consequently refusing to share his power with Ovitz from the very beginning (Deutschman, 2005). This shows that top managers can pay lip service to change management programme.

19.3.9 Communication and culture

Communication is vital if a change is going to be successful (Price and Chahal, 2006). Communication with employees can eliminate their resistance and avoid confusion too.

The assumption behind the use of communication (i.e. keeping employees updated with the need for change) and consultation is that this approach for effecting change will be more successful than an imposed leader-driven one.

In each organisation there are corporate and personal cultures and these will impact on how the organisation operates. Apparently, the organisations' culture is the main determinant of whether change will be adopted while the extent of participation of the individuals will depend on the extent of power they hold in the eyes of the management. In this regard, a strong trade union will be given a more detailed hearing in the consultation process.

In order to make a change objective successful, employees should not be made to take for granted the need for change. They should rather be made to be in agreement with the top management's agenda for their negotiations with the consultant or the planning team (Freidman, 1977).

19.3.10   Consultation

Sometimes, an outside consulting organisation can be used to bring about change, if for instance, current competency is deemed inadequate. Consultants can be used where their added expertise is needed for a short time, wherein short-term consultation would prevail over long-term hiring.

Practice shows that the outcome with the use of consultants can be less than expected. According to Smid *et al.* (2006), the hiring of consultants can create a feeling of over confidence and complete handing-off to these consultants. The advice is not to limit your role and contribution despite the presence of consultants.

The role of management consultants as change-agents has been criticised as a 'cynical charade' where they are often blamed for creating a crisis out of a given situation in order to create a demand for their services, albeit, change management. Further, consultants are often blamed for resorting to management fads, overuse of buzzwords and adopting management plans which are difficult to execute but sound nice when explained to the management. These mismatches are more harmful than helpful to organisations (Burrello, 2006).

19.4   Conclusion

No doubt the world is changing and it will benefit an organisation if it is able to respond to the changes in its external and internal environment. The concept of change is easier said than done. Effort and resources must be devoted to a change programme to enhance success. Co-operation from the workforce and adoption of an implementable change programme can be identified as keys to success (Paton and McCalman, 2000). For example, corporate giants such as Toyota and Mercedes Benz constantly engage in numerous change endeavours and some of them have been successful.

The undermining of the workforce as human assets of the organisation has led to almost 70% failure rates in transformation programmes. This highlights the need for adopting human resource (HR) development as a strategic function of the top management, whether this is done through an HR department or a single person responsible for looking after the well-being of its employees. Ironically HR plays a less optimal role in change management.

Most organisational change endeavours have not been completely successful and this is a growing problem in the business world for which no solution has yet been found.

Claims about change and change management have massive rhetorical underpinnings. Practitioners of change management continue to relegate the transformation process to simple step-by-step procedures aimed at changing employees' attitudes and behaviours.

Our understanding of the myriad reactions to change management can be explained by Sir Isaac Newton's (1666 in Beiser 2003:25) Third Law of Motion:

Any object at rest tends to remain at rest. An object in motion tends to remain in motion. Every action has an equal and opposite reaction. Force equals mass times acceleration.

In line with the above observation, if employees feel comfortable in their environment, they usually tend to resist change in it. But if the change will benefit them there will be a large-scale acceptance to it which will in turn lead to improved job satisfaction and organisational performance. On the contrary, if the proposed change tends to disturb their equilibrium then it is met with equal amount of resistance, the counter-reaction to the proposed change and the total chaos created in the organisation out of this resistance will lead to negative performance. One failed initiative leads the organisation to adopt another one which ensures 'apparent success' – such an organisation will continue to be in a state of motion.

The basis that change management lies on the three-stage process of unfreezing-changing-refreezing is now obsolete. The unpredictability of human nature contradicts the unassuming reaction cycle explained in the model. The big question is: how can one plan an orderly approach to something which keeps changing all the time (albeit 'the constantly changing world')? The further question of the sustainability of the benefits from change initiatives is also one which is usually overlooked by managers in the quest of remaining abreast with the 'changing times'. To sum up in the words of Alphonse Karr (1849):

Plus ça change, plus c'est la meme chose. (The more things change, the more they stay the same.)

Endnotes

1. Process-map: Process-maps set out the correct sequence of activities and decision points within a process. It is usually used to illustrate internal departmental processes but it does not show who carries out the tasks or how they are carried out (Stafylarikis and Eldridge, 2005).
2. Strategic Candidate Games – played to effect change in an organisation; individuals/groups try to promote through political means their own favoured changes of a strategic nature (Mintzberg, 1998).
3. Mass Movements – range from protest to rebellion and are collective attempts to express grievances to promote/resist change (Mintzberg, 1998).

References

Beiser, A. (2003) *Applied Physics*. New York: McGraw-Hill Trade.

Burrello, N.K. (2006) The Consultant's Role, available at www.diversitydtg.com (Accessed on 15 March 2006).

Civilisation: Modern Culture, available online at www.ecotao.com.

Clemmer, J. (2006). Change Management is an Oxymoron, available at http://proquest.umi.com (Accessed on 15 March 2006).

Dawson, M. and Jones, M. (2005) Herding Cats: Human Change Management, (available online): www.pwcglobal.com/extweb/pwcpublications.nsf/. (Last accessed on 12 March 2006).

Deutschman, A. (2005) 'Change or Die', *Fast Company*, Issue 94, p. 53, available at http://www.fastcompany.com (Accessed on 15 March 2006).

Found, P., Beale, J., Sarmiento, R. and Francis, M. (2005). An exploratory study of the keys to successful and sustainable change management, Cardiff University, IMRC Working Paper for ICPR 2005, available at www.cuimrc.cf.ac.uk (Accessed on 16 March 2006).

Freidman, A. (1977) *Industry and Labour*. London: Macmillan.

Grey, C. (2003). Fetish of change. *Journal of Critical Postmodern Organization Science*, pp. 1–10.

Heckscher, C. (1994) Defining the post-bureaucratic type. In: Heckscher, C. and Donnelon, A. (eds) *The Post-Bureaucratic Organization: New Perspectives on Organizational Change*. Thousand Oaks, CA: Sage Publications.

Hirst, P. and Thompson, G. (1996) *Globalization in Question*. Cambridge, UK: Polity Press.

Jimmieson, N.L., Peach, M. and White, K.M. (2008) Utilizing the theory of planned behavior to inform change management. *Journal of Applied Behavioral Science*, **44**(2):237–262.Joseph, B.D. and Janda, R.D. (2003) *The Handbook of Historical Linguistics*. Oxford: Blackwell Publishing.

Karr (1849) *Les Guêpes*. In Joseph, B.D. and Janda, R.D. (eds) *The Handbook of Historical Linguistics*. Oxford: Blackwell Publishing.

Kubr, M. (2002) *Management Consulting: A Guide To the Profession*. Geneva: International Labour Office.

Markham, C. (1991) *Practical Management Consultancy*. London: Accountancy Books.

Mintzberg, H. (1998) *Strategy Safari: The Complete Guide through the Wilds of Strategic Management*. London: Prentice Hall Financial Times.

Morgan, G. (1997) *Imaginization: New Mindsets for Seeing, Organizing, and Managing*. London: Sage Publications.

Nichelson, N. (1998) How hard-wired is human behaviour. *Harvard Business Review*, Jul/Aug98, **76**, Issue 4.

Paton, A.R. and McCalman, J. (2000) *Change Management: A Guide to Effective Implementation*. London: Sage Publications.

Pendlebury, J., Grourard, B. and Meston, F. (1998) *The Ten Keys to Successful Change Management*. Chichester: John Wiley & Sons.

Price, A.D.F. and Chahal, K. (2006) A strategic framework for change management. *Construction Management and Economics*, **24**(March):237–251.

Robbins, H. and Finley, M. (1998) *Why Change Doesn't Work*. London: Orion Business.

Sadler, P. (2001) *Management Consultancy: A Handbook For Best Practice*. London: Kogan Page.

Senior, B. (2002) *Organisational Change*. London: Prentice Hall Financial Times.

Smid, G., Van Hout, E. and Burger, Y. (2006) Leadership in organisational change: rules for successful hiring in interim management. *Journal of Change Management*, **6**(1):35–51.

Smircich, L. (1983) Concepts of culture and organizational analysis. *Administrative Science Quarterly*, **28**:339–358.

Stafylarikis, M. and Eldridge, D. (2005), *Course Notes*, MA HRD and Consulting, Lancaster University.

Stewart, A. (1993) Reengineering: the hot new managing tool. *Fortune*, **128**(4):32–37.

The American Heritage Dictionary of the English Language (2000) Houghton Mifflin Company (available online): www.dictionary.com (Accessed on 12 March 2006).

20 Case Studies

Ezekiel Chinyio

20.1 Introduction

This chapter provides several case studies. Having discussed different aspects of stakeholder management hitherto, the present chapter is a supplement that supports the main discussions in the book. The case studies also demonstrate the practicality of the concepts of the subject matter. It must be pointed out that these case studies are not exhaustive and are not meant to cover every aspect of stakeholder management.

The case studies have been provided by several contributors. Each case study is attributable to the author(s) associated with it. In terms of referencing, therefore, the respective contributors of the case studies should be cited as the authors. For instance, the first case study below should be cited as:

Chanley, G. (2009) Stakeholder Dynamism – Baltimore Community Environmental Partnership (US EPA 1999). In: Chinyio, E. and Olomolaiye, P. (eds) *Construction Stakeholder Management*. London: Blackwell-Wiley, pp 351–352.

Table 20.1 provides a list of the case studies.

The case studies are not listed in any particular order and details of each of these now follows.

Table 20.1 List of case studies

Case study no.	Title	Author(s)
1	Stakeholder dynamism – Baltimore Community Environmental Partnership (US EPA 1999)	Dr Gail Charnley
2	Co-producing Space and Value: End-User Involvement in the Nokia Cable Factory Building Renewal Project	Dr Ritsuko Ozaki
3	Stakeholder Conflict: A Case Study of the African Cherry: *Prunus africana*	Dr Robert Kowalski
4	Family life cycle and planning of multifamily housing	João Alberto da Costa Ganzo Fernandez and Roberto de Oliveira
5	The Natural Environment as a Stakeholder of the Built Environment: A Case Study of Residential Development and Forests in South Africa	Johan J. Bester and I. J. van der Merwe
6	Lessons from a railway project	Dr Stefan Olander
7	Managing stakeholders by strategically influencing the distribution of power within the stakeholder network	Dr Athena Roumboutsos
8	Lessons from a housing project	Stefan Olander
9	Economic Analysis of Housing Designs	Renato da Silva Solano and Roberto de Oliveira
10	Let's save lives	Ron Rosenhead

20.2   Case Study No.1

Stakeholder Dynamism

Baltimore Community Environmental Partnership (US EPA 1999)

By Gail Charnley

This case study is culled from Charnley (2000) and concerns the diversity of stakeholders and their often conflicting interests. Sometimes the interest of stakeholders could be tacit and means of understanding these fully should be sought. This case study from Baltimore in the USA demonstrates this complex dimension of stakes.

Southern Baltimore is an industrialised area with a large concentration of industrial, commercial, and waste treatment and disposal facilities. Major facilities include chemical manufacturers, petroleum storage facilities, a medical waste incinerator, the city landfill and a municipal wastewater treatment plant, 11 of which report air emissions to the EPA Toxics Release Inventory. Additional facilities, such as the city waste incinerator, a large steel mill and two utility power plants, are located nearby. Altogether, more than 175 chemicals are emitted from facilities in the area, which led residents to rank air quality first on their list of concerns at a community priority-setting meeting. In particular, the community residents were concerned about the possible public health consequences of exposure to the combined emissions from all the industrial, commercial, and waste treatment and disposal facilities located in and around their neighbourhoods. A Community Environmental Partnership was started in southern Baltimore as a community-based approach to environmental protection and economic development. A subcommittee of the partnership comprising representatives of different community sectors was formed to address air quality, while a separate subcommittee was formed to address community health. The goals of the air quality subcommittee, co-chaired by one resident and one industry representative, were to determine whether levels of air toxics resulting from industrial emissions in partnership neighbourhoods had an effect on community health and to recommend actions to improve air quality. All decisions were made by consensus.

The air quality subcommittee chose to use a risk-based screening method to help provide information on the potential health risks associated with airborne chemicals in partnership neighbourhoods. The approach used standard methods to identify chemicals from air pollution sources that might pose the greatest health risks. Three successive screens of the original 175 chemicals of potential concern identified four chemicals as being of most concern to the partnership neighbourhoods. Of those four, only benzene emissions were estimated to result in airborne concentrations above the subcommittee's screening level, suggesting that local industrial emissions did not pose a threat to public health in that area. Petrochemical storage facilities in one neighbourhood were identified as the primary source of the modelled benzene, but contributed only 12% of the measured ambient benzene concentrations in the area. Mobile sources were thought to account for most of the ambient benzene concentrations but mobile sources were not considered in the screening exercise, which looked only at point-source emissions.

The limited scope of the subcommittee's investigation produced a dilemma. The subcommittee wanted to focus on facility-related point-source chemical emissions and to develop concrete recommendations to improve community health. As it turned out, the study found that the point sources evaluated were not likely to be a significant contributing factor to community health concerns. By not including a potentially important source of air pollution – mobile sources – in the study, the subcommittee did not have enough information to develop the most effective recommendations. Thus it was possible that poor air quality did contribute to public health problems in South Baltimore, but by failing to look at the whole picture, the study could not answer the question in full. The relationship between the limited scope of the subcommittee's work and its ability to make recommendations for improving community air quality and health was not adequately discussed, understood and agreed to at the beginning of the effort.

When the participants realized that the results of the study were not going to be able to show what some expected – that industrial air emissions posed risks to their health – the environmental advocacy group representatives resigned from the subcommittee. In a letter to EPA (timed to be released one day before the study results were made public), those who resigned (and others who had not been involved in the project at all) stated that they were 'deeply committed to the Partnership's ultimate goal: the discovery of more effective ways to reduce pollution through the reinvention of traditional regulatory programs.' That goal had not, in fact, been articulated and agreed to at the start of the effort. The letter's authors went on to say that what they had sought by participating in the project was 'a real opportunity [to develop] a new and deeper understanding of the environmental conditions *that threaten us* and [to debate] the best way to address those problems' [emphasis added]. Thus those who resigned had started with the assumption that the environmental conditions they were addressing posed risks to their health. When that assumption was not borne out by the results of a process they had agreed to and participated in from the start, they resigned in an attempt to discredit the process and findings and to maintain their adversarial position. In this way, the conflict became one less about what science was relevant and more about whether science was relevant. Scientific legitimacy was appealing when it suited the needs of the environmental advocacy participants; scientific information was sought as a means to buttress their beliefs, not to answer a question or solve a problem.

Looking back, Chapter 7 of this book has discussed stakeholders' varying priorities. Much more, stakeholders' interests are not static but fluid. In any undertaking, therefore, all stakeholders must be considered and the dynamism of their expectations monitored.

References

Charnley, G. (2000) *Enhancing the Role of Science in Stakeholder-Based Risk Management Decision-Making*. A report for the American Industrial Health Council and the American Chemistry Council [Online report] Available at http://www.healthriskstrategies.com/pub.html (Accessed on 28 November 2008).

US Environmental Protection Agency (EPA) (1999) *Baltimore Community Environmental Partnership Air Committee Technical Report. Community Risk-Based Air Screening: A Case Study in Baltimore, MD*. Unpublished draft document. Prepared by the US EPA Office of Pollution Prevention and Toxics and by Versar Inc. Washington, DC (1999).

20.3   Case Study No.2

Co-producing Space and Value: End-User Involvement in the Nokia Cable Factory Building Renewal Project

By Ritsuko Ozaki

With an awareness of the environmental impacts of construction activities, we are forced to consider how we can revitalise old building stock rather than demolish, which is traditionally practised. Utilisation of old building stock requires consideration to meet end-users requirements and to ensure that various features of renewed buildings are met so that value can be added on the users' terms. Technological and engineering innovations will encourage the effective renewal of old buildings; however, it is also crucial for construction professionals to identify what users expect of such rehabilitated buildings. Literature claims that end-user involvement in the product development and design process provides a sense of ownership and leads to success in businesses. Nonetheless, it is not easy to identify potential users and their complex and diverse requirements in construction. With this in mind, this case study, taken from a successful renovation project of a former Nokia Cable Factory building in Helsinki, Finland, discusses the ways in which end-users can be involved in construction projects and design processes.

The Cable Factory (Nokia Kaapeli) is an example which shows a certain respect for the industrial past. A balance between existing old structures and new interventions was achieved by both contrasting and blending the two. With the expansion of business and shortage of space in the 1950s and new town planning which moved industrial buildings farther from the city centre in the 1960s, Nokia Kaapeli decided to relocate. During the last few years of ownership, Nokia Kaapeli invested very little in the maintenance of the building. As the industry was moving from the area, Nokia Kaapeli started renting the premises at very affordable rates, and many artists and small businesses moved to the Cable Factory, due to the fact that they were able to secure peaceful working spaces. There were also spaces suitable for performances and exhibitions. The potential of the factory and its ideological and philosophical starting point was proved to be effective in practice before any official decisions were made. The administrative decision-making took 4 years. In 1987, the city of Helsinki and Nokia agreed on the procedures for the transitional period and formed a delegation to plan the future use of the factory in the ownership of the city. Plans were made to build schools, hotels, museums and even a car park to the former factory. The concerned tenants of the Cable Factory founded an association, Pro Kaapeli. Architects who had worked at the Factory also created a parallel plan to save the building and the activities that were prevalent at the post-industrial Cable Factory. They considered what kind of space was available in the building and the ways internal spaces could be used by different users (artists and visitors). Pro Kaapeli pointed out deficiencies in the planning of the area and even got the media involved. Pro Kaapeli was featured in the leading national newspapers and national TV and managed to dissolve deeply rooted prejudices against house squatters and artists who were often considered as 'shady'. Along the lines with Pro Kaapeli, a commissioned report described the identity of the Cable Factory and was used as a guideline for the future development, so that the unique atmosphere would not be destroyed and the building would act as a cultural symbol. For Helsinki the value of the Cable Factory lies on an emotional level. Those who were involved in this renovation project felt that the building and its newly found artistic community were too unique and valuable to be wiped away. A new agreement was made with Nokia, the city council decided to protect the

Cable Factory and its milieu and an estate company was founded. Almost all tenants were allowed to stay. Now, the new Cable Factory hosts theatres, museums, art schools and many small businesses and art workshops, and is regarded as a renowned art establishment in Europe. The usable surface area is 53 348 square meters, of which 40 000 have been rented, and 99% of the work spaces are in use. More than 200 000 people annually attend the events taking place in the halls, museums and dance theatres. The Cable Factory finances its own operations; the turnover surpassed 3.5 million Euros in 2005. The new Cable Factory was – albeit initially born accidentally – was realised with tenants' proactive involvement, mainly Pro Kaapeli.

Acknowledgement

The author is grateful to the architect on the renovation team, Jan Verwijnen, for allowing to interview him and for showing the Cable Factory building.

References

Högström, H. Finnish Cable Factory. http://www.kaapelitehdas.fi/.
Verwijnen, J. (1996) The Cable Factory: a story of another kind of interior renovation. In: I. Helkama (ed) *Helsinki Interiors*. Helsinki: Rakennustieto.

20.4 Case Study No.3

Stakeholder Conflict: A Case Study of the African Cherry: *Prunus africana*

By Robert Kowalski

In the tropical forests that clothe the mountain slopes of the West African Republic of Cameroon grows a tall tree measuring 30 metres or more that is the source of both global concern and community conflict. *Prunus africana* is the provider of traditional medicine in Africa. But it is now traded on the international market for the manufacture of products used to treat a number of prostate conditions that currently affect more than 50% of men over the age of 50. The bark of *Prunus* and its extracts are traded on a scale larger than those of any other wild African tree. The retail value of the trade in *Prunus africana* is estimated at over US$220 million a year.

However, since 1995, its international trade has been regulated by the Convention on Trade in Endangered Species (CITES). Commercial exploitation, habitat loss and unsustainable harvesting have led to a decline in the tree and unsustainable extraction methods, involving excessive debarking or the felling of entire trees (it is easier to strip the bark from a felled log), are threatening the species. Nevertheless the commercial demand has been so high that these unfavourable practices had become common.

However, there is one harvesting technique which can sometimes be used sustainably. This involves the removal of opposite quarters of the bark on the lower part of a trunk, with harvesting of the other quarters possible after 4 or 5 years. But all the sanctions and incentives have legislated against the adoption of these methods.

Although, in the long run, the assurance of the conservation of the species and the introduction of sustainable harvesting systems will be to the benefit of the pharmaceutical companies that manufacture products from *Prunus*, commercial imperatives around gaining adequate supplies have often dictated the demonstration of less enlightened behaviours.

In these circumstances I had the privilege of working with one particular group of villages on a biodiversity conservation project that had to focus upon conflict management. The central stakeholders involved were the village elders – the traditional male authority figures; the young men of the villages who undertook most of the harvesting of the bark, using mostly unsustainable methods and operating outside of legal controls; the representatives of a French pharmaceutical company that were the major purchasers of the bark from these communities; and the local and regional officers of the Ministry of Forests who were charged with the conservation of the species and it protection from illegal harvesting.

The monies that flowed into the local economy from the illicit trade in the bark were having a number of effects, including the undermining of the traditional authority structures, the adoption of behaviours by the young men that were portrayed by the elders as 'squandering' the communities' resources, and increase in rent seeking behaviours by some officials.

There were three levels of stakeholder interests that were supporting the conflict and the devastation of the local population of African Cherry trees. The primary interests were around power and authority within the community and the economic imperative of the company to obtain bark. The secondary interests involved on the one hand the economic resources that underpinned the primary interest and the continuity of supply of bark in an increasingly unpredictable situation on the other. The tertiary interests revolved around the individuals' personal and community histories (e.g. scores and rivalries) and engagement in corrupt practises.

The approach that we took, as external agents of change, was not to be judgemental and to accept that the primary and secondary interests were substantially legitimate, and in any event the business of the stakeholders to manage, but that the tertiary interests were illegitimate and had to be avoided, so as not to provide perverse incentives for maintaining the conflict.

Using an approach and a set of tools for community wide situation analysis known as Participatory Learning and Action (PLA), and with staff from the Ministry of Forests seconded to the project as facilitators, we engaged the various community stakeholders in an appraisal of the prevailing circumstances. In a series of group meetings, focus groups and individual interviews we explored the local history, aspirations, needs, perceptions and priorities of as many different individuals and groups as we practically could, over a period of several months. The first product of this process was the emergence of the voice of local women whose views, up to this point, had been essentially ignored.

As the project proceeded it was this strand that became stronger, and eventually presented itself to all parties as a potential solution. Working with the pharmaceutical company the idea of a certification scheme was developed that involved the identification of legitimately harvested bark through the use of seals. Each sack of bark was to have two seals, one that was held and applied by the traditional elite authority and a second that was in the hands of a newly formed women's council. Both seal holders kept tallies of the number of sacks authorised – and these were regulated through the Ministry officials to ensure compliance with agreed sustainable quotas. The young men, now trained by the Ministry of Forests in sustainable harvesting techniques, collected the bark, transported the sealed sacks to the company and received payment, part of which was returned as a remittance to the seal holding authorities for use in developing the community.

For us the lessons that came from this experience revolved around the empowering approaches of participatory analysis with the stakeholders that enabled the solutions to emerge from those directly involved in the conflict.

Acknowledgement

Dr Kowalski wishes to thank Dr P Scott Jones for his contributions to the development of the project from which this case study is taken.

20.5   Case Study No.4

Family Life Cycle and Planning of Multifamily Housing

By João Alberto da Costa Ganzo Fernandez and Roberto de Oliveira

In the Brazilian context, literature identifies the attributes of a multifamily housing apartment as: number of bedrooms, playground, two parking spaces, service bathroom, service entrance, maid's room, two ensuite bedrooms, at most two apartments per floor, a good view from the living room, pool, hobby box, hot-water system, ample kitchen with table, sophisticated security system, Jacuzzi, small wash basin, individual water meter, etc. Meanwhile, the family life cycle is very important in choosing a type of house. Also, the lot or land which is usually bought in advance because of market circumstances, and its location, serve as a starting point for the architectural programming (or briefing, as it is called in some other places) and the segmentation process of a multifamily housing product. The cost of the lot will be a decisive factor in the price per unit in addition to the quality and size. A primary segmentation by income emerges almost naturally because of the price and characteristics of the location of the lot. Since the possibilities of the finishing standards are restricted to those considered coherent with the value of the location (primary segmentation based on income), morphological characteristics such as the number and type of compartments and equipment for general use, are decision variables among the various project possibilities, which should be defined by secondary segmentation. The key question here is: *how to correctly segment a housing arrangement?* Well, knowing the preferences of occupants at each stage of the family life cycle provides a basis for establishing a worthwhile secondary segmentation.

In this case study, the important architectural attributes of middle-class potential apartment buyers in each phase of the family life cycle were observed in Florianópolis (SC/ Brazil) from 2002 to 2005. Citywide Real Estate Exhibitions were used for this purpose. Five stages of the classical life cycle were considered, i.e. childless couples, families with small children, families with teenagers, families with adult children living at home and couples in the empty nest. In this categorisation, couples in the empty nest are older and have children who have married and moved out of their parents' home; while childless couples are relatively young and yet to have children.

Our observation in Florianópolis showed that the priorities of different family types differ. While childless couples prioritise the living room, barbecue facilities on the balcony, and the pool; those who have children prioritise the playground and the service bathroom. When the children grow older, families start to value attributes related to space and privacy, such as the maid's room, a hobby box, two parking spaces, two ensuite bedrooms and acoustic insulation. Buyers in the last phase of the family life cycle – the empty nest – differ from others by the importance they give to an independent service entrance and a nice view from the living room. The desired number of bedrooms is greater during the phases of family expansion and smaller during the childless and contraction phases.

If a designer is to get the secondary segmentation right, he/she needs to engage the buyers and users of buildings who are a subset of stakeholders. The appropriate form of communication must also be used therein. The priorities of users can be established by using, for example, in-person intercept surveys, which can be conducted at Real Estate Exhibitions (with potential apartment buyers), as we have done in Brazil. The questions to use therein must be worded simply and easy to answer. This kind of survey is low-cost and can properly define the potential family life cycle stages for each market area (neighbourhood) as well as identify the most important architectural attributes in each phase. It can also provide a basis to improve and better define the architectural programming.

Chapter 7 discussed stakeholder mapping. An understanding of stakeholders' expectations informs their mapping. This case study has used an effective approach to decipher the interests of a set of stakeholders in Brazil. The particular approach to use with other stakeholders will depend on who they are. The bottom line is to be effective in the understanding of your stakeholders and their expectations. That way, you can serve them more effectively.

20.6   Case Study No.5

The Natural Environment as a Stakeholder of the Built Environment: A Case Study of Residential Development and Forests in South Africa

By Johan J. Bester and Izak J. van der Merwe

Introduction

Should one consider the natural environment as a stakeholder when developing the built environment? The natural environment can be severely affected, but does it have any influence? Although feedback mechanisms may exist between the natural and the built environment, the time lag may be long before the effects become manifested and these effects may vary in magnitude and significance (for humans). While we as humans are only one species among the multitude of living things on earth, our actions can have profound impacts on other species. Does our society provide adequate measures for balancing its potential impacts, or do we mainly respond to the concerns and demands of other humans?

Case background

The forests
Natural forests in South Africa are limited in extent, highly fragmented and rich especially in plant species diversity (Geldenhuys, 2000; Vermeulen, 2000). They occur mainly along the humid southern and eastern coast and escarpment where they contribute significantly towards scenic quality of the landscape, they play an important role in soil protection and represent a small albeit important carbon reservoir and sink. Besides environmental functions the forests also provide forest goods and cultural services for rural people in their vicinity and in more distant urban areas (e.g. Lawes *et al.*, 2004; von Maltitz and Shackleton, 2004). These forest ecosystems are ancient and have persisted and evolved through fluctuations in global climate (Geldenhuys, 2000; Lawes *et al.*, 2007). In terms of South Africa's National Forests Act of 1998 (NFA), all natural forests are protected and may not be disturbed or destroyed unless there is very good justification. Because they

are small and fragmented, each forest patch is of great importance as a link in the network of forest habitats across the landscape (Geldenhuys, 2000; Vermeulen, 2000).

The arid Camel thorn savannah of the Northern Cape Province provides another case example. Savannah in general is less fragmented although equally significant for biodiversity and aesthetic reasons (which presents a strong basis for tourism). The savannah also provides a variety of other environmental functions (Shackleton *et al.*, 2007). Savannah of the Kalahari is characterised by the Camel thorn tree, *Acacia erioloba*, which is regarded as a keystone species. It has been a protected species under forestry legislation since the 1940s. All savannah woodlands are not protected by law as is the case with natural forests. However, an exceptionally dense and scenic area of Camel thorn savannah in the Northern Cape, the case example cited below, was protected as State forest for several decades until 1956.

Socio-economics
During the 1990s and into the new millennium the residential property market thrived in South Africa. With prevailing perceptions of crime and insecurity, a strong demand has emerged for residential developments that provide security infrastructure and access control. Towards the exclusive upper end of the market, such developments are often linked to specific lifestyle themes, for example golf estates or eco-estates. The latter presents marketing utility in an era when the general public begins to take note of environmental concerns. Residential property development within the setting of a natural forest or the savannah environment and associated with wild animals provides an ideal package of attributes for 'eco-estate' labelling. This may however be quite misleading.

Environmental impacts
Habitat destruction, fragmentation and modification, among other due to urban expansion, are major problems with respect to already fragmented forest ecosystems such as the natural forests (Seydack, 2000). Regardless of their labelling as 'eco-estates', residential developments that infringe on natural forests and woodland can impose severe long-term burdens on forest ecosystems. Within and around human settlements fire regimes are changed; forest canopies and margins are opened up resulting in alteration of the microclimate inside the forest; compaction of soils affect the long-term vitality of trees; human residence considerably elevates the risk of introduction of pathogens, alien species and pollution; some natural fauna, especially larger vertebrates are displaced or excluded, some taking flight from the area at the very first signs of development activity (Figure 20.1).Security fencing also isolates these forests so that any remaining terrestrial fauna are effectively contained. So unless their populations are actively managed, problems with nutrition, overpopulation, social structure, inbreeding, etc. affect the viability of some animal species. Abstraction of underground water may lead to salinisation or desiccation of aquifers while use of fertilisers may cause eutrophication. These are some general impacts on which one could elaborate (Department of Water Affairs and Forestry, 2008).

Two case examples – different scenarios
The Kalahari Estate, Northern Cape: The developers of this exclusive residential complex anchored their development on the character of the arid Kalahari savannah, well illustrated by Figure 20.2, combined with the usual theme of golf. EIA procedures were followed in compliance with National Environmental Management legislation, which prompted objections to the development among others from non-government environmental interest groups. Further, an application for licenses under the NFA was required to cut some Camel thorn trees for example for installation of services if the development would be authorised. The National

Figure 20.1   Disturbance associated with building activity in coastal forests. Building material introduced onto the site present a source of pollutants. A large opening was created in the forest canopy, which will remain indefinitely, and access roads with associated soil compaction will also affect tree root systems over the long term. These are examples of impacts of varying intensity and temporal scale.

Figure 20.2   Camel thorn savannah in undisturbed condition in the case study area provides an attractive view as well as rich diversity of habitat elements. Large mammals that occur naturally in these arid savannahs require extensive areas to meet all their survival needs, especially for viable populations.

Department of Water Affairs and Forestry (Branch: Forestry) sub-divided the Camel thorn forest into three zones as illustrated in Figure 20.3. The core habitat, Zone 1, was the highest priority for conservation and Zone 3 was a buffer zone in which some development could be considered. A task team comprised of the Branch: Forestry, provincial conservation authorities, the municipality, non-governmental organisations and the developer negotiated conflicts

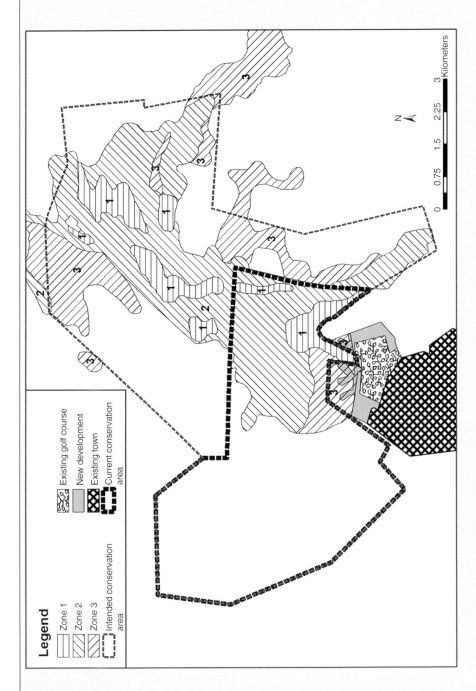

Legend

Zone 1
Zone 2
Zone 3
Intended conservation area

Existing golf course
New development
Existing town
Current conservation area

Figure 20.3   Map showing the new development and Camel thorn forest (*Source*: Department of Water Affairs and Forestry, South Africa; Used with permission). The portions of highest conservation priority (Zone 1) occur as fragmented patches. Zone 2, is an extension of core habitat, which is important for connectivity. Limited developments were allowed in Zone 3, the buffer zone.

of interest. Through an offset agreement the developer agreed to provide additional land for conservation of the main Camel thorn forest, they reduced the number of residential units in the proposed new development, which would be restricted to the buffer zone, and reconfigured the lay-out of infrastructure and property boundaries to minimise the impact of the development. In view of these concessions made by the developer, the development was approved and a licence was granted for the removal of 113 small and six large specific Camel thorn trees. This enabled the development to proceed while maintaining a reasonable level of care for the ecosystem and compliance with legislation.

Coastal Forest Estate, KwaZulu-Natal: This proposed development of about 140 residential units was originally designed in a grid layout that would affect one of the most threatened coastal forest types in South Africa – a forest type that had already been reduced by more than a third of its original extent through various incremental impacts including agriculture and urban expansion. Umdoni Forest, earmarked for the development, provides habitat for some IUCN red data species, including the blue duiker antelope and six plant species. Altogether 198 plant species were identified in this 56 hectare area and over 200 bird species, some of which are very rare, occur here either as resident species or migrants.

Although the specific site of this development was designated to remain natural within the local Government's open space planning framework, the developers were allowed to proceed with the planning of the Coastal Forest Estate. The land was private property and had no formal conservation status. There were few economically viable and ecologically compatible land use alternatives in this case. The EIA process elicited objections from the non-government environmental interest groups, provincial government as well as the public. Although some forestry professionals were opposed to the development, the official response was contradictory. Environmental authorities in the Province eventually granted an environmental authorisation. Appeals followed. One environmental consultant argued that the forests on this land constitute fairly new forest expansion, thus not old primary forest. Some ecologists opposed this view in the appeals. Conflicting responses from the various authorities caused considerable delays in the decision process.

Elsewhere along this coastline similar developments were being planned. In some cases opportunistic developers opened illegal clearings in forests for residential development. Others attempted to circumvent the requirements of environmental legislation by presenting their developments under the framework of legislation that is intended to accelerate low cost housing development. In one tribunal case involving a proposed development in Hawaan Forest, contradictory scientific opinions about the environmental impact were presented. In this test case the tribunal ruled against the developer and set a 40-meter buffer between development and the forest as a requirement.

Analysis of some key elements

1. Both case examples can demonstrate that residential development has significant impacts on the forests as ecosystems or on specific species associated with these forests, especially rare and localised species.
2. The preferential selection of forest sites (or any other pristine natural site) for residential developments suggest that the natural environment can have a positive influence on the development of the built environment. The presence of natural trees and forests for example are thus seen as an attribute that enhances the appeal of the site for

residential development. The result is an unavoidable conflict. Such preferential selection of sites of special natural quality is ironic because development on such sites often compromises the very qualities about the site that constitutes its basis of attraction.

3. When one considers the ability of the natural living environment to influence or oppose a new development directly, it is obvious that the ecosystem is a stakeholder with no voice. It is entirely vulnerable and dependent on the ethical disposition of the developer or intercession by the public, civil society or Government.

4. Over time, as the affected sites lose their original pristine character and sense of place, it may be interpreted as a way through which the natural environment asserts its influence. Most probably, however, such changes will go undetected except to the most informed property owners and may even be preferred by the residents. In more serious cases local environmental changes may render a site more susceptible to hazards such as erosion, flooding, forest fire, etc. Such delayed influences are often unforeseen (by the developer) and when they occur, damage to the ecosystem may be irreversible. Mitigation measures then need to be devised.

5. Market-based approaches, for example financial compensation, substitution or even replacement, implies various further questions. It may be possible to compensate forest dependent local people for lost consumptive benefits, however, can one pay back to the natural environment what a property developer takes away from it? It can be said that forest ecosystems are undervalued and thus vulnerable in a world governed by market considerations. While it may be possible to attach a financial value to a tree or a natural forest, should one accept such values as relevant instruments in all cases? Refer also to McCauley (2006) and Sapontzis (1995). When people obtain ownership of land, should that give them absolute right over everything on that land? If not, how are they compensated for lost opportunities such as income from development? Sometimes developers propose replacement of natural features on a site. Is it possible, however, to replace a natural forest or woodland ecosystem which resulted from centuries of evolution and succession? It is obviously not possible, even within the time span of several human generations, to recreate a forest that consists of living elements several centuries old. Furthermore forest ecosystems are more than random assemblages of trees and other species. Their existence within the landscape is often the unique result of complex interactions of many different variables. Forests are dynamic although the rate of change and development often far exceeds the short time frames within which most people think and conduct their lives. Do we as humans consider the impact of our short-term decisions on the long-term dynamics of forests?

Concluding observations

Humans have been part of the natural environment for millennia, using its resources and co-existing with other living things. However, during the last few centuries we have acquired substantial technological abilities compared to our ancestors, enabling a human population of greater numbers and capable of inflicting disproportionate impacts on the natural environment. Our ability to extract materials and construct persistent structures for our own accommodation is one example.

In a world of ever fewer and smaller remnant refugia of pristine nature, humans still tend to view the environment from the perspective of self-benefit. Hence developers perceive forests as vacant sites, not recognising that they are indeed occupied by non-human inhabitants; plants and animals, many of which are unable to co-exist closely with humans. Existing forests or trees can even be seen as attractive attributes that render sites desirable for human residence, thus an opportunity to convert natural capital to physical capital for the benefit of making profit.

The natural environment finds itself vulnerable against this onslaught. Unable to articulate their needs and the consequences of proposed developments, the ecosystem and its constituent parts depend on human ethics, the ability of humans to care for anything other than their own direct short-term benefit, and on the institutions of society to intervene with adequate knowledge and appropriate mechanisms to provide for the ecosystem and its natural dynamics (Miller, 1996).

Note

The views and opinions expressed in this case study are of the authors and do not necessarily reflect the views or policy of the Department of Water Affairs and Forestry or the South African Government

References

Department of Water Affairs and Forestry (2008) *Policy Principles and Guidelines for Co-operative Control of Development Affecting Natural Forests.* Pretoria: Department of Water Affairs and Forestry, 19pp.

Geldenhuys, C.J. (2000) Classification and distribution of forests and woodlands in South Africa. In: Owen, D.L. (ed) *South African Forestry Handbook.* Menlo Park: South African Institute of Forestry, pp 591–599.

Lawes, M.J., Obiri, J.A.F. and Eeley, H.A.C. (2004) The use and value of indigenous forest resources in South Africa. In: Lawes, M.J., Eeley, H.A.C., Shackleton, C.M. and Geach, B.G.S. (eds) *Indigenous Forests and Woodlands in South Africa – Policy, People and Practice.* Scottsville: University of KwaZulu-Natal Press, pp. 227–273.

Lawes, M.J., Eeley, H.A.C., Findlay, N.J. and Forbes, D. (2007) Resilient forest faunal communities in South Africa: a legacy of paleoclimatic change and extinction filtering. *Journal of Biogeography,* **34**(7):1246–1264.

McCauley, D.J. (2006) Selling out on nature. *Nature,* **vol 443**(7107 ~ 7 September 2006) pp. 27–28.

Miller, G.T. (1996) *Living in the Environment.* Belmont, California: Wadsworht Publishing Company, Pp. 110–126.

Sapontzis, S.F. (1995) The nature of the value of nature. *The Electronic Journal of Analytic Philosophy,* **3** (Spring1995). Available online at http://ejap.louisiana.edu/EJAP/1995.spring/sapontzis.abs.html (Accessed last on 10 July 2009).

Seydack, A.H.W. (2000) Larger forest mammals; ecology and conservation. In: Owen, D.L (ed) *South African Forestry Handbook.* Menlo Park: South African Institute of Forestry, pp. 659–665.

Shackleton, C.M., Shackleton, S.E., Buiten, E. and Bird, N. (2007) The importance of dry woodlands and forests in rural livelihoods and poverty alleviation in South Africa. *Forest Policy and Economics,* (9):558–577.

Vermeulen, W.J. (2000) Management of natural forests. In: Owen, D.L. (ed) *South African Forestry Handbook.* Menlo Park: South African Institute of Forestry, pp. 601–612.

von Maltitz, G.P. and Shackleton, S.E. (2004) Use and management of forests and wood-lands in South Africa: stakeholders, institutions and processes from past to present. In: Lawes, M.J., Eeley, H.A.C., Shackleton, C.M. and Geach, B.G.S. (eds) *Indigenous Forests and Woodlands in South Africa – Policy, People and Practice.* Scottsville: University of KwaZulu-Natal Press, pp. 109–135.

20.7   Case Study No.6

Lessons from a Railway Project

By Stefan Olander

Background

Project location: Lund, Sweden
Type of project: Railway project, the construction of a two-way railway track through the town centre
Project cost: Approximately 1.2 billion Swedish Kronor (about 130 million EUROs)
In the late 1980s, the Swedish government decided to expand the west coast railway from a single- to a double-track railway. The route passes through a number of communities, one of which is Lund. The developer (the national railway administration) proposed an expansion of the railway along the existing route through the town centre and some popu-lated areas, with the argument that it was the most rational alternative. Moreover, it was asserted that residents along the existing railway would enjoy a better living environment from noise reduction measures that would have to be introduced. The basic issue was that no alternative routes were properly examined.

Since the railway was routed through a built-up area, the municipality would need to issue final approval for the project. Nearby residents formed an interest group with the pur-pose of forcing the municipality and the developer to reroute the railway through a more sparsely populated area. This action resulted in demands on the developer to investigate alternative routes for the railway. After several investigations, all of which showed that the most rational route was to expand along the existing railway, the municipality gave approval. However, the residents were not satisfied and started an extensive campaign through the media in order to coerce the decision-makers to relocate the railway. After an appeal that took 6 years to process, final approval to expand along the existing route was given.

Outcome

Demands by the municipality and residents in the vicinity led to further study of the most appropriate route for the double track railway. Indeed, a number of studies were con-ducted which all showed that the best alternative was to build the railway along the exist-ing route. On that basis, the municipality approved the proposed route. In response, the residents in the vicinity organised themselves with the aim of trying to influence deci-sion-makers to propose a different route. This case of inadequate communication with residents in the earlier stages of the project had ruined any basis for their trust and so they appealed the approval decision, but lost in both instances. The national government took 6 years to consider the appeal with the eventual result that it required some changes to address concerns over noise and safety.

Consequences

The main consequence for the project was that it was delayed by approximately 7 years, primarily because of adverse public opinion expressed by local residents. This led to a time-consuming inquiry. In addition, media coverage was almost always based on the views of residents, which constituted 'bad press' and generated a generally bad reputation for the project.

20.8 Case Study No.7

Managing Stakeholders by Strategically Influencing the Distribution of Power within the Stakeholder Network

By Athena Roumboutsos

Project description

In June 1998 within the framework of the European Structural Funds (ESF) 1994–99, the University of the Aegean was assigned the pilot project 'IALYSOS' by the Greek Ministry of Education. This was a turnkey contract with a scope to develop 40 school libraries on the islands in the Aegean Archipelagos. The project involved surveying approximately 250 schools on 20 islands, identifying the 40 schools that could accommodate libraries by reverting existing classrooms into library space, whilst justifying the respective investment in terms of student numbers or the social impact of overcoming isolation; design, procurement of construction work, procurement of furniture, computers, other tutorial equipment and 6000 book titles per library; provision of library services and librarian training. The project budget was approximately 7.5 million euros and the project duration was less than 30 months as financial closing (ESF framework period closing) had to be achieved by 31 December 2000. From an operational point of view, as it was a pilot project, each task constituted a milestone as it requested ministerial (client) approval prior to implementation. Considering the time limitations, budgetary restrains, decision-making process and geographical dispersion, it was a uniquely complex project.

Key project implementation issues

The project team identified four discrete project phases: school selection; design and procurement of construction works and supplies (furniture, equipment, books), as well as preparation of training material; implementation and project transfer. While developing the initial project implementation plan and project schedule, the project management team identified the following major sources of project risk:

1. Authority and Authorisation issues
   a. Authorisation to access and intervene in school property. School buildings, in Greece, are owned by the respective Municipality, which receives school funding by the respective Province Authority (Prefecture). Public technical services are responsible for public buildings and, in this case, the respective prefecture technical services would be, primarily, responsible for the selected school buildings.
   b. Authority over school staff and moreover the school principal and teacher responsible for the library, as they belonged to a totally different hierarchy, which composed of the school principal, the secondary school prefecture director, the Ministry of Education Secondary School Directorate. The latter is responsible for teachers' placements and the application of the respective employment regulatory framework and independent to the Directorate funding the project.

The hierarchical structure depicted (Figure 20.4) meant that the project would be directly dependent on a large number of stakeholders (40 school principals and 40 teachers assigned to the school library, 5 secondary school prefecture directors, 5 prefecture technical services and a large number of mayors – 34 after school selection).

2. Pressure was anticipated from local communities wishing to have their school selected as one of the 40 project sites, their major influence being towards the Municipality and the school staff.
3. Dispersed geographical project locations (Map of Greek Islands and project locations (▲ – Figure 20.5) had a significant impact on project organisation, monitoring and control. Principle liabilities considered were:
   a. On-site project progress reporting and quality control, as budgetary restrains would not allow for permanent on-site staff.
   b. Misuse of facilities before the project transfer phase.

It was obvious to the project management team that the project was directly dependent on too many dispersed stakeholders with unpredictable interests. Their number had to be reduced and dependency reversed. But how?

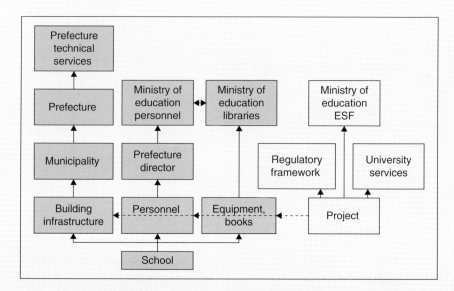

Figure 20.4 Hierarchical dependencies within school environment and project.

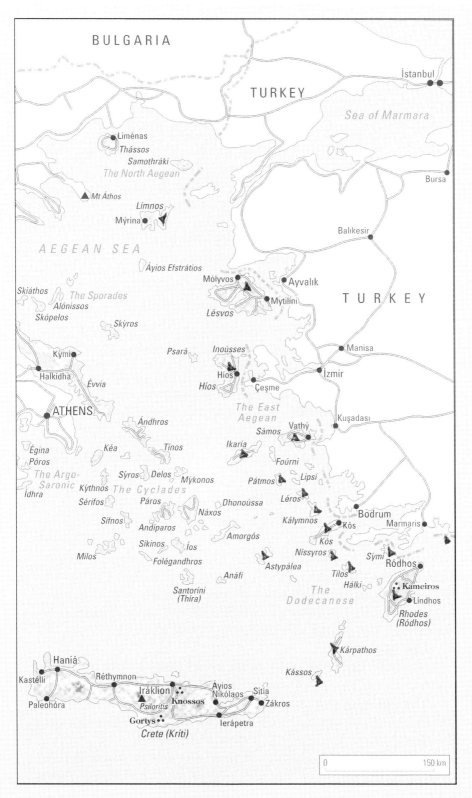

Figure 20.5   Map of Greek Islands and project locations.

Mapping power distribution within the stakeholder network and strategic planning

The project team decided to study the problem from a stakeholder management point of view. Due to the dispersion of the project locations, the tasks involved in project completion and the lines of command involved in the schools, the number of immediate project stakeholders was, inevitably, large. However, it was noticed that each stakeholder was not only characterised by its individual predisposition towards the project (or particular project outcomes/activities) but also its power dependency on other stakeholders. Therefore, *dominant* and *dependent* stakeholders needed to be identified. The agents formulating these dependencies were none other than contractual agreements, legislations, regulations, organisational structures, etc. as well as the political environment linking the stakeholders. In that context, the relations between stakeholders and the project were viewed as a distribution of power within the project – stakeholder network.

Hence, the problem could be reduced by focusing on dominate stakeholders, identifying their disposition towards the project and developing a respective strategy.

Mapping the power distribution within the project stakeholder network was the way forward concerning the identification of the *dominant* stakeholders. The project team decided to carry out the task per project phase as different stakeholders would be involved in each phase and the dependencies altered according to the expected outcomes.

The school selection phase was the first the project team concentrated on. Mapping of hierarchical dependencies accompanied by brainstorming within the project team, were used to produce the power distribution map of Figure 20.6.

The power dependency arrows led to three *dominant* groups of stakeholders: The Ministry of Education Personnel and Libraries Directorate, the prefecture technical services and the local communities.

Public project promotion and campaigning, a prerequisite of European Commission funding, would influence positively local communities but during the initial project phase would increase pressure over school selection. *The project team concluded: Timing wasn't right for the public campaign.*

The prefecture technical services were considered a true bottleneck, as no pronounced dependency was identified that might secure their undivided collaboration with the project team. Moreover, five (5) prefecture technical services would be involved in the project.

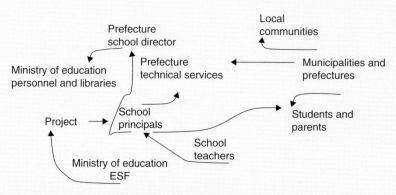

Figure 20.6   Power distribution within the stakeholder network – school selection phase.

The project team reverted to scope management. What was needed from the prefecture technical services? Design and construction work approvals of a *public* technical unit. *Public* was the keyword. The prefecture technical services could be replaced by those of the University. Collaboration was not guaranteed, however, the number of stakeholders would be reduce from five (5) to one. In addition, it was noted that during the school selection phase no 'approvals' were required for surveying the schools.

The last dominant stakeholder group, that is the directors of the Ministry of Education Personnel and Library Directorates had to be interviewed in order to identify their interests in the project. Prior to interviewing, the project team considered the options. Stakeholder interests might impose constrains on an otherwise independent selection process. *Timing* was again an issue. The project team decided to acquire permission to survey schools from the five prefecture school directors whose disposition towards the project was estimated to be either neutral or positive. Stakeholder interviews would be carried out following the completion of the selection phase and before the ministerial decision on the selected schools, so as any intervention to be addressed towards the ESF Directorate and not towards the project team.

Mapping power distribution proved to be a useful tool, and timing the influence on dominate stakeholders was highlighted as, equally, important.

The timing was also right for school selection. The summer school holidays were, in fact, the right time to survey schools with minimum disruption of school operation and minimum dissemination of project objectives, hence, minimising the probability of local community intervention.

With the successful and timely completion of the school selection phase, the project team concentrated on mapping the next phases. Emphasis was placed on the design and procurement phase leading to the implementation phase of the project. Power distribution and dependencies were mapped as shown in Figure 20.7. The decision made to seek the cooperation of the University Technical Services was included. However, monitoring and control of implementation on a maximum of 20 (17 after selected schools were announced) isolated (island) locations was not a simple task. Regular monitoring was foreseen but for day-to-day updates, the project team had to depend on school principals and staff. These dependencies were, also, mapped.

While studying the dependencies during the two successive phases, two stakeholders (the University and the Ministry of Education Personnel and Libraries) remained *dominate*, while the relation between the project and the (potential) subcontractors and the project

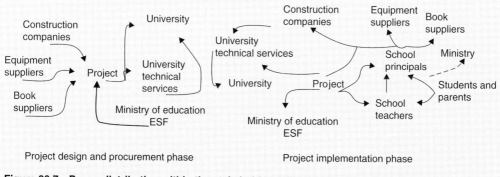

Project design and procurement phase   Project implementation phase

Figure 20.7   Power distribution within the stakeholder network – design and implementation phase.

and the Ministry ESF Directorate indicated an inversion in dependencies, as during the implementation phase the project would, for example, be dependent on the Ministry for a regular flow of payments.

Prior to procurement, designs had to be approved. The collaboration of the University Technical Services was of paramount importance to the project. Only the University chief administration could secure that. *But why would the University attach more importance to this project than any other carried out by the services? On whom is the University dependent? To whom does the University provide services to?* These were brainstorming questions the project team addressed. The answer was simple: Students and future students potentially coming from these very schools! Therefore, the campaign envisaged during the school selection mapping phase had to highlight the involvement of the University and promote University work to the local communities who, in essence, had an obvious interest in the project. The local communities' positive dependency on the project was considered to bear the potential of enhancement and the project was promoted as a 'scarce commodity' and a unique opportunity to improve school standards. Local communities bought into the project objective and became dependent on the project management team to deliver the project to their communities. Consequently, Mayors and Municipalities were positioned dependently on the project, as this promoted their position within the local community. Additionally, the motivation of the local community had the anticipated positive effect on the University, as it presented itself as an agent open to community needs and supporting community interests in the field of education. In the end the positive image of the University was dependent on the timely and quality progress of the project and University Heads pressured their Technical Services, by using their power of hierarchy, to support by all means the project.

The Ministry of Education Personnel and Libraries Directorates were, once again, identified as *dominant* stakeholders. A possible negative disposition towards the project would influence their dependent staff (Library teachers and School Principals), who, in turn, would not be cooperative or their cooperation would be obstructed or undermined. There was a possibility that all stakeholders in their line of command involved in the project would have negative dependencies. In-depth interviews with the respective directors highlighted the fact that their basic 'need' centred on information, as there was no official link to the Ministry Directorate funding the project (ESF). Moreover, in their perception, they were deprived of their entitled hierarchical power by the ESF Directorate's fund control power. The answer was simple: insertion of a reporting task and the project was positioned favourably towards these *dominant* stakeholders and secured the collaboration of school principals and staff.

One important feature highlighted by the comparison of the power distribution maps of Figure 20.7, was the project's funding agent (Ministry of Education, ESF Directorate), which, though, presented dependent on the project during the project design and procurement phase became a *dominant* stakeholder in the project implementation phase. The project team knew that all subcontracts had to be signed before the end of 1999, according to the ESF regulation framework. Therefore, the funding agent would support the management team in any way possible in order to achieve that goal. Following this, the smooth implementation of the signed subcontracts would depend highly on a regular and timely cash flow provided by the Ministry. Hence, the project, during the implementation phase, would be dependent on the Ministry. The solution would be to involve the Ministry, contractually, in the subcontracts, highlighting it as the source of funding and the project team as the intermediate. As contract negotiation was carried out during the procurement phase

(Ministry dependent on project), and at the time, the Ministry was happy to be involved in the proposed trilateral agreements.

Finally, the inversion of dependence between the project and the subcontractors during project procurement and implementation phase is a normal and anticipated relation, safeguarded by the terms of reference in any contract. However, in this particular case, the individuals anticipated to be the projects day-to-day reporters (i.e. the school principals) had to be provided with a minimum of power. Again, inclusion of their activities in the contracts' terms of reference provided the necessary status and reinforced the cooperation with the Ministry of Education Personnel and Libraries Directorates.

The analysis of the power distribution maps and the strategic planning of the project that followed eliminated most of the sources of project risk initially identified. Prior to proceeding with the design and procurement phase, the project management team carried out the mapping of the final stage of the project to verify the expected evolution. Figure 20.8 presents the project transfer stage following the anticipated changes in stakeholder dependencies.

The project team was satisfied that, through the planned interventions all project stakeholders would be ultimately dependent on the project and its success. However, there was one final risk to be addressed: Misuse of procured equipment and materials prior to project transfer. Again the solution was found to be the inclusion of specific clauses in the assigned subcontracts: all equipment and materials would be the property of the schools. The proposed solution had dual benefits: released the University of any liability concerning misuse and enhanced the position of the school principals vis-à-vis the subcontractors.

Concluding remarks

By developing a strategy to reverse the power dependency within the described project stakeholder network, the risks associated with project implementation, that is on-site control, authorisation and authority issues, as well as property ownership issues were

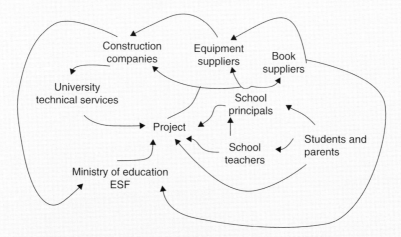

Figure 20.8   Power distribution within the stakeholder network – project transfer phase following strategic interventions.

overcome in a very geographically dispersed project (Map of Greek Islands and project locations (▲ in Figure 20.5).

Mapping of power distribution within the project stakeholder network was primarily based on brainstorming and observations made by the project management team, which provided useful insight concerning relations and dispositions towards the project and its outcome. It was highlighted that relationships between stakeholders are dynamic and multiple feedback is required to monitor these changes.

The successful completion of the described project is promising for the methodology illustrated herewith.

20.9   Case Study No.8

Lessons from a Housing Project

By Stefan Olander

Background

Location: Lund, Sweden
Type of project: A housing project consisting of 60 apartments
Project cost: Approximately 60 million Swedish Kronor (about 7 million EUROs)

The property, which was purchased by a developer in the late 1980s, was an old residence located in a large park. It had been left vacant for several years and was in a state of disrepair. Nonetheless, both the residence and the park were considered by the community to have some cultural value; in addition, the park also had a recreational value. The proposal by the developer was to build two 9-storey buildings, with senior citizens as the main target group. The old residence was to be preserved and used as a joint facility by residents.

The proposal met opposition from neighbours and from groups that had an interest in the cultural values of the city. The main criticism was that the proposed buildings were too high and that the park would lose its recreational purpose. Furthermore, the neighbours objected to a foreseen deterioration in their environment. The proposal was stopped after a 5-year planning process, following an appeal from the neighbours. The project was then redefined. A nearby gas station was moved, which made it possible to preserve a larger area of the park. In addition, the height of the proposed buildings was reduced, whilst the number of buildings would be allowed to increase from two to five. At that point, objections from the neighbours remained the only stumbling block and these were based on the same arguments as before.

Outcome

With the removal of the gas station, the cultural and recreational values were better preserved. The municipality thus approved the construction of the two 9-storey houses, despite their being amongst the highest buildings in the city and affecting the city skyline within which the cathedral was a significant characteristic. The residents in the vicinity appealed twice against the decision and obtained approval on the second attempt on a legal technicality. The developer was forced to start all over again and presented a new proposal consisting of five 4–6-storey houses, with a large part of the park preserved as a public space. Concerns about the city skyline and the intrinsic cultural and recreational values were consequently no longer considered to be major issues. The residents living nearby were, however, still concerned about shade, traffic and the potential decrease in their property values. Once again, they appealed against the proposed construction, but this time they lost, and construction proceeded. In the late 1990s, the project was finally completed after a planning process spanning 10–12 years.

Consequences

The main consequence was the time delay; the project took 12 years from the first ideas to completion. This delay and the extensive changes to the project also had the effect that a significant amount of committed resources were wasted, representing approximately 5–10% of the total project cost. The developer did, however, identify at least one important positive effect, which was that the final project outcome was better than the first proposal of two 9-storey houses. This case study demonstrates the importance of considering all stakeholders' views in the course of a project. If not, the project could be delayed or faced other downsides.

20.10   Case study No.9

Economic Analysis of Housing Designs

By Renato da Silva Solano and Roberto de Oliveira

This case study concerns an economic analysis of the design of two multifamily housing projects in the city of Porto Alegre in the Brazilian state of Rio Grande do Sul. Both projects concern rectangular high-rise buildings, which have indentation on the façade in order to add value to the building. The designs were analysed prior to approval to ensure that costs were within budget as well as matching the desired profit.

The Compacity Index (CI) was used as a decision criterion; and the major objective in both projects was the identification of possible interventions that could improve economic performance while adhering to architectural solutions. Also, the aim of the designer and the construction firm was to keep the quality of the façade within a reasonable cost and on the same construction area. In this regard, the external characteristics of the finishings in both projects were kept constant as they were initially proposed by the designers while the effect of changing the geometric shape was used to evaluate the CI effect on the Construction Global Cost (CGC). The shape option is indicated by the number of edges whose number directly affects CC/m^2 and the results were set in a table.

Project No.1 is for the staff of a State Firm Foundation and its design was based on the requirements of 5000 typical users. The product is a condominium with 96 housing units, 125 parking spaces, two elevators and the dimensions of the rooms were in compliance with local codes. The lot has three multistorey buildings over columns, underground parking spaces; residences are from the second to the ninth floor and water reservoir and elevator installations over these spaces. Each tower can be erected in a

'15.75 x 27.50 meters' rectangle without constraining the lot. It also has semi-built-in balcony on the building core, a Compacity Index (CI) of 46.46% and estimated cost ranging from R$5.415.486 34 to R$5.447.072 09 (1£Sterling = R$4.00). According to Brazilian norms this design can be classified as H8/3N (i.e. having eight floors and 3rd class type of finishings; meaning that it is not of high quality in terms of the specification of materials).

In this project, eight steps named as 'ACTION' were employed in the decision analysis. This dynamic interaction involved the construction firm, engineer and its estimator. The analysis sought for an optimal solution by a step-wise variation of the building shape, that is changing the number of edges while seeking to increase the Compacity Index (CI). At the end, a saving of 4.5% was made on total construction cost.

Project No. 2 was designed by a realtor who is based in the city of Porto Alegre after consulting their brokers. The design was constrained by the lot. The dwellings in this project are on the first floor and consist of 24 one-bedroom and 48 two-bedroom apartments. This residential floor has balconies that are integrated with respective living rooms. The building has 144 parking spaces from the second to the thirteenth floor and its estimated cost was R$6.31M. Its CI is 64%. According to Brazilian Norms it is an H12/2N building (height of 12 floors with 2nd class type of finishings).

In its design analysis, three options were considered wherein the Realtor and designer chose the option that represented the saving of R$210 thousands on construction costs: the main reason being that this option provided minor effects on design and more results in decreasing construction costs. The number of the edges was reduced initially to 24, then to 12, and finally to 4; which reflected savings of 2.06%, 3.32% and 4.65% respectively. Modifying the façade had an intangible effect on cost on this occasion. The last option was chosen due to cost and not as a perfect decision. A deterministic approach might have identified an optimal solution however a cost-based decision was acceptable to the decision makers.

In the two projects, the analyses provided the designer and owner with a clear vision of forecasting budget conflicts that can arise between certain stakeholders, i.e. designers and estimators. The foregoing illustrates how stakeholder interactions can bring about interventions in design phase decisions that can increase the economic performance of designs. Chapter 13 discusses other benefits of stakeholder management.

20.11 Case Study No.10

Let's Save Lives

By Ron Rosenhead

Why is stakeholder management so important?

We regularly run project management events for our clients and find all too often that projects frequently impact upon people who have not been involved in any consultation or communications within the project. This has led to:

- severe delays in projects
- increased costs to bring stakeholders on board
- belated development of a communications plans for the project

The following two examples have all come from our work in training professionals to deliver projects on time and to budget. All names have been omitted for obvious reasons.

Case study 1: Let's save lives

A local council in the UK reviewed its accident statistics. It identified its worst areas in the Borough for accidents and decided on its priority areas for action. One stretch of road had claimed the lives of several people and there had also been quite a number of serious accidents. This road was seen as a priority area for action.

The council surveyed the road and involved the local residents and businesses in the area. Several schemes were developed and comments received and welcomed. Eventually, the council agreed on the scheme to reduce accidents on this stretch of road.

On the day that the work started the council received an injunction to stop work. The injunction said that the council has not involved a key stakeholder; the railway authority. They had a bridge not too far away from the work that was being suggested and they were worried about the impact the work would have on this structure.

Legal proceedings commenced and the council lost! It had to pay £¼ million in legal costs and amend the scheme – after discussions with the railway authority and the local population. The overall scheme was delayed for slightly over 12 months and who knows how many accidents happened in that time period.

The case study shows that it takes time to formally sit down and review who will be affected. I always suggest if in doubt include them; you can always exclude them later on if you discover they are not a stakeholder. Omissions can be costly, monetarily or at an extreme, loss of life.

Case Study 2: Drive On

During a project management course with a large manufacturing company we were discussing stakeholder management and its benefits to project management. One person suddenly shouted out: 'Oh no, the unions!'

His story was that as he was Transport Manager for a very large fleet of trucks all based in the UK. The company was trying to deliver goods all over the UK and he wanted to ensure that deliveries were done not only effectively but were managed well and considered the 'green environment.' He commissioned the installation of an electronic device into each vehicle which he said the unions would suggest is a spy in the cab as it:

- tells a central control room where each vehicle is
- gives information on how long each journey has taken
- identifies fuel consumption per vehicle including spending time at the side of the road with engine ticking over – a big problem!

He rushed out from the course and through his staff arranged to meet with the unions as soon as possible.

We never heard any more about this issue however, this transport manager was really worried about the lack of involvement with the trade unions and even suggested the drivers could go on strike.

The key lesson is to look at not only who is involved but which of the key stakeholders is likely to have the *most* impact on your project, if they are not 'involved', they could interfere with your project at some stage.

Conclusion

Projects involve planning where all eventualities are considered and catered for upfront. Failure to plan is a plan to fail. One aspect that planning should not ignore is the potential impact of stakeholders. A project would pass through several phases and its stakeholders would have varying levels of influence along the line. A good plan would identify who is who and what they can do at each stage. The foregoing two case studies demonstrate that stakeholders can impact on a project negatively. However, a pre-plan can forestall most of, if not all, the negative impacts which stakeholders can bring.

20.12 Chapter summary

The case studies illustrate the consequences of dealing with stakeholder issues in an inadequate manner. Chapter 1 did signify that stakeholders abound in construction projects and other types of projects. The levels of power and interest of these stakeholders would be different and these attributes are dynamic. The challenge is to monitor the changing profile of your stakeholders and to be ready to address their demands. It is not an insurmountable challenge but a demand for due diligence. If you can track and pay attention to your stakeholders the chance is: you will likely satisfy them all the time.

The case studies have emphasised the potential downsides of stakeholders. However, we must not forget the flip side: the benefits. Chapters 1 and 15 have discussed some of the benefits which stakeholders can bring to an organisation. The main concept of stakeholder management is to maximise the benefits which can be derived from stakeholders while minimising the downsides. In this regard, the intention of the case studies is to flag up the downsides, provoke a thought and enhance a means of planning for these upfront. That way, the objective of stakeholder management will be achieved. On this note, the following chapter wraps-up the discussions.

21 Conclusion

Ezekiel Chinyio and Paul Olomolaiye

This concluding chapter is a call to excellence in stakeholder management. Our expectation is that stakeholder management will be a core competence in construction. The aim of stakeholder management is to maximise the benefits that can be derived from stakeholders while minimising the possible downsides that can arise by associating with them. Stakes are dynamic and thus must be monitored. Stakeholders also have dynamic levels of power and urgency. Stakes and stakeholders must be tracked at all times just like, air traffic control, would track and monitor aircrafts. The scope of the subject matter is large and the intention of this book is not to drive fear into individuals and organisations that apply it but to advise that one should not drop guard.

There are gains and pains in managing stakeholders. Some of the gains of stakeholder management include (Roome and Wijen, 2006):

- increased process and organisational efficiency;
- waste reduction and lower costs;
- stronger market positioning;
- reduced risk of prosecution;
- identification of new business opportunities;
- having a good public and local community image;
- having foresight of upcoming issues;
- lower insurance premiums;
- easier access to financial support;
- enhanced organisational learning.

Failure to manage stakeholders can be detrimental to an organisation. Some of these detriments include (Ismodes, 1997; Carroll and Buchholtz, 2006):

1. conflicts with the local community;
2. complicated decision-making processes;
3. time delays and associated cost overruns while assessing and responding to claims;
4. negative publicity for the companies involved;
5. difficulty in prioritising and thus responding to stakeholders' claims.

These detriments warrant that attention is paid to stakeholder management. A full recognition of stakeholder management as a core organisational function would boost its implementation. Every organisation does manage its stakeholders in one way or the other. However, many organisations have practised this at an informal level with their departments and font-line operatives allowed to use their initiative to implement it. A placid approach to stakeholder management is no longer worthwhile especially as activities in the global economy have heightened and the level of competition in all sectors. Organisations and individuals may manage stakeholders unconsciously. However, this vital function needs to be formalised in organisations. This should start by recognising the contributions of stakeholder management.

When a function is formalised, there is a demand for resources to be committed to it. A spin-off from this is that the function will then be accountable wherein someone will bear ultimate responsibility for it. In contrast, when a function is not formalised, then some employees who are especially laid-back may opt not to implement it.

The gains and pains of stakeholder management should inform practice. When the contributions of stakeholder management are valued, its implementation will be upheld and sustained.

21.1   A corporate issue

Sometimes an organisation may wish to act alone; and sometimes it will want to partner with others. This fluid situation is applicable to stakeholder management as well. The decision to coalesce with whom and when should be made at top level. Management should provide guidance in terms of stakeholder management.

Stakeholder management should also flow from strategy, i.e. it should enable an organisation to actualise its strategies. Thus, a strategy-statement will boost stakeholder management in organisations. While 'top-down' guidance is required for effective stakeholder management, a 'bottom-up' feedback from those who implement it is equally vital. That way, the practise can be updated continuously.

21.2   Also a people issue

Stakeholder management is a 'people-issue'. People represent organisations as well as spearhead their stakeholding functions. An understanding of how to deal with people and their concerns is a big step towards good stakeholder management practice. In this regard, stakeholder management involves appropriate communication which involves the use of diverse channels to supply relevant information.

A lot of issues can be resolved if communication can be forthcoming at the right time and manner. If only the community had been kept informed, their opposition could have been forestalled. If only the supply chain had been provided the relevant information, the project would not have experienced a delay. There can be so many if-only's. The point is that a stitch in time saves nine.

Stakeholder management is also about making and managing relationships. This calls for an understanding of individual and corporate behaviour and how to respond to these. As stakeholder management involves dealing with several people, one has to be adaptable. A flexible approach helps in dealing with different scenarios. However, flexibility is not tantamount to weak principles. When guidance is in place, stakeholder management can be implemented in a deem-fit approach while relying on underlying principles.

Sometimes problems will manifest while people interact. Some problems can escalate into major disputes. The ideal is to be proactive with stakeholder management such that disputes are avoided. When this is not possible, and such situations should be very few, then any resulting dispute should be resolved. The ability to resolve disputes is part of the stakeholder management drive and is discussed more in Chapter 17.

21.3   Preparation

The task of stakeholder management may seem daunting. However, when it is implemented proactively and with appropriate guidance, it will yield the benefits. Preparation may be needed in stakeholder management, as in other functions. Preparation is needed at two fronts:

corporate and individual. At the corporate front, each organisation needs to prepare how they will implement stakeholder management. This calls for guidance from the top. The management side of an organisation should provide leadership on how to implement stakeholder management. Individuals should be made to know what they need to do and be empowered to carry it out. Without empowerment, people can be confused, laid-back or frustrated. A stakeholder management champion should be visible in an organisation to drive through the implementation and act as custodian. Matters arising in the course of implementation can be discussed with this champion – who can, for instance, be the human resource manager. That way, the stakeholding efforts of individuals in an organisation will be coordinated.

A stakeholder management champion can provide an oversight function. This individual will be able to identify areas where employees need support and to arrange that to be forthcoming. Training and development will also be the remit of this champion. Some organisations encourage or even offer training on risk management, presentation skills, budgeting, etc. This package can be enlarged to include stakeholder management.

When an organisation feels it is not at its best, it can opt for training and development. Employees at different levels should be encouraged to excel in stakeholder management. From operational activities to strategic decisions, stakeholder management should be visible, and training can be used to enhance the practice.

21.4   Research potentials

Stakeholder management in construction is under-researched and offers a huge research potential. There are lots of attributes to be studied. For instance, the impact of stakeholder management on motion study can be investigated; likewise the cost of implementing stakeholder management. Other researchable aspects are the effects of stakeholder management on sustainability, profit, risks, etc.

21.5   Summary

Stakeholder management is about managing diverse stakes. It is a worthwhile endeavour that can yield several tangible and intangible benefits to individuals and organisations. In contrast failure to manage stakeholders can impact negatively on individuals and organisations. It is thus worthwhile, to practice stakeholder management proactively. Firms should be able to identify their stakeholders and nurture and sustain relationships with these. The aim is to optimise by maximising the benefits while minimising the downsides. Several tactics for engaging stakeholders have been discussed in this book which individuals and organisations will find useful.

References

Carroll, A.B. and Buchholtz, A.K. (2006) *Business and Society: Ethics and Stakeholder Management* (6th edn). Mason: Thomson South-Western.

Ismodes, A. (1997) *Socio-economic Aspects in Water Resources Development*. Stockholm: Avdelningen för Vattenbyggnad, Kungliga Tekniska Högskolan.

Roome, N. and Wijen, F. (2006) Stakeholder power and organizational learning in corporate environmental management. *Organization Studies*, **27**(2):235–263.

Index